Philosophical Applications
of Free Logic

Philosophical Applications of Free Logic

With an Introduction and Edited by
KAREL LAMBERT

New York   Oxford
OXFORD UNIVERSITY PRESS
1991

Oxford University Press

Oxford   New York   Toronto
Delhi   Bombay   Calcutta   Madras   Karachi
Petaling Jaya   Singapore   Hong Kong   Tokyo
Nairobi   Dar es Salaam   Cape Town
Melbourne   Auckland

and associated companies in
Berlin   Ibadan

Copyright © 1991 by Oxford University Press, Inc.

Published by Oxford University Press, Inc.,
200 Madison Avenue, New York, New York 10016

Oxford is a registered trademark of Oxford University Press

All rights reserved. No part of this publication may be reproduced,
stored in a retrieval system, or transmitted, in any form or by any means,
electronic, mechanical, photocopying, recording or otherwise,
without the prior permission of the publisher.

Library of Congress Cataloging-in-Publication Data
Philosophical applications of free logic / with an introduction
and edited by J. Karel Lambert.
p. cm.   Includes bibliographical references.
ISBN 0-19-506131-4
1. Logic, Symbolic and mathematical.
I. Lambert, Karel J., 1928E
BC135.P64   1991   160—dc20
90-36714

1 3 5 7 9 8 6 4 2
Printed in the United States of America
on acid-free paper

*For Kal, Kathryn and Christopher
who, in due course, will
also enjoy the
solemn duties and privileges
of
The Colonel's Club*

Acknowledgments

Thanks are due to the authors of the various articles in this book for permission to include their efforts, and also to the various journals and publishing houses for their permission. The latter are detailed in the prefaces to Sections II, III, and IV. I would also like to thank Ted Pasternak for his help in compiling the bibliography and Cynthia Read of Oxford University Press for her good-natured and helpful attitude during all phases in the preparation of this anthology.

K.L.

Laguna Niguel, Calif.
May 1990

Contributors

Ermanno Bencivenga
Department of Philosophy, University of California, Irvine, Irvine, CA 92717

Tyler Burge
Department of Philosophy, University of California, Los Angeles, Los Angeles, CA 90024

Nino Cocchiarella
Department of Philosophy, Indiana University, Sycamore Hall 026, Bloomington, IN 47405

James Garson
Department of Philosophy, University of Houston–University Park, Houston, TX 77004

Richard Grandy
Department of Philosophy, Rice University, Houston, TX 77251

Jaakko Hintikka
Department of Philosophy, Boston University, Boston, MA 02215

Karel Lambert
Department of Philosophy, University of California, Irvine, Irvine, CA 92717

William Mann
Department of Philosophy, University of Vermont, Burlington, VT 05405

Storrs McCall
Department of Philosophy, McGill University, 855 Sherbrooke St. West, Montreal, Quebec, Canada H3A 1G5

Carl Posy
Department of Philosophy, Duke University, Durham, NC 27706

Contributors

Ronald Scales
c/o Karel Lambert at UC–Irvine

Dana Scott
Department of Computer Science, Carnegie Mellon University, Pittsburgh, PA 15213

Peter Simons
Department of Philosophy, Salzburg University, Residenplatz 1, 5020 Salzburg, Austria

Bas van Fraassen
Department of Philosophy, Princeton University, Princeton, NJ 08544

Contents

I. Introduction

    The Nature of Free Logic, 3
        KAREL LAMBERT

II. Logic and Language

    1. A Theory of Definite Descriptions, 17
        KAREL LAMBERT

    2. Existence and Description in Formal Logic, 28
        DANA SCOTT

    3. A Free IPC is a Natural Logic: Strong Completeness for Some Instuitionistic Free Logics, 49
        CARL J. POSY

    4. Singular Terms, Truthvalue Gaps, and Free Logic, 82
        BAS C. VAN FRAASSEN

    5. Free Semantics, 98
        ERMANNO BENCIVENGA

    6. Applications of Free Logic to Qantified Intensional Logic, 111
        JAMES W. GARSON

III. Knowledge and Truth

    7. *Cogito, Ergo Sum*: Inference or Performance?, 145
        JAAKKO HINTIKKA

    8. A Definition of Truth for Theories with Intensional Definite Description Operators, 171
        RICHARD E. GRANDY

    9. Truth and Singular Terms, 189
        TYLER BURGE

10. Presupposition, Implication, and Self-reference, 205
BAS C. VAN FRAASSEN

11. A Russellian Approach to Truth, 222
RONALD SCALES

IV. Metaphysics

12. Abstract Individuals, 229
STORRS MC CALL

13. Quantification, Time, and Necessity, 242
NINO B. COCCHIARELLA

14. Definite Descriptions and the Ontological Argument, 257
WILLIAM E. MANN

15. Predication and Ontological Commitment, 273
KAREL LAMBERT

16. Free Part—Whole Theory, 285
PETER M. SIMONS

Bibliography, 307

PART I
Introduction

The Nature of Free Logic

Karel Lambert

1. Genesis and Leading Principles

Free logics are logics devoid of existence assumptions with respect to their terms but whose quantifiers have existential force. Despite anticipations in the first half of the twentieth century (for example, by Rosser)—and even earlier (e.g., by Punjer)—concentrated philosophical and technical study dates only from the mid-1950s.[1] Now in abundant supply, they are an important development in modern logic. Their genesis and leading principles may be explained as follows.

In the Port Royal theory of immediate inference (1662), which derived ultimately from Aristotle, inferences from a statement of the form

A: All S are P

to a statement of the form

I: Some S are P

and from a statement of the form

E: No S are P

to a statement of the form

O: Some S are not P

where 'S' and 'P' are placeholders for general terms, were counted valid. General terms are terms that are true (or false) of each of possibly many objects. Moreover, inferences from an A statement to the negation of the corresponding O statement (and vice versa), and from an E statement to the negation of the corresponding *I* statement (and vice versa), also were counted valid.

It is well known that when *A* and *E* statements are interpreted as universal conditionals and *I* and *O* are interpreted as existential conjunctions, both *A* to *I* and *E* to *O* inferences break down unless the placeholder '*S*' is restricted to nonempty general terms—general terms true of at least one (existent) object.[2] Corroboration is easily obtained by letting '*S*' be the empty general term 'planets between the Earth and the Moon' and '*P*' be the nonempty general term 'in rotation around the Sun'. In traditional language, the inferences described earlier are preserved by requiring that all statements of the four basic forms have *existential import* (with respect to their constituent *general terms*).

The drastic policy requiring that all replacements of '*S*' in the four basic forms be nonempty has adverse consequences. First, it restricts the scope of the Port Royal theory of immediate inference and thus precludes its use in assessing the validity of inferences containing empty general terms in subject position. For instance, the Port Royal theory, under the current interpretation of the four basic statement forms, cannot be applied to many inferences containing statements of physical law. The statement 'All bodies on which no external forces are acting move uniformly in a given direction' is such a statement because, many maintain, it contains the empty general term 'bodies on which no external forces are acting'. Second, the restricted Port Royal theory does not allow one to distinguish between inferences whose validity depends on no assumption that its constituent general terms in subject position are nonempty from those whose validity does require such an assumption. The mutual inference between the appropriate *A* and the negation of *O* statements is an example of the former, and the inference from the appropriate *A* to *I* statements is an example of the latter.

In the modern logic dating from Frege (1893-1903), object language counterparts of the emptiness and nonemptiness of general terms became available; a general term (or predicate) '*S*' is nonempty (or true of at least one (existent) thing) just in case there exists an object x such that x is S; otherwise it is empty. *A* to *I* and *E* to *O* inferences are modified to hold, in general, on the addition assumption that there is an object x such that x is S, but the mutual inferability of an *A* statement and the negation of the corresponding *O* statement and of an *E* statement and the negation of the corresponding *I* statement require no such additional assumption. The result, apparently, is an unrestricted quantifier logic with respect to its general terms capable of distinguishing between inferences in which existence assumptions with respect to their constituent general terms are crucial and those in which they are not.

Many scholars have characterized the connection between the Port Royal theory of immediate inference and the Frege-inspired modern theory of general inference as in part the making explicit in the latter of what was implicit in the former. Given the object language counterparts of nonempty general terms, it is now customary to say that the modern theory, in contrast to the Port Royal theory, is "free of existence assumptions with respect to its general terms."

There is, however, an important exception to the modern treatment of both A to I and E to O inferences, an exception that reduces the impact of the claim often made in favor of the modern theory of general inference over its medieval counterpart that it has rid itself of existence assumptions with respect to its general terms. Let 'S' be a general term of the form '(is) the same t', where 't' is a singular term. Then both A to I and E to O inferences hold unconditionally in the modern treatment despite the fact that 'There exists something the same as Vulcan (the putative planet)' is false because 'Vulcan' refers to no (existent) object. So unless instances of 't' in the relevant schemata of modern logic are restricted to singular terms that refer to an (existent) object, the special cases of the unmodified A to I and E to O inferences contemplated earlier will fail. Indeed, this very course of action is reflected in the modern logics' policy that anything of the form

    SE: There exists something the same as t

the object language counterpart of 't refers to an (existent) object', is logically true. In traditional language, the special cases of the unmodified A to I and E to O inferences are preserved in the modern theory of general inference by requiring that all statements of the appropriate special forms of A, E, I, and O have existential import (with respect to their constituent *singular terms*).

This policy, however, is prima facie no more palatable than the restrictive policy discussed earlier. First it, also, restricts the scope of the modern theory of general inference and thus precludes its use in assessing the validity of inferences containing statements with special empty general terms—for instance, inferences involving statements such as 'Everything the same as Vulcan is in rotation around the Sun', which contains the special empty general term '(is) the same as Vulcan'. Second, the modern theory of general inference does not allow one to distinguish between inferences whose validity requires the instances of the special general term scheme '(is) the same as t' in subject position to be nonempty (such as in A to I inferences) from those whose validity requires no such assumption (such as in A to the negation of O inferences).

In the modification of the modern theory of general inference called "free logic" not all instances of the statement form *SE* are logically true, and, indeed, some may be false. In the latter case, the general term '(is) the same as t' is empty and, in fact, this is equivalent to asserting that the singular term 't' does not refer to an (existent) object. So in free logic the special cases of *A* to *I* and *E* to *O* inferences discussed above are modified to hold on the condition that there exists an object x such that x is the same as t; no such condition is needed for *A* to the negation of *O* inferences (and vice versa), or for *E* to the negation of *I* inferences (and vice versa).

The principle of the indiscernibility of identicals yields, via quantifier distribution and quantifier confinement,

> *St*

from

> All objects x are such that x is *S*

on the condition that

> There exists an object x such that x is the same as t

This restricted form of universal instantiation is a characteristic feature of a free logic. Indeed, it is easily shown that *SE* is validating just in case unrestricted universal instantiation is validating.

Many philosophical logicians have characterized the relation between the modern theory of general inference and free logic as in part the making explicit in the latter what was implicit in the latter. Given the object language counterparts of nonempty general and singular terms, it is common to say that free logic, in contrast to the modern theory of general inference stemming from Frege, is "free of existence assumptions with respect to its general *and* its singular terms."

2. Proof Theory of Free Logic

Proof theoretical developments of free logic typically are of two sorts depending on whether the primitive predicate of singular existence, usually symbolized by Russell's '$E!$', is available. Axiomatic formulations of both kinds are provided here. (There are also many natural deduction versions of free logic whose unique character is straightforwardly reflected in the ensuing axiomatic treatments.)

Except for the possible presence of '$E!$' (exists), the *vocabulary* of free first-order predicate logic is not essentially different from that of classical first-order predicate logic. There is a list of variables, a list of individual constants (singular terms), a list of n-placed predicates; the connectives '$\sim$' and '$\rightarrow$', for example; the quantifier '$\forall$', for example, the comma; and the parentheses '(' and ')'. An *expression* is any finite sequence of signs from the vocabulary. A sign is *foreign to an expression* if it does not occur in the expression, and *foreign to a set of expressions* if it is foreign to every member of the set. Predicates are referred to by 'P', 'Q', 'R', ... with or without subscripts, singular terms are referred to by 'a', 'b', 'c', ... with or without subscripts, and variables are referred to by 'x', 'y', 'z', ... with or without subscripts. Statements, soon to be defined, are referred to by 'A', 'B', 'C', ..., but 't', 's', and 'r', with or without subscripts, are reserved for variables and singular terms. $A(t_1/s_1, \ldots, t_n/s_n)$ refers to the result of substituting simultaneously $t_1, \ldots, t_n$ for $s_1, \ldots, s_n$ in A.

The *formation* rules are also conventional. A *statement* is any of the following forms:

(i) $P(a_1, \ldots, a_n)$, where P is an n-place predicate;
(ii) $\sim A$ where A is a *statement*;
(iii) $(A \rightarrow B)$, where A, B are *statements*;

and

(iv) $\forall x A(x/a)$, where A is a *statement* and x is foreign to A; a may, but need not occur in A.

If the primitive vocabulary contains '$E!$', then the statements include also those of the form

(v) $E!a$.

A *quasi-statement* is any statement, and any expression of the form $A(x/a)$, where A is a quasi-statement and x is foreign to A.

From now on, unless otherwise stipulated, 'A', 'B', 'C', ... refer to quasi-statements, and 'S' refers to sets of statements. The contextual definitions of $(A \& B)$, $(A \vee B)$, $(A \equiv B)$, and $\exists x A$ are the conventional ones.

The *transformation rules* for the system without '$E!$' in its primitive vocabulary (FS_1) are as follows. An *axiom* of FS_1 is a tautology or any statement of the following forms:

MA_1 $A \rightarrow \forall x A$
MA_2 $\forall x(A \rightarrow B) \rightarrow (\forall x A \rightarrow \forall x B)$
MA_3 $\forall y(\forall x A \rightarrow A(y/x))$
MA_4 $\forall x \forall y A \rightarrow \forall y \forall x A$
MA_5 $\forall x A(x/a)$ if A is an axiom

The only rule of inference is

   D: From A and $A \to B$ infer B

A *derivation* of statement A from a set of statements S is a sequence $\langle A_1, \ldots, A_n \rangle$ such that

(i) $A = A_n$
(ii)(a) $A_i \in S$ or
(ii)(b) A_i is an axiom or
(ii)(c) A_i is consequence of previous members of the sequence by D.

If S is empty, A is a *theorem* of FS_1 and the sequence $\langle A_1, \ldots, A_n \rangle$ constitutes a *proof* of A.

If identity ('=') is added to the vocabulary of FS_1, the formation rules expanded to allow

   $a = b$

to count as a statement, and the transformation rules supplemented by at least

   $MA_6 \quad a = b \to (A \to A(b//a))$

where '$A(b//a)$' is the result of replacing a at one or more places in A by b, if any at all, then, as indicated earlier, the restricted version of universal specification characteristic of free logic is derivable, namely,

   RUS $\forall x A \to (E!a \to A(a/x))$

with the help of the following definition:

$E!a = \mathrm{df}\ \exists x(x = a)$

In the formulation of free predicate logic in which '$E!$' is a primitive one-place predicate (FS_2), the significant differences from FS_1 not already alluded to above lie in the transformation rules. In FS_2, MA_1, MA_2, and MA_5 are retained but MA_3 and MA_4 are replaced, respectively, by

   $MA'_3 = \mathrm{RUS}$

and

   $MA'_4 = \forall x E!x$

All else remains the same. When identity is added to FS_2, and also *both* of the classical meta-axioms for identity, then, as Hintikka first showed, the earlier definition of $E!a$ is justified by the theorem

$E!a \equiv \exists x(x = a)$

Hintikka's theorem, and hence the preceding definition of $E!a$ assumes even greater importance in light of a recent result by Meyer, Bencivenga, and Lambert that $E!a$ is indefinable in FS_2.[3] Their result also suggests that the bald assertion in FS_2 of Meyer's existence, or the calumny that he does not, can only be effected, sans identity, by use of the general term 'exist' much tradition to the contrary notwithstanding.

3. Model Theory for Free Logic

Model theoretic developments for free logic are motivated in part by the following informal considerations. First, there are competing world pictures, one consisting of both a (possibly empty) set of existing objects and some set of nonexisting objects, and the other consisting only of a (possibly empty) set of existing objects. Using Scales's useful terminology, the first kind of world picture is called the *Meinongian world picture*, and the second kind, the *Russellian world picture*.[4] Typically, when the Meinongian world picture prevails, the interpretation function in the formal model is *total*; it always assigns to a singular term either an existent or nonexistent object as value. In short, all singular terms refer. When the formal model theory is based on the Russellian world picture, the interpretation function defined on the singular terms is *partial*; some singular terms do not refer.

Second, matters are complicated by differing philosophical views about the truthvalues, if any, of at least the simple (atomic) statements containing singular terms not referring to an existent object. Three positions may be distinguished. The first—*negative free logic*—holds all simple statements containing at least one singular term not referring to an existent to be false. The second—*positive free logic*—holds at least some such statements to be true. The third—*neuter free logic*—holds all such statements, with the exception, perhaps, of simple statements of the form 'a exists', to be truthvalueless.

In the remainder of this section, two common kinds of model theory will be outlined. The first of them yields a positive free logic based on a Meinongian world picture; the second yields a negative free logic based on a Russellian world picture. It ought not be inferred, however, that these associations are invariable; there are, for example, positive free logics based on a Russellian world picture.[5] (The only neuter free logics seem to be those developed by Smiley and Skyrms. The latter, however, exploits one of the procedures used in developing a positive free logic based on a Russellian world picture, namely, van Fraassen's technique of supervaluations).[6]

The first kind of model theory, FM_1, consists of a *model structure* $\langle D_o, D_I, f \rangle$, where D_o (the outer domain) is a (possibly empty) set of objects (individuals), D_I (the inner domain) is a (possibly empty) set of objects such that D_I and D_o are disjoint (but their union is nonempty), and f (the interpretation function) is a function such that

(a) $f(a)$ is a member of the union of D_o and D_I;
(b) $f(P)$, where P is an n-adic predicate, is a set of n-tuples of members in the union of D_o and D_I;

and

(c) Every member of the union of D_o and D_I has a name.

Intuitively, D_I is the set of existent objects, and D_o is the set of nonexistent objects.

A statement of FS_1 is *true or false in* FM_1 under the following conditions:

(a) If A has the form of $P(a_1, \ldots, a_n)$, then A is true in FM_1 just in case $\langle f(a_1), \ldots, f(a_n) \rangle \in f(P)$; otherwise A *is false in* FM_1.
(b) If A has the form $\sim B$, then A *is true in* FM_1 just in case B *is false in* FM_1; otherwise A *is false in* FM_1.
(c) If A has the form of $B \to C$, then A *is true in* FM_1 just in case B *is false in* FM_1 or C *is true in* FM_1; otherwise A *is false in* FM_1.
(d) If A has the form $\forall xB$, then A *is true in* FM_1 just in case $B(a/x)$ is true in FM_1 for all a such that $f(a) \in D_I$; otherwise A *is false in* FM_1.

In this kind of semantic development the truth definition is entirely conventional except in clause (d) in which $\forall xB$ is defined just on the singular terms having values in D_I (the existents). That FS_1 is a positive free logic is evident in the case where $f(P) = D_0$, $f(a) = d$ such that $d \in D_o$ and D_I is empty. Then *Pa is true in* FM_1. Intuitively, *Pa* could be 'Hafner is a firebreather'. FS_1 is semantically adequate in the sense that A is a *theorem* of FS_1 just in case A *is true in all* FM_1 models.[7]

To accommodate FS_2, an additional clause must be added to the truth definition, namely,

(a$_2$) If A has the form $E!a$ then A *is true in* FM_1 just in case $f(a) \in D_I$; otherwise A *is false in* FM_1.

FS_2 is both a positive free logic and is semantically adequate.[8] To accommodate classical identity in either theory, an additional clause must be added to the truth definition:

(a_3) If A has the form $a = b$ then A *is true in* FM_1 just in case $f(a)$ is the same as $f(b)$.

Again, both FS_1 with identity and FS_2 with identity are positive free logics—for example,

$$a = a$$

is true in both even when $f(a) \notin D_1$ (a does not refer to an existent). And they are also semantically adequate.[9]

FM_1 models allow the domain of existents to be empty. But this is not an essential feature of free logics. One could exclude empty domains, but then to secure semantic adequacy a meta-axiom such as

$$\exists x(A \to A)$$

would have to be added to FS_1 or FS_2. This remark simply reinforces the idea that free logic simpliciter has to do with the existential import of terms, singular and general.

Turning now to the second kind of model theory, an FM_2 model structure is an ordered pair $\langle D, f \rangle$ where D is a (possibly empty) set of (existent) objects and the interpretation function f is a function such that

(a) $f(a)$ is a member of D if $f(a)$ is defined;
(b) $f(P)$, where P is an n-adic predicate, is a set of n-tuples of members of D;
(c) Every member of D has a name.

A statement of FS_1 *is true or false in* FM_2 under the following conditions:

(a) If A has the form $P(a_1, \ldots, a_n)$ then A *is true in* FM_2 if each of $f(a_1), \ldots, f(a_n)$ are defined and $\langle f(a_1), \ldots, f(a_n) \rangle \in f(P)$; if any of $f(a_1), \ldots, f(a_n)$ are undefined (do not refer to an existent object), then A *is false in* FM_2.
(b)–(d) duplicate the clauses in the truth definition based on FM_1.

FS_1 is a *negative free logic* under these conditions because any simple statement containing at least one singular term is false. The same is true of FS_2. In FM_2 models the interpretation function is partial as clause (a) in the definition of the FM_2 model structure makes clear and the truth definition is also unconventional in the clause for simple statements.

Neither FS_1 nor FS_2 is semantically adequate as they stand. But they can easily be made so. For instance, it suffices to add to FS_2 the meta-axiom

$$A(a/x) \to E!a, \text{ provided } A(a/x) \text{ is simple (or atomic)}[10]$$

Moreover, when identity is considered, the clause (a_3) in the truth definition must be modified as follows

(a′₃) If A has the form $a = b$, then *A is true in* FM_2 provided $f(a)$ and $f(b)$ are defined and are the same; if either $f(a)$ or $f(b)$ is undefined, *A is false in* FM_2.

As a result, neither FS_1 with identity nor FS_2 with identity contains the meta-axiom

$$a = a$$

and thus the identity theory in negative free logic is nonclassical. Indeed it is a theorem in FS_2 with identity, for example, that

$$a = b \rightarrow (E!\,a\ \&\ E!\,b)$$

The upshot is that though

$$A(a/x) \rightarrow \exists x A$$

fails in general in negative free logics, it holds if $A(a/x)$ is a simple statement. But, as in positive free logic,

$$A(a/x) \rightarrow (E!\,a \rightarrow \exists x A)$$

holds in general (and indeed is a theorem).

4. Applications of Free Logic

Because free logics treat expressions such as 'the planet causing the perturbations in the orbit of Mercury' as genuine singular terms (contra Russell) even though they do not refer to existent objects (contra a reformist Frege), they have yielded new foundations for the theory of definite descriptions (see Chapter 1), for set theory (see Chapter 2), for the theory of partial functions,[11] and for modal logic (see Chapter 6). They have been applied in philosophical areas as diverse as the philosophy of religion (see Chapter 14) and the philosophy of mathematics (see Chapter 3). They have proved especially useful in the analysis of topics in epistemology (see Chapters 7 to 11) and in the analysis of topics in ontology such as existence, predication (see, for example, Chapter 15), the theory of objects,[12] the part–whole relation (see Chapter 16), time and possibility (see Chapter 13), and abstract individuals (see Chapter 12). The applications described are far from exhaustive but constitute a representative sample of the breadth of philosophical uses of free logic. Finally, as is clear from the earlier remarks on the model theory of free logics, these logics are compatible with often dramatically opposed world pictures—for instance, the Meinongian and Russellian world pictures. The usefulness of free logic as a neutral instrument in much ontological argument thus seems assured.

Notes

1. See "Dialog mit Punjer über Existenz. I. Der Dialog," in Frege, G., *Nachgelassene Schriften*, Felix Meiner Verlag, Hamburg (1969) and Rosser, J. B., "On the consistency of Quine's *New Foundations for Mathematical Logic*," *Journal of Symbolic Logic*, vol. 4 (1939). But the pivotal paper in the development of free logic is H. S. Leonard's, "The logic of existence," *Philosophical Studies*, vol. 7 (1956) pp. 49–64. The other pioneering studies are Jaakko Hintikka's, "Existential presuppositions and existential commitments," *The Journal of Philosophy*, vol. 56 (1959) pp. 125–137; Hugues Leblanc's and Theodore Hailperin's "Nondesignating singular terms," *Philosophical Review*, vol. 68 (1959) pp. 129–136; Timothy Smiley's, "Sense without denotation," *Analysis*, vol. 20 (1960) pp. 125–135; Karel Lambert's "Existential import revisited," *Notre Dame Journal of Formal Logic*, vol. 4 (1963) pp. 133–144 and his "Notes on E!III: A theory of descriptions," *Philosophical Studies*, vol. 13 (1963) pp. 51–59 and "Notes on E!IV: A reduction in free quantification theory with identity and definite descriptions," *Philosophical Studies*, vol. 15 (1964) pp. 85–88; Rolf Schock's "Contributions to syntax, semantics and the philosophy of science," *Notre Dame Journal of Formal Logic*, vol. 5 (1964) pp. 241–290; Bas van Fraassen's "The completeness of free logic," *Zeitschrift für Mathematische Logik u. Grundlagen der Mathematik*, vol. 12 (1966) pp. 219–234; and Nino Cocchiarella's "A logic of possible and actual objects," *Journal of Symbolic Logic*, vol. 31 (1966) pp. 688–689 (Abstract). The label "free logic" is due to Lambert.
2. For a full account see Eaton, R. M., *General Logic*, Charles Scribners' Sons, New York (1931) pp. 223–226.
3. Meyer, R., Bencivenga, E., and Lambert, K., "The ineliminability of E! in free quantification theory without identity," *Journal of Philosophical Logic*, vol. 11 (1982) pp. 229–131.
4. Ronald Scales, *Attribution and Reference*, University of Michigan Microfilms (1969).
5. See, for example, Bas van Fraassen's "Singular terms, truthvalue gaps and free logic," and Ermanno Bencivenga's "Free semantics," in this volume (Chapters 4 and 5, respectively).
6. Ibid., and see the first footnote for the reference to Smiley's essay. Brian Skyrms' essay is entitled "Supervaluations: Identity, existence and individual concepts," *The Journal of Philosophy*, vol. LXV (1968) pp. 477–483.
7. See, for example, Leblanc, H., and Meyer, R. K., "On prefacing $\forall x A \supset A(Y/X)$ with $(\forall y)$: A free quantification theory without identity," *Zeitschrift für Mathematische Logik u. Grundlagen der Mathematik*, vol. 16 (1970) pp. 447–462.

8. See, for example, Meyer, R. K., and Lambert, Karel, "Universally free logic and standard quantification theory," *The Journal of Symbolic Logic*, vol. 33 (1968) pp. 8-26. This essay is a variant of the outer-domain approach in which the members of the outer domain are expressions. It is undoubtedly a case in which Scales' label "Meinongian semantics" has been stretched beyond useful application.
9. Ibid.
10. See, for example, Burge, T., "Truth and singular terms" (Chapter 9).
11. See, for example, Karel Lambert's and Bas van Fraassen's *Derivation and Counterexample*, Dickenson, Encino, Calif. (1972), for a brief discussion of the basic idea.
12. See, for example, Karel Lambert's *Meinong and the Principle of Independence*, Cambridge University Press (1983).

PART II
Logic and Language

Lambert's essay "A Theory of Definite Descriptions" is a slightly amended fusion of his two essays, respectively, "Notes on E! III: A Theory of Descriptions," *Philosophical Studies*, vol. 13 (1962) pp. 51–59, and "Notes on E! IV: A Reduction in Free Quantification Theory with Identity and Descriptions," *Philosophical Studies*, vol. 15 (1964) pp. 85–88. (The segments of those essays included here are reprinted by permission of Kluwer Academic Publishers.) It contains the first consistent (and complete) published version of a free theory of definite descriptions, often referred to in the literature as the system FD_2. Scott's later essay, "Existence and Description in Formal Logic," a reprint of the essay of the same title in *Bertrand Russell: Philosopher of the Century*, edited by Ralph Schoenman, Little, Brown and Co., Boston (1967) pp. 181–200, contains an independently discovered close cousin of FD_2, and a treatment of virtual class theory based on a bivalent positive free logic. Van Fraassen's essay, "Singular Terms, Truthvalue Gaps, and Free Logic," a reprint of the essay of the same title in *The Journal of Philosophy*, vol. 63 (1966) pp. 481–95, introduced the novel and now widely exploited semantical technique of supervaluations; it is reprinted here with the kind permission of *The Journal of Philosophy*. Bencivenga's "Free Semantics," amended for this volume, augmented and amplified the technique of supervaluations; the original essay of the same title appeared in *Italian Studies in the Philosophy of Science*, edited by M. L. Dalla Chiara, D. Reidel Publishing Co. (1980) pp. 31–48. (The segments of the original used here are reprinted by permission of Kluwer Academic Publishers.) Both van Fraassen's and Bencivenga's essays reflect, and, indeed, have stimulated, distinctive views in the philosophy of language. Posy's essay, "A Free IPC is a Natural Logic," is reprinted in its entirety from *Topoi*, vol. 1 (1982) pp. 30–43, and is a defense of the view that free intuitionistic predicate logic is a natural logic for mathematical intuitionism. (It is reprinted here by permission of Kluwer Academic Publishers.) Finally, Garson's study, "Applications of Free Logic to Quantified Intensional Logic," though dependent on an earlier essay entitled "Quantification in Modal Logic" published in *Handbook of Philosophical Logic* (edited by D. Gabbay and F. Guenther) by D. Reidel Publishing Co. (1984) pp. 249–307, is new in this volume, and contains a vigorous defense of the conviction that the quantificational foundations of modal logic are free. (The segments of the original essay included here are reprinted by permission of Kluwer Academic Publishers.)

1
A Theory of Definite Descriptions

Karel Lambert

PART I

For the purpose of explaining ordinary singular inference it is desirable to have a logic that is free of existence assumptions with respect to its argument constants.[1] In this kind of logic *any* singular term purporting to refer to an existent, whether it actually succeeds in so referring or not, can replace the free argument variables (parameters) $x, y, z, \ldots$ in the logical formulas of the system. The key step in securing what I have called elsewhere a *free logic*[2] is to replace the rule of universal instantiation (or the law of specification) in an otherwise "standard" first-order predicate calculus by a rule (or a law) that makes the tacit existence assumption in the associated reasoning (or declaration) explicit. For example, a free logic can be obtained from *Principia* by replacing

(1) $(x)(\phi x) \supset \phi y$

by

(2) $((x)(\phi x) \mathbin{\&} E!y) \supset \phi y$

A theory of definite descriptions based on free logic demands certain changes in Russell's theory of descriptions. For example, from Russell's definition of '$\psi(\imath x)\phi x$' in *Principia* *14.01, the theorem

(3a) $\psi(\imath x)\phi x \supset (\exists x)(\phi x)$

is derivable. In effect, (3a) says that any sentence containing a description containing 'ϕ' logically implies the sentence that there are ϕ's. From (3), letting 'ψ' be '$(\lambda y)(y = (\imath x)\phi x)$', with the help of the axiom of self-identity, one can obtain

(3a₁) $(\exists x)(\phi x)$

that is, any predicate 'ϕ' is exemplified. But (3a₁) is not always true as is shown by letting 'ϕ' be 'is a headless horseman'. Another consequence of *14.01 is the theorem

(3b) $\psi(\imath x)\phi x \supset (\exists y)((x)(\phi x \supset x = y))$

In effect, (3b) says that any sentence containing a description containing 'ϕ' logically implies the sentence that there is at most one thing that is ϕ. (3b) is also defective. The theorem

(3b₁) $(\exists y)((x)(\phi x \supset x = y))$

can be deduced from (3b) in precisely the same way that (3a₁) was deduced from (3a). Theorem (3b₁) is not always true as is shown by letting 'ϕ' be 'is a writer'. Rejection of (3a) and (3b), first, demands rejection of, or at least a change in, Russell's definition in *Principia* *14.01 and, second, supports the intuition of some ordinary language philosophers (e.g., Strawson) that, in general, descriptional statements logically imply neither existence nor uniqueness.

Moreover, free description theory can at least entertain the unconditional acceptance of certain "natural" theorems that, in *Principia*, are assertable only under restriction. Cases in point are the propositions that the such and such is (a) such and such, that is, in symbols,

(4) $\phi(\imath x)\phi x$

and

(4₁) $(\imath x)\phi x = (\imath x)\phi x$

In *Principia*, (4) and (4₁) are not independently assertable; they are assertable if and only if the descriptum exists. That is, for instance,

(5) $E!(\imath x)\phi x \equiv \phi(\imath x)\phi x$[3]

In Hintikka's[4] recent revision of description theory (4) is unconditionally assertable. It is a direct consequence of the axiom

(6) $y = (\imath x)\phi x \equiv (x)(\phi x \supset x = y \ \& \ \phi y)$

The proof sketch is as follows:

(a) From (6), $y = (\imath x)\phi x \supset (x)(\phi x \supset x = y \ \& \ \phi y)$: by Definition of '$\equiv$' and Simplification.

(b) From (a), $y = (ix)\phi x \supset \phi y$: by Distribution and Simplification.
(c) From (b), $\phi(ix)\phi x$: by Substitution $(ix)\phi x/y$ and Self-identity.

However, Henry Leonard[5] has noted that if (4), and Russell's definition in *Principia* *14.02, that is,

(7) $E!(ix)\phi x = \text{Df} (\exists y)((x)(\phi x \equiv x = y))$

are accepted, the following difficulty arises. From (4), one can obtain by substitution,

(8) $(\exists y)(y = (ix)(\exists y)(y = x))$

Given the biconditional (acceptable both to Leonard and to Hintikka)

(9) $E!x \equiv (\exists y)(y = x)$

from (8), one gets

(10) $E!(ix)E!x$

Performing like operations on the biconditional licensed by Russell's definition in (7) one can obtain

(11) $E!(ix)E!x \equiv (\exists y)((x)(E!x \equiv x = y))$

Propositions (10) and (11) yield

(12) $(\exists y)((x)(E!x \equiv x = y))$,

a proposition that says there is exactly one existent!

In his own free modal theory of descriptions, Leonard avoids (12) by putting a restriction on (4), namely,[6]

(13) $\sim(x)(\sim \Diamond \sim \phi x \supset E!x) \supset \phi(ix)\phi x$

Proposition (13) does not allow deduction of the bothersome $E!(ix)E!x$.

The problem posed by (12) arises in free description theory, whether it be modal in character as is Leonard's theory, or nonmodal in character as is preferred by both Hintikka and myself, on the assumption of the essential correctness of '$\phi(ix)\phi x$'. But '$\phi(ix)\phi x$' is defective in other ways that are more important than merely replacement of 'ϕ' by '$E!$'.

1. The Description Paradox

On the face of it, nothing could be more natural than the truth of 'The such and such is (a) such and such'. In symbols, again, this is

(14) $\phi(ix)\phi x$

But naturalness is not always a good measure of truth; witness the difficulties that ensue from an unwary acceptance of the natural method of identifying classes specified in an unrestricted version of the principle of class abstraction, that is, in

(15) $(x)(x \in \hat{y}(\phi y) \equiv \phi x))$

Substitution of '$\sim(x \in x)$' into 'ϕ' in (15) produces a straightforward contradiction, the famous paradox of classes, as Russell saw. Similarly with (14), let us replace 'ϕ' by the predicate '$(\lambda x)(\phi x \ \& \sim \phi x)$' in that formula. By concretion one concludes from (14), under this interpretation of 'ϕ', that

(16) $\phi(ix)(\phi x \ \& \sim \phi x) . \sim \phi(ix)(\phi x \ \& \sim \phi x)$,

which is a contradiction. It follows that Hintikka's nonmodal theory of descriptions is inconsistent. Leonard's modal theory, as the reader can verify by a consideration of (13), avoids this result.

One of the virtues of Russell's theory of descriptions, then, lies in the fact that the description paradox is not derivable there. The most that can be deduced in Russell's theory is

(17) $\sim E!(ix)(\phi x \ \& \sim \phi x)$

The blemish, of course, must also be removed from the otherwise more plausible free description theory. Accordingly,

(18) $\phi(ix)\phi x$

and, hence,

(19) $y = (ix)\phi x \supset \phi y$

and, hence, the axiom (6) in Hintikka's theory must be rejected. We are now free to entertain Russell's definition of '$E!(ix)\phi x$' (in the more usable form),

(20) $E!(ix)\phi x = \text{Df } (\exists y)((x)(\phi x \supset x = y) \ \& \ \phi y)$

Some important problems now are these: Can '$\phi(ix)\phi x$' be asserted under some restriction without leading (1) to contradiction or (2) to the problem posed by (12)? How much, if any, of Hintikka's key biconditional in (6) can we accept? What conditions, if any, are necessary so that descriptional expressions in general can be eliminated in favor of the materials of quantification theory with identity?

A Theory of Definite Descriptions

Let us take these problems in the reverse order. In Hintikka's theory, the biconditional[7]

(21) $\quad (E!(ix)\phi x \ \& \ \psi(ix)\phi x) \equiv (\exists y)((x)(\phi x \supset x = y) \ \& \ \phi y \ \& \ \psi y)$

is acceptable. In fact, (21) is deducible with the help of (6). Proposition (21) seems quite reasonable. It allows us to eliminate a description in favor of quantificational materials only under the condition that the description exists. Notice that the unwarranted (3a) and (3b) are not derivable from (21); the most that can be obtained are

(22) $\quad (E!(ix)\phi x \ \& \ \psi(ix)\phi x) \supset (\exists x)(\phi x)$

and

(23) $\quad (E!(ix)\phi x \ \& \ \psi(ix)\phi x) \supset (\exists y)((x)(\phi x \supset x = y))$

which appear entirely reasonable. Of additional importance is the fact that there can be deduced from (21),

(24) $\quad E!(ix)\phi x \equiv (\exists y)((x)(\phi x \supset x = y) \ \& \ \phi y)$

a proposition that justifies acceptance of the definition in (20).

Let us now turn to the second question posed at the end of the preceding section.

It has already been established that (4) follows from the now rejected axiom in (6). Specifically, it follows from the already rejected part [see (19)] of the half of (6) that reads

(25) $\quad y = (ix)\phi x \supset (x)(\phi x \supset x = y \ \& \ \phi y)$

But how is it with the other part of (25), the part that reads

(26) $\quad y = (ix)\phi x \supset (x)(\phi x \supset x = y)$?

Taking our cue from the fallacious (3b), we ought at least to be suspicious of (26). Indeed, this attitude is justified; (26) is, in fact, defective. Substituting '$(ix)\phi x$' into 'y' in (26), and with the help of the axiom of self-identity, one gets

(27) $\quad (x)(\phi x \supset x = (ix)\phi x)$

Proposition (27), in turn, yields, with the help of (2),

(28) $\quad (E!y \ \& \ \phi y) \supset y = (ix)\phi x$

But (28) is not always true as is shown by letting 'y' be 'Russell' and 'ϕ' be 'is a man who is a writer'. So (26), and hence, all of (25), must be rejected in their present form.

Consider the other half of the axiom in (6), which reads:

(29) $(x)(\phi x \supset x = y) \& \phi y) \supset y = (ix)\phi x$

Is (29) acceptable? The answer is, "Almost!" But from (29), it follows that

(30) $\sim (\exists x)(\phi x \& \phi y) \supset y = (ix)\phi x$

But (30) can be false as is shown by letting 'ϕ' be 'is a flying horse' and 'y' be 'the flying horse captured by Bellerophon'. It appears then that none of the rather direct consequences of (6) are unconditionally assertable.

Notice that (30) would not fail on the assumption that y exists. The same would hold for (29). I believe the same condition applies to the other half of (6) [see (25)]. Accordingly, let us adopt the following revision of Hintikka's axiom:

(31) $E!\,y \supset (y = (ix)\phi x \equiv (x)(\phi x \supset x = y \& \phi y))$

In effect (31) says that y is $(ix)\phi x$ if and only if y is uniquely characterized by ϕ, provided that y exists.

In free logic, with the help of universal generalization, the identity theorem, '$(x)(\exists y)(x = y)$' and the biconditional '$E!\,x \equiv (\exists y)(y = x)$', one can obtain from '$E!\,y \supset \phi y$', '$(y).\phi y$'. Further, with revised specification [see (2)], one can obtain from '$(y)\phi y$', '$E!\,y \supset \phi y$'. Accordingly, (31) may be rewritten as[8]

A_1 $(y)(y = (ix)\phi x \equiv ((x)(\phi x \supset x = y) \& \phi y)))$

Adoption of this axiom has the following desirable consequences. First, the two important biconditionals in Hintikka's system described here in (21) and (24) are deducible from Axiom 1. Second, as suggested earlier, we can no longer deduce the faulty '$y = (ix)\phi x \supset \phi y$' or the faulty '$y = (ix)\phi x \supset (x)(\phi x \supset x = y)$' but only the reasonable

(32) $(E!\,y \& y = (ix)\phi) \supset \phi y$

and

(33) $(E!\,y \& y = (ix)\phi x) \supset (x)(\phi x \supset x = y)$

Third, from (32) and (33) we can deduce the legitimate

(34) $E!\,(ix)\phi x \supset (\exists x)\phi x$

and

(35) $E!\,(ix)\phi x \supset (\exists y)((x)(\phi x \supset x = y))$

A Theory of Definite Descriptions

but not the defective

(36) $\quad \psi(\imath x)\phi x \supset (\exists x)\phi x$

or

(37) $\quad \psi(\imath x)\phi x \supset (\exists y)((x)(\phi x \supset x = y))$

Propositions (34) and (35) suggest that the descriptional context '$E!(\imath x)\phi x$' does logically imply both existence and uniqueness. This analysis also provides the answer to the first question posed at the end of the last section.

From (32) we can deduce, with the help of the axiom of self-identity,

(38) $\quad E!(\imath x)\phi x \supset \phi(\imath x)\phi x$

That is, the such and such is (a) such and such, provided that the such and such exists. Proposition (38) neatly avoids the too-weak '$\phi(\imath x)\phi x$' and the too-strong '$E!(\imath x)\phi x \equiv \phi(\imath x)\phi x$' of Russell; the half of the Russell biconditional that reads

$\phi(\imath x)\phi x \supset E!(\imath x)\phi x$

is sometimes false, as is shown by letting 'ϕ' be 'is identical with the headless horseman'. On the other hand, (38) is strong enough to preclude the description paradox while yet avoiding the bothersome '$E!(\imath x)E!x$'. Concerning the former point, the present theory has the Russellian virtue that

(39) $\quad \sim E!(\imath x)(\phi x \ \& \sim \phi x)$

Let me close this part with a recommendation. An important theorem in traditional description theory (which, also, is deducible in Hintikka's theory) is

(40) $\quad y = (\imath x)(x = y)$

In the present version of free description theory, (40) is not deducible without restriction. But (40) appears to be unconditionally assertable. Therefore, let us add, as our second axiom in free description theory,

$A_2 \quad y = (\imath x)(x = y)$

PART II

In the previous part, a free description theory based on the axioms

$A_1 \quad (y)(y = (\imath x)\phi x \equiv ((x)(\phi x \supset x = y) \ \& \ \phi y)))$

and

$$A_2 \quad y = (\imath x)(x = y)$$

was presented.

An unnoticed result of the theory is that, when appended to identity theory, the identity principle

(41) $\quad x = x$

gets reduced to the status of a derived principle. Replace A_2 by the variant

$$A_2 \quad (\imath x)(y = x) = y$$

Replace 'x' by '$(\imath x)(y = x)$' and 'ϕ' by '$(\lambda x)(x = y)$' in

(42) $\quad x = y \supset (\phi x \supset \phi y)$

and (41) follows with the help of A_2.[9]

Is it possible to reduce the present version of free description theory without risk of the untoward results mentioned in the previous part? Yes; such a theory can be formulated with a single axiom. Further, variants of A_1 and A_2 are rather direct consequences of this theory. The result is a very elegant free quantification theory with identity. For example, Hintikka has shown that his version of free quantification theory with identity is simpler by one principle than the traditional version.[10] If the present suggestion for treating descriptional contexts of the form '$y = (\imath x)\phi x$' is appended to Hintikka's theory, his theory is reducible by two principles, viz. (in his notation) the principles

(4a) $\quad f \to x = x$, where f contains a free occurrence of x

and

(4b) $\quad f \to a = a$, where a occurs in f

are reducible to the status of derivable principles.

Assume the following variant of A_1:

$$A_1 \quad (y)((\imath x)\phi x = y \equiv (\phi y \,\&\, (x)(\phi x \supset y = x)))$$

In free identity theory we have the theorem

(43) $\quad (\imath x)\phi x = z \supset (y)(z = y \equiv (\imath x)\phi x = y)$

A_1 and (43), by quantification theory, yield

(44) $\quad (\imath x)\phi x = z \supset (y)(z = y \equiv (\phi y \,\&\, (x)(\phi x \supset y = x)))$

The suggestion of the present paper is that the converse of (44) is also

A Theory of Definite Descriptions

plausible, for example,

(45) $\quad (y)(z = y \equiv (\phi y \ \& \ (x)(\phi x \supset y = x)))) \supset (ix)\phi x = z$

If the converse of (43) were deducible in free quantification theory with identity, (45) would also be deducible. But in free quantification theory with identity, the converse of (43) is deducible only under the condition that $E!x$ [i.e., $(\exists z)(x = z)$]. This shows the neutrality of free quantification theory with identity on the question of the truthvalue to be assigned to identity contexts containing nonreferential singular terms. Note especially that if follows from the converse of (43) that

(46) $\quad (\sim(\exists y)(y = z) \ \& \ \sim(\exists y)(ix)\phi x = y) \supset (ix)\phi x = z$

that is, if the substitution instances of 'z' and '$(ix)\phi x$' have no designata any statement of the form '$(ix)\phi x = z$' is true. Such cases of what Quine calls "don't cares" can be decided arbitrarily in a way consistent with the rest of the theory.[11] For example, in the present theory no principle can be condoned that allows '$(ix)\phi x = (ix)\phi x$' to turn out false where '$(ix)\phi x$' is nonreferential. For the truth of '$(ix)\phi x = (ix)\phi x$', even where '$(ix)\phi x$' is nonreferential, follows from the unexceptionable identity principle '$x = x$'.[12] Accordingly, I shall adopt (45), noticing particularly that where '$\sim(\exists y)(y = z)$' and '$\sim(\exists x)\phi x$', '$(ix)\phi x = z$' is deducible, and that (44) and (45) yield the biconditional justifying the axiom

$A_3 \quad (ix)\phi x = z \equiv (y)(z = y \equiv (\phi y \ \& \ (x)(\phi x \supset y = x)))$

The common sense of A_3 is explicit in its verbal rendition. For what A_3 says is this: 'The so and so is z just in case everything is z if and only if it and it only is so and so'.

It remains to be proved that A_3 yields (41) (and, hence, A_1 and A_2) when appended to free quantification theory with identity where the only identity axiom is (42). A consequence of A_3 is

(47) $\quad (ix)(z = x) = z \equiv (y)(z = y \equiv (z = y \ \& \ (x)(z = x \supset y = x)))$

Now, as the reader may verify, the right-hand side of (47) requires only (42) for its proof. Hence, we obtain

(48) $\quad (ix)(z = x) = z$

But it has been pointed out earlier that (48) and (42) yield

(49) $\quad x = x$

An instance of (49) is

(50) $\quad (ix)\phi x = (ix)\phi x$

A_3 and (50) yield A_1. And this completes the task, for A_1 and A_2 easily follow.

The proof that '$x = x$' can be removed from among the basic assumptions of identity theory offers some additional support for free quantification theory with identity, vis-à-vis simplicity.[13] For in traditional identity theory, '$x = x$' is not derivable without additional assumptions.

Notes

1. The definitive papers are by Henry S. Leonard, "The logic of existence," *Philosophical Studies*, vol. 4 (June 1956) pp. 49–64; Hugues Leblanc and Theodore Hailperin, "Non-designating singular terms," *Philosophical Review* (April 1959) pp. 239–44; and by Jaakko Hintikka, "Existential presuppositions and existential commitments," *Journal of Philosophy*, vol. 3 (Jan. 1959) pp. 125–37.
2. See the abstract of my address "The definition of E(xistence)! in free logic" in *Abstracts: The International Congress for Logic, Methodology and Philosophy of Science*, Stanford University Press (1960).
3. Another example is the faultless '$(ix)\phi x = (ix)\phi x$', which, in *Principia*, is not unconditionally assertable.
4. K. J. J. Hintikka, "Towards a theory of definite descriptions," *Analysis*, vol. 19 (1959) pp. 79–85.
5. "The logic of existence," p. 62.
6. Ibid.
7. The symbolic language in the present paper departs from that of Hintikka's and is similar to Leonard's. But this difference is not essential to my arguments.
8. The relationship between Axiom 1 and (31) is one of the key points of difference between free and classical predicate logic. In the latter, given 'ϕx' one can always obtain '$(x)\phi x$', and vice versa. Among the consequences of this difference between the two logics is that a completeness proof for free logic based on the universal closures of the formulas of free logic will no longer work. In free logic, it is not true that to every wff ϕ containing free argument variables there corresponds a wff ϕ^c containing no free argument variables (= the closure of ϕ) such that ϕ is a theorem if and only if ϕ^c is a theorem, and such that ϕ is valid if and only if ϕ^c is valid. For example, '$(y)((x)(\phi x) \supset \phi y)$' is an example in free logic that violates the requisite condition.
9. In general, given a formula of the form '$x = y$', where 'x' and 'y' are distinct and 'y' is a variable, (41) is deducible with the help of (42).
10. Jaakko Hintikka, "Existential presuppositions and existential commitments," *Journal of Philosophy*, vol. 56 (no. 3) (1959) pp. 126–27. Hintikka's system is truly one without existential presuppositions, both in the sense that the system does not require that its argument constants have bearers and in the sense that its principles hold in every domain including the empty one (i.e., in *all* possible worlds!).

11. W. V. Quine, *Word and Object*, Wiley, New York (1960) p. 259.
12. This is the case because, in free logic, nonreferential terms are admitted to the position of the argument variables.
13. The usefulness of free logic in ontological questions is shown by Hintikka (cf. note 3). He shows that an intuitive object language counterpart of Quine's thesis "To be is to be the value of a variable" is provable in free logic. It is not possible to do so in traditional symbolic logic, since the thesis has no suitable equivalents in that logic. Further, as I hope to show in the near future, free logic has important implications for the theory of classes; for example, '$(x)(x \in \hat{y}(\phi) \equiv \phi x)$', a proposition not assertable without restriction in class theories based on traditional symbolic logic, is assertable in free logic.

Historical Note. The theory of definite descriptions outlined in this essay, later called FD_2, was the first sound and complete free theory of definite descriptions ever devised. This was known by me in the early 1960s based on a semantics which later was to be called "outer domain semantics" (and independently discovered by Nuel Belnap). In that semantics a model structure intuitively was a triple consisting of an outer domain of nonexistents, an inner domain of existents and an interpretation function which assigned all individual constants to members in the union of the inner and outer domains, all n-adic predicates to set of n-tuples of members in the union of those two domains, the individual in the inner domain of which the basis of a definite description is uniquely true to that definite description and an arbitrary member of the outer domain to all definite descriptions that did not meet the previous condition. Truth in a model was absolutely conventional except for the quantifiers which were defined only on the inner domain to give them their requisite existential force. This outer domain semantical approach, which I elaborated in logic classes in the late 1950s and early 1960s, did not then appeal to me because of its Meinong-like cast. Indeed it was this concern which helped to motivate Bas van Fraassen, who, as an undergraduate, did logic with me, to develop his own later anti-Meinongian supervaluational approach to free logic.

2
Existence and Description in Formal Logic

Dana Scott

The problem of what to do with improper descriptive phrases has bothered logicians for a long time. There have been three major suggestions of how to treat descriptions usually associated with the names of Russell, Frege, and Hilbert-Bernays. The author does not consider any of these approaches really satisfactory. In many ways Russell's idea is most attractive because of its simplicity. However, on second thought one is saddened to find that the Russellian method of elimination depends heavily on the scope of the elimination. Further, the semantical meaning of Russell's transformation is not all that clear, although it could be made quite precise. Frege's use of a null entity for the denotation of an improper description has of course an immediate semantical interpretation, but the arbitrary choice of a null entity in each domain is really not very natural. In many axiomatic theories, Euclidean geometry for example, the choice of a distinguished point is not possible to even very desirable. Bernays (Bernays and Fraenkel 1958) used Fregean descriptions with a kind of 'local' null entity carried along in the notation itself. This idea, though clear and workable, is not very elegant in the author's opinion.

It is curious that in ordinary mathematical practice having undefined function values, a situation close to using improper descriptions, does not seem to trouble people. A mathematician will often formulate conditionals of the form

if $f(x)$ exists for all $x < a$, then ...

and will not give a moment's thought to the problem of the meaning of $f(a)$. More careful authors never use a description or a function value unless it has been previously proved that its value exists. This style led

Hilbert and Bernays (1934) to the point of requiring such a proof before a formula containing the description can be considered as well formed. This suggestion is to be rejected on many grounds. As has often been pointed out, the class of well-formed formulas will hardly ever be recursive. Also the class of well-formed formulas will change upon the introduction of additional axioms. More serious is the fact that it is quite natural to employ descriptions *before* they have been proved to be proper. In axiomatic set theory in the discussion of recursive definitions, it is very tempting to give an explicit definition of the required function by means of a description and then prove a theorem of the form:

for all a in a set well-ordered by $<$,
if $f(x)$ exists for all $x < a$, then $f(a)$ exists

It will then follow by transfinite induction that $f(a)$ exists for all a in the well-ordered set. Only a logician would have objections to this use of the 'exists'. It is the purpose of this paper to lay these objections to rest by presenting a formal theory of descriptions that corresponds quite faithfully to such natural modes of reasoning.

After the author had explicitly formulated his plan (December 1963), he discovered that around 1959–1960 several other logicians had come to nearly the same idea: notably Hailperin and Leblanc (1959), Hintikka (1959a, b), Rescher (1959), and Smiley (1960). These papers have not received the attention they deserve; thus a complete exposition including a full discussion of the semantics required seems desirable. Further, the author wishes to show how the idea can be applied to a theory like Quine's system of virtual classes (Quine 1963). Quine, following Russell, employs contextual definitions that avoid giving an independent meaning to the virtual classes. The author will replace Quine's definitions by axioms and present a simple semantical interpretation for the theory. The paper will conclude with a model-theoretic discussion of eliminability of notions by contextual definitions. There is an interesting problem here that is left open.

The author is indebted to Professors Hintikka, Kaplan, Kreisel, Mostowski, Quine, Robinson, Suszko, Suppes, and Tarski who were kind enough to comment on earlier versions of this paper.

Note added in proof (1990). Since the writing of this paper, the author worked for many years on intuitionistic logic. In such a framework the Law of the Excluded Middle is not available, and the problem of what to do about nondenoting descriptions is not quite obvious. A particular solution was found by the author in 1975 and published under the title "Identity and Existence in Intuitionistic Logic" for the 1977 Durham

Conference on Applications of Sheaves in the Springer-Verlag Lecture Notes in Mathematics, vol. 753 (1979), pp. 660–696. Many authors have remarked on the need for partial elements in models for intuitionistic logic and have given semantical interpretations—and these interpretations agree with the approach advocated in the 1979 paper. When intuitionistic logic is specialized to classical logic (say, by adding Excluded Middle), then the result is exactly as suggested here in this 1967 paper. The author feels that the additional experience with the nonclassical theories shows that principles advocated are both workable and sound.

1. Descriptions

To simplify matters let us consider a first-order logic with just one nonlogical constant: a binary predicate symbol R. The logical symbols are ¬ (for negation), → (for implication), ∀ (for universal quantification), = (for equality), and **I** (for description). Note that **I** is an *inverted* capital I, which the author prefers to Russell's inverted iota. The other propositional connectives ∧, ∨, ↔, and the quantifier ∃ should be considered as introduced by definition, or better, the formulas involving them may be taken as abbreviations of formulas containing only the basic symbols. The individual variables are $v_0, v_1, \ldots, v_n, \ldots$. In the metalanguage x, y, z, w are metavariables ranging over the individual variables of the object language.

We define the notion of *term* and *formula* in the usual way:

(i) All variables are terms.
(ii) If α and β are terms, then $\alpha = \beta$ and α R β are (atomic) formulas.
(iii) If Φ and Ψ are formulas and x is a variable, then ¬Φ, $[\Phi \to \Psi]$ and $\forall x \Phi$ are formulas; while $\mathbf{I}x\Phi$ is a term.

The precise definition of *free* and *bound* variables need not concern us here, and it may be assumed as known.

To give a semantical interpretation of this language one first gives a structure $\langle A, R \rangle$, where A is a set (the domain of individuals) and R is a binary relation (the interpretation of the predicate symbol R). Then relative to the given structure one defines the *values* of formulas (they will be truthvalues) and terms (they will be objects) corresponding to the values given to the free variables. Before presenting this definition, it will be wise to consider some informal, motivating principles that have guided the choice of our precise formulation. Above all we wish to follow:

Principle 1. *Bound individual variables should range only over the given domain of individuals.*

The author will not attempt to define what he means by "range over" since he is sure everyone understands this statement. In case someone does not, he should wait to see the formal definition of value for the quantified formulas and for the descriptive phrases. The second principle is not so important, but the author wishes to include it with an eye to future applications:

Principle 2. *The domain of individuals should be allowed to be empty.*

Finally, and very important for the basic idea of the paper, we have:

Principle 3. *The values of terms and free variables need **not** belong to the domain of individuals.*

To see in a simple example the usefulness of Principle 3, consider the question raised by Mostowski (1951) in connection with the empty domain. Namely, it is "clear" that the formula

$$x \mathrel{R} x \to x \mathrel{R} x$$

is valid in all domains including the empty one. Similarly, the sentence

$$[x \mathrel{R} x \to x \mathrel{R} x] \to \exists y [y \mathrel{R} y \to y \mathrel{R} y]$$

is valid, because if x is given a value in the domain, then there is some value of y to satisfy the formula within the quantifier. On the other hand, the formula

$$\exists y [y \mathrel{R} y \to y \mathrel{R} y]$$

is *not* valid in the empty domain; hence, the valid formulas are not closed under the rule of modus ponens when the empty domain is included. The fallacy (or better inconvenience) here lies in allowing the second formula to be valid. The first formula is completely valid no matter what value we assign to x. The second formula will fail in the empty domain, however, if we recognize Principle 3. To have a valid formula we must modify the implication to read:

$$[x \mathrel{R} x \to x \mathrel{R} x] \land \exists y [x = y] \to \exists y [y \mathrel{R} y \to y \mathrel{R} y]$$

We can make this point better after the precise definition of value is given.

Before presenting the definition of value, we must still decide what to do with the improper descriptive phrases. Under the guidance of Principle 3 we are no longer required to give such terms values *within* the given domain. Indeed, it seems much better to give an improper description a value definitely *outside* of the domain, thereby emphasizing its impropriety. The way to do this is to assign to each domain A a null entity $*_A$ such that $*_A \notin A$. This is much easier than trying to make the null entity belong to A as Frege wished (especially when A is empty!). Assuming a reasonable set theory, we could let $*_B$ be the set of all sets belonging to A which are nonself-members. Thus $*_A \subseteq A$ but $*_A \notin A$. Assuming the so-called Axiom of Regularity we could even take $*_A = A$. The exact choice is quite irrelevant as long as we agree $*_A \notin A$. Now we are ready for the definition of value.

In the following we shall write $\mathfrak{A} = \langle A, R \rangle$ for short; while s will denote an *assignment* which is simply a function whose domain is the set of integers $N = \{0, 1, 2, \ldots\}$. For $i \in N$, $s(i)$, or simply s_i, is the value we wish to assign to the variable v_i. We define

$$s(i/a) = (s \sim \{\langle i, s_i \rangle\}) \cup \{\langle i, a \rangle\}$$

in other words $s(i/a)$ is like s except the ith value s_i has been replaced by a. We shall read

$$\vDash_\mathfrak{A} \Phi[s]$$

as: the (truth) value of the formula Φ is *true* for the assignment s relative to the structure $\mathfrak{A}$, or better, s *satisfies* Φ in $\mathfrak{A}$, or also, Φ is *true* at s in $\mathfrak{A}$. The symbol

$$\|\alpha[s]\|_\mathfrak{A}$$

is read: the (object) value of the term α for the assignment s relative to the structure $\mathfrak{A}$, or better, the value of α at s in $\mathfrak{A}$. (Maybe the use of "in" is bad here, because in view of Principle 3 $\|\alpha[s]\|_\mathfrak{A} \in A$ need not be so.) The exact clauses of the recursive definition of these notions are as follows:

$\|v_i[s]\|_\mathfrak{A} = s_i$

$\vDash_\mathfrak{A} \alpha \mathbf{R} \beta[s]$ iff $\langle \|\alpha[s]\|_\mathfrak{A}, \|\beta[s]\|_\mathfrak{A} \rangle \in R$

$\vDash_\mathfrak{A} \alpha = \beta[s]$ iff $\|\alpha[s]\|_\mathfrak{A} = \|\beta[s]\|_\mathfrak{A}$

$\vDash_\mathfrak{A} \neg \Phi[s]$ iff not $\vDash_\mathfrak{A} \Phi[s]$

$\vDash_\mathfrak{A} [\Phi \to \Psi][s]$ iff if $\vDash_\mathfrak{A} \Phi[s]$, then $\vDash_\mathfrak{A} \Psi[s]$

$\vDash_\mathfrak{A} \forall v_i \Phi[s]$ iff for all $a \in A$, $\vDash_\mathfrak{A} \Phi[s(i/a)]$

$$\|\mathbf{I}v_i \Phi[s]\|_\mathfrak{A} = \begin{cases} a & \text{if } a \text{ is the unique element of } A \text{ such that} \\ & \vDash_\mathfrak{A} \Phi[s(i/a)]; \\ *_A & \text{if there is no such element} \end{cases}$$

We say that Φ is *valid* in $\mathfrak{A}$ and write $\vDash_\mathfrak{A} \Phi$ to mean that $\vDash_\mathfrak{A} \Phi[s]$ holds for all assignments s. We say that Φ is *universally valid* and write $\vDash \Phi$ to mean that $\vDash_\mathfrak{A} \Phi$ holds for all structures $\mathfrak{A}$.

The question now is to find an axiomatization of the universally valid formulas. Note first that these two rules are correct:

(MP) If $\vDash \Phi$ and $\vDash [\Phi \to \Psi]$, then $\vDash \Psi$

(UG) If $\vDash [\Phi \to \Psi]$ and x is not free in Φ, then $\vDash [\Phi \to \forall x \Psi]$

Next note that these schemata comprise only valid formulas:

(S0) Φ, if Φ is a tautology

(S1) $\forall x[\Phi \to \Psi] \to [\forall x \Phi \to \forall x \Psi]$

(S2) $\forall y \exists x[x = y]$

(S3) $\alpha = \alpha$

(S4) $\Phi(x/\alpha) \wedge \alpha = \beta \to \Phi(x/\beta)$

where $\Phi(x/\alpha)$ is the result of substituting α for all the free occurrences of x in Φ rewriting bound variables if necessary.

Inasmuch as $\exists x \Phi$ abbreviates $\neg \forall x \neg \Phi$, it is easy to see that the rule (UG) includes the rule

(EG) If $\vDash [\Psi \to \Phi]$ and x is not free in Φ, then $\vDash [\exists x \Psi \to \Phi]$

Also using the schemata, especially (S1), we can show that

$$\vDash \forall x \Phi \wedge \exists x \Psi \to \exists x[\Phi \wedge \Psi]$$

and

$$\vDash \forall x[\Phi \to \Psi] \to [\exists x \Phi \to \exists x \Psi]$$

Using these together with (S4) we establish easily the validity of the schema

(UI) $\forall x \Phi \wedge \exists x[x = \alpha] \to \Phi(x/\alpha)$

where x is not free in the term α. This is the correct version of the law of universal instantiation which is valid not only when the domain is empty but also when the values of terms are allowed to be outside the domain. Using (S2) we can also show

$$\vDash \forall y[\forall x \Phi \to \Phi(x/y)]$$

which some authors would take as an axiom but which is superfluous when principles of equality are available. On the other hand, (S2) is practically a special case of this last schema. Replace Φ by the formula $\neg x = y$, obtaining

$$\vDash \forall y [\forall x [\neg x = y] \rightarrow \neg y = y]$$

from which we derive

$$\vDash \forall y [y = y \rightarrow \exists x [x = y]]$$

and then

$$\vDash [\forall y [y = y] \rightarrow \forall y \exists x [x = y]]$$

In view of (S3), we can now easily obtain (S2). So it is really just a matter of taste as to which schemata are chosen as the fundamental ones.

To understand better what is going on here, consider the meaning of

$$\exists x [x = \alpha]$$

under our semantical rules. When x is not free in α, then

$$\vDash_\mathfrak{A} \exists x [x = \alpha][s]$$

holds if and only if $\|\alpha[s]\|_\mathfrak{A} \in A$. Let us call the elements of A the (properly) existing individuals (of the particular structure $\mathfrak{A}$). Then to say that $\exists x [x = \alpha]$ is true means that the value of α exists (properly). Is that not exactly what $\exists x [x = \alpha]$ ought to mean? Thus if $\forall x \Phi$ is true, it is not correct to conclude that $\Phi(x/\alpha)$ is true *unless* the value of α exists. Again, is that not quite reasonable? "To be is to be the value of a bound variable," as Quine would say.

Turning now to the descriptive operator we have first of all this valid schema:

(I1) $\quad \forall y [y = \mathbf{I} x \Phi \leftrightarrow \forall x [x = y \leftrightarrow \Phi]]$

where y is not free in Φ. In words: an existing individual is the value of a descriptive phrase if and only if it is indeed the unique individual satisfying the formula of the phrase. As a consequence of (I1) we have at once:

$$\vDash \exists y [y = \mathbf{I} x \Phi] \leftrightarrow \exists y \forall x [x = y \leftrightarrow \Phi]$$

That is, proper phrases are the only ones whose values exist. What of improper phrases? According to our definition of value they are all given the same value $*_A$. Now the term $\mathbf{I} v_0 [\neg v_0 = v_0]$ clearly is an improper descriptive phrase; call it $*$ for short. The rest of the definition of value for descriptive phrases can be expressed by the schema:

Existence and Description in Formal Logic

(I2) $\quad \neg \exists y[y = \mathbf{I}x\Phi] \to \ast = \mathbf{I}x\Phi$

The converse of this implication already follows from (I1).

One important reason for insisting that improper descriptions all assume the same improper value is to have this highly useful law of extensionality:

$\vdash \forall x[\Phi \leftrightarrow \Psi] \to \mathbf{I}x\Phi = \mathbf{I}x\Psi$

This would not be valid if one wanted 'the golden mountain' and 'the round square' to have different values. While making unkind remarks about 'the golden mountain', Russell also rejected this law of extensionality, which this authors considers an unfortunate choice. Of course, Russell was particularly interested in eliminating descriptions altogether, and we now must discuss that question.

Using (I1) and (I2) we can almost completely eliminate descriptions, because we have the schema of elimination:

(IE) $\quad \Psi(y/\mathbf{I}x\Phi) \leftrightarrow \exists y[\forall x[x = y \leftrightarrow \Phi] \wedge \Psi] \vee$
$$[\neg\exists y \forall x[x = y \leftrightarrow \Phi] \wedge \Psi(y/\ast)]$$

where the variable y is not free in Φ. Several applications of (IE) will confine all occurrences of the descriptive operator to the following contexts:

$v_i \, \mathsf{R} \, \ast \qquad \ast \, \mathsf{R} \, v_i \qquad \ast \, \mathsf{R} \, \ast$

$v_i = \ast \qquad \ast = v_i \qquad \ast = \ast$

Now, assuming that a formula Θ has no free variables, the equality formulas can be eliminated, because the last one is true, and the first two are always false in contexts where the variable v_i is bound. To be able to eliminate descriptions completely we would have to add a new schema such as:

(I3) $\quad \ast = \alpha \vee \ast = \beta \to \neg \alpha \, \mathsf{R} \, \beta$

However, this schema is not valid with our present semantics. It *is* valid in those structures $\mathfrak{A} = \langle A, R \rangle$, where $R \subseteq A \times A$. This restriction that the relation of a structure should be confined to existing individuals is not at all desirable, as we shall see when we discuss Quine's virtual classes.

We could have validated (I3) by choosing $\ast_A$ to lie outside the field of the relation R. Again this is not too desirable, because it is often felt that the valid formulas should be closed under substitution of formulas for predicate symbols. Clearly (I3) becomes invalid when R is replaced by $=$. So for pure logic we reject (I3). When giving axioms for a theory on the other hand, a schema like (I3) might be very reasonable. Then in that theory complete elimination of descriptions from sentences would be possible.

In summary the author feels that it is fair to say that the theory of descriptions presented here combines the best features of Russell's and Frege's theories. With Frege, we preserve the laws of identity and the extensionality of the descriptive operator without giving improper descriptions an unintended proper designation. Assuming the very reasonable (I3), we would be in complete agreement with Russell in nonequality atomic contexts, for from (I3) we could derive

$$\vDash \alpha \mathrel{\mathsf{R}} \mathbf{I}x\Phi \leftrightarrow \exists y[\forall x[x = y \leftrightarrow \Phi] \wedge \alpha \mathrel{\mathsf{R}} y]$$

This possibility is of course excluded by Frege.

We shall not pause here to give the proof that every valid formula can be derived from (S0)–(S4), (I1)–(I2) by the rules (MP) and (UG), because a more general completeness proof will be presented in full in a later section.

2. Virtual Classes

In *Set Theory and Its Logic* Quine (1963) makes thorough use of what he calls *virtual classes* to simplify the development and comparison of various systems of set theory. For example, the different kinds of existential assumptions about the real classes can be presented in a uniform manner in Quine's notation. More than that, with mild assumptions on real classes, the reduction of arithmetic to class theory can be very conveniently described in Quine's style; so the device has considerable appeal.

In the author's opinion, the only thing missing in Quine's presentation is a semantical analysis of the notion of virtual classes. No doubt Quine feels no need for such an analysis, since his class symbols are all eliminable by design. Virtual classes function mainly as an aid in condensing long formulas; the program is successful owing to the transfer of standard set-theoretical notions from the real to the virtual. Nevertheless, semantical insights can be helpful in understanding a formal system; especially when one can check formulas without having to first eliminate the contextually defined notions. In the presentation to be given here, virtual classes will be treated axiomatically, Quine's contextual definitions will be proved as theorems, and the model theory for the system will naturally suggest itself along the lines of what we did for descriptions.

Our language will be much like the first-order language of Section 1, except that we replace the predicate symbol R by the symbol ∈ for membership. Further we drop the descriptive operator for the time being and use instead the operator of class abstraction; thus, the terms are now either single variables or expressions of the form

$\{x:\Phi\}$

where x is a variable and Φ a formula. The construction of compound formulas proceeds as before.

As axioms and rules of inference for the theory we use (MP), (UG), and (S0)–(S4) as before, except the notions of terms and formulas must be understood in the new sense. In addition we employ three principles governing the behavior of membership and abstraction:

(Q1) $\forall y[y \in \alpha \leftrightarrow y \in \beta] \to \alpha = \beta$

(Q2) $\alpha \in \beta \to \exists y[y = \alpha]$

(Q3) $\forall y[y \in \{x:\Phi\} \leftrightarrow \exists x[x = y \wedge \Phi]]$

where the variable y is not free in α, β, or Φ. The last schema could also have been written in the form

(Q3′) $\forall y[y \in \{x:\Phi\} \leftrightarrow \Phi(x/y)]$

Combining (Q2) and (Q3) note that

(Q3″) $\alpha \in \{x:\Phi\} \leftrightarrow \exists x[x = \alpha] \wedge \Phi(x/\alpha)$

is a consequence, where x is not free in the term α.

It is quite easy to see that every theorem provable in the present theory is provable in Quine's theory. Note that (S3) and (S4) are Quine's 6.4 and 6.6. Next (Q1) follows from Quine's 2.7; (Q2) follows from 6.9 and 6.12; and (Q3) is given in Mostowski (1951, p. 17). Conversely, except for our allowing the empty domain of (real) individuals, all of Quine's theory can be deduced from ours. In particular, the contextual definitions 2.1, 2.7, and 5.5 are provable at once as biconditionals, and the lone axiom 4.1 is a special case of (S4). We are not concerned here with the additional axioms on the existence of real classes.

To discuss the models of Quine's theory, we first remark that having virtual class forces us to contemplate *many* improper individuals and not just *one* as was the case for descriptions. However, a virtual class is completely determined by its real members. So let us take as models structures of the form $\mathfrak{A} = \langle A, E \rangle$, where $E \subseteq A \times A$, and where the relation E is *extensional* in A. In other words the structure $\mathfrak{A}$ must safisfy the sentence

$$\forall x \forall x'[\forall y[y \in x \leftrightarrow y \in x'] \to x = x']$$

in the usual sense. The *elements* of A will correspond to the real classes; while the *subsets* of A will correspond to the virtual classes—well, not quite. We must identify the real classes with the virtual classes having the same members. This is best done by making the values of terms *always* be subsets of A. To get the correspondence between the elements of A and the subsets of A we define a function $\dot{E}$ on A such that for $a \in A$,

$$\dot{E}(a) = \{b \in A : \langle b, a \rangle \in E\}$$

By virtue of the extensionality of E, this is a one–one correspondence between elements of A and certain subsets of A. Next in the definition of value we make these changes:

$\vDash_{\mathfrak{A}} \alpha \in \beta[s]$ iff for some $a \in \|\beta[s]\|_{\mathfrak{A}}$, $\dot{E}(a) = \|\alpha[s]\|_{\mathfrak{A}}$

$\vDash_{\mathfrak{A}} \forall v_i \Phi[s]$ iff for all $a \in A$, $\vDash_{\mathfrak{A}} \Phi[s(i/\dot{E}(a))]$

$\|\{v_i : \Phi\}\|_{\mathfrak{A}} = \{a \in A : \vDash \Phi[s(i/\dot{E}(a))]\}$

All the other clauses remain the same. In the definition of validity we make the restriction that assignments should have *only* subsets of A as values. Even if A is empty there is one subset of A; so this restriction does not cause any trouble.

Note that if $a, b \in A$ and $s_0 = \dot{E}(a)$ and $s_1 = \dot{E}(b)$, then

$\vDash_{\mathfrak{A}} v_0 \in v_1[s]$ iff $\langle a, b \rangle \in E$

and

$\vDash_{\mathfrak{A}} v_0 = v_1[s]$ iff $a = b$

Hence, if Φ is a formula without free variables and without the abstraction operator, then Φ is true in $\mathfrak{A}$ in the new sense if and only if it is true in the sense of Section 1.

It is easily checked that all the schemata (S0)–(S4) and (Q1)–(Q3) are valid in all extensional structures. The converse, that all formulas valid in all extensional structures are provable in the present theory, will follow from the completeness theorem of Section 3. Thus we have a full explanation of the model theory for Quine's system.

When he introduced the descriptive operator into his system, Quine, the modern-day champion of contextual definition, abandoned the Russell approach in favor of Frege's idea. That seems a bit odd, does it not? The explanation is probably this: with Russell's elimination the formula $\alpha = \alpha$

is not always valid, whereas Quine wants this law of equality. Besides, it is a waste of effort to introduce new operators by contextual definition when an explicit definition is at hand. Quine chose this definition (more or less):

$$\mathbf{I}x\Phi = \{x\colon \exists y[\forall x[x = y \leftrightarrow \Phi] \wedge z \in y]\}$$

Thus when $\neg\exists y\forall x[x = y \leftrightarrow \Phi]$ holds, $\mathbf{I}x\Phi$ denotes the empty class. Since Quine wants the empty class to be a real class, we see that the improper description is behaving in the Fregean manner.

Another definition was open to Quine, however, namely:

$$\mathbf{I}x\Phi = \{z\colon \exists y[\forall x[x = y \leftrightarrow \Phi] \wedge z \in y] \vee \\ [\neg\exists y\forall x[x = y \leftrightarrow \Phi] \wedge \neg z \in z]\}$$

If we let Δ be the term $\{z\colon \neg z \in z\}$, then by the argument of the Russell paradox we can prove that

$$\neg\exists y[y = \Delta]$$

Hence, with the revised definition we prove exactly (I1) and (I2) with $*$ replaced by Δ. Of course $* = \Delta$ is at once provable, so the former theory is recaptured. The author strongly feels that this path to descriptions is much more in harmony with the concept of virtual classes than is the version adopted by Quine. (See, however, Professor Quine's remarks quoted at the end of this section.)

Let us see now what happens to the elimination of descriptions with the definition just proposed. In view of (Q2) we have first:

$$\mathbf{I}x\Phi \in \beta \leftrightarrow \exists y[\forall x[x = y \leftrightarrow \Phi] \wedge y \in \beta]$$

where y is not free in Φ or β. That would please Russell. In the other argument place we have:

$$\alpha \in \mathbf{I}x\Phi \leftrightarrow \exists y[\forall x[x = y \leftrightarrow \Phi] \wedge a \in y] \vee \\ [\neg\exists y\forall x[x = y \leftrightarrow \Phi] \wedge \exists y[y = \alpha \wedge \neg y \in y]]$$

That would probably confound followers of Russell or Frege; the author hopes it is not too displeasing to the Quine school, however.

Parallel to the revision of the definition of descriptions, the author would like to also suggest a revision of Quine's definition of *function value*. Let us assume along with Quine enough axioms to guarantee the existence of ordered pairs of real classes. In the author's notation, the definition of function value will read:

$$\phi(\xi) = \mathbf{I}y\exists x[x = \xi \wedge \langle y, x \rangle \in \phi]$$

where x and y are variables not free in ϕ and ξ. Quine gave as his definition:

$$\phi(\xi) = \mathbf{I}y[\langle y, \xi \rangle \in \phi]$$

This is "defective" not only because the wrong kind of description was used, but because when ξ does not exist (i.e., $\neg \exists x[x = \xi]$), then

$$\langle y, \xi \rangle = \{\{y\}, \{y, \xi\}\} = \{\{y\}\} = \langle y, y \rangle$$

By chance there might be a unique y with $\langle y, y \rangle \in \phi$, and we are uncomfortable. Now Quine avoids this unpleasantness by restricting attention to the class

$$\arg \phi = \{x: \exists z \forall y[y = z \leftrightarrow \langle y, x \rangle \in \phi]\}$$

Using the proposed *new* definition we can simplify this last equation to

$$\arg \phi = \{x: \exists y[y = \phi(x)]\}$$

Further, there is no need to avoid unintended function values, for we can prove quite generally

$$\exists y[y = \phi(\xi)] \leftrightarrow \xi \in \arg \phi$$

This last biconditional reads so well that it seems justification enough for the revised definition. (This discussion of function value is improved over an earlier version at the suggestion of David Kaplan.) The general principle to be applied to such questions is this: things should exist only when it is intended that they exist. It seems quite remarkable to the author that there is a flexible enough formalism that actually allows us to follow this principle.

In connection with these suggestions, Professor Quine wrote to the author on May 3, 1965, as follows:

> The redefinition of description and of function value that you propose sacrifices an advantage that I had gone out of my way for: the freedom to substitute descriptions for bound variables without regard to special existence premisses. This freedom, touched on in pp. 58, 68, and 107 of *Set Theory and Its Logic*, covers a lot, since so many notations are defined as function values and ultimately as descriptions. It even covers function values where function and argument are rendered by Greek letters, without presumption of existence; cf. p. 68. Also it covers arithmetical expressions containing Greek letters; cf. p. 107. Without this freedom the book would be appreciably more labored. Perhaps you could devise alternative conventions, on your basis, that would work smoothly too; but then I'd want to see some trial runs for comparison.

Professor Quine is quite justified in asking for trial runs for comparison, and the author will try to apply these comparisons in future publications. For the time being the reader is asked to consider the merits of the proposal on the grounds of "naturalness" as indicated above. He should also imagine having to make all existential assumptions explicit, and ask himself whether unrestricted substitution is to be preferred over the gain of information obtained by using formulas with explicitly displayed assumptions.

3. General Operators

The system that we shall treat here will be of the same type as the systems of Sections 1 and 2. The language will involve a binary predicate symbol R and a variable binding operator O of the same syntactical category as the operators of description and abstraction. Thus,

$$Ox\Phi$$

is a term when x is a variable and Φ is a formula. As before we shall not assume that the values of terms are necessarily in the range of the individual variables. A convenient way to express this is to consider structures of the form

$$\mathfrak{A} = \langle A, A_*, R, O \rangle$$

where A is a set (the domain of properly existing individuals), A_* is a *nonempty* superset of A (the domain of "improper" individuals), R is a binary relation where $R \subseteq A_* \times A_*$, and O is a function defined on *subsets* of A taking values in A_*. The definition of value is now modified in this particular:

$$\|Ov_i\Phi[s]\|_\mathfrak{A} = O(\{a \in A : \vDash_\mathfrak{A} \Phi[s(i/a)]\})$$

Further, the definition of validity is changed so that $\vDash_\mathfrak{A} \Phi$ means that $\vDash_\mathfrak{A} \Phi[s]$ only for assignments s where the values of the function s are *all* in the set A_*. When the set A_* is explicitly mentioned it is not reasonable to allow the values of assignments to the free variables to be completely arbitrary.

If $\mathfrak{A} = \langle A, R \rangle$ is a structure in the sense of Section 1, then we can correlate with it a structure in the new sense, namely:

$$\mathfrak{A}_* = \langle A, A_*, R, I \rangle$$

where

$$A_* = A \cup \{*_A\} \cup \text{field}(R)$$

and where I is defined on subsets $X \subseteq A$ so that

$$I(X) = \begin{cases} a & \text{if } X = \{a\}, \\ *_A & \text{if } X \neq \{a\} \text{ for all } a \in A \end{cases}$$

It is then easy to prove that for an assignment s with values in A_*, $\vDash_\mathfrak{A} \Phi[s]$ holds in the old sense if and only if $\vDash_{\mathfrak{A}^*} \Phi[s]$ holds in the new sense, and that $\|\alpha[s]\|_\mathfrak{A} = \|\alpha[s]\|_{\mathfrak{A}^*}$. Of course the symbol O should be replaced by I to make sense of this last statement.

If $\mathfrak{A} = \langle A, E \rangle$ is a structure in the sense of Section 2 then the correlated structure is

$$\mathfrak{A}_0 = \langle A_0, A_*, E_0, J \rangle$$

where $A_0 = \{\dot{E}(a): a \in A\}$, A_* is the set of all subsets of A, J is the identity function on A_*, and the relation E_0 is defined for $X, Y \in A_*$ so that

$$X E_0 Y \text{ iff for some } a \in Y, \dot{E}(a) = X$$

Again, for assignments with values in A_*, the old and new definitions of value agree completely. Therefore, the structures considered here do properly generalize those used in earlier examples.

Aside from (S0)–(S4), which are all valid in the present sense, we have also

(O1) $\quad [\forall x[\Phi \leftrightarrow \Psi] \rightarrow Ox\Phi = Ox\Psi]$

and

(O2) $\quad Ox\Phi = Oy\Phi(x/y)$

where the variable y is not free in Φ.

These kinds of schemata were not needed explicitly in Sections 1 and 2 because the required formulas were in each case deducible from the others given. In the general case they are the only schemata required that involve the operator O in a special way. We want now to show that a formula is universally valid if and only if it is deducible from (S0)–(S4), (O1), (O2) by the rules (MP) and (UG). Let $\vdash \Phi$ mean that the formula Φ is so deducible. All we need to prove is that if not $\vdash \Phi$, then there is a structure $\mathfrak{A}$ and an assignment s such that not $\vDash_\mathfrak{A} \Phi[s]$.

To this end let Φ_0 be a particular formula such that not $\vdash \Phi_0$. Let

$$\Psi_0, \Psi_1, \ldots, \Psi_n, \ldots$$

be a list containing *every* formula at least once such that Ψ_0 is $\neg \Phi_0$, and if Ψ_n is of the form

$$\neg \forall v_i \Phi$$

then Ψ_{n+1} is of the form

$$\exists v_{j+1}[v_{j+1} = v_j] \wedge \neg \Phi(v_i/v_j)$$

where v_j is the first variable not free in $\Psi_0, \Psi_1, \ldots, \Psi_n$. It is easy to show that such a sequence exists. We define by recursion the sequence of formulas

$$\Psi'_0, \Psi'_1, \ldots, \Psi'_n, \ldots$$

where Ψ'_n is $\neg \Psi_n$ or Ψ_n according as

$$\vdash [\Psi'_0 \wedge \cdots \wedge \Psi'_{n-1} \to \neg \Psi_n]$$

or not.

We let

$$M = \{\Psi'_n : n \in N\}$$

Clearly $\Psi_0 \in M$ and the set M of formulas has these properties:

(i) if $\vdash \Phi$, then $\Phi \in M$
(ii) $\neg \Phi \in M$ iff $\Phi \notin M$
(iii) $[\Phi \to \Psi] \in M$ iff $\Phi \notin M$ or $\Psi \in M$
(iv) $\forall v_i \Phi \in M$ iff for all j if $\exists v_{j+1}[v_{j+1} = v_j] \in M$, then $\Phi(v_i/v_j) \in M$

So far the details of the proof are just as in any standard version as the completeness proof for first-order logic based on the method due to Henkin.

Let T be the set of all terms, and define an equivalence relation $\equiv$ on the set T by the condition that

$$\alpha \equiv \beta \text{ iff } [\alpha = \beta] \in M$$

The equivalence class of a term α is denoted by $\alpha/\equiv$. We define

$$A = \{v_j/\equiv \, : \, \exists v_{j+1}[v_{j+1} = v_j] \in M\}$$

and

$$A_* = \{\alpha/\equiv \, : \, \alpha \in T\}$$

The relation $R \subseteq A_* \times A_*$ is defined by the equation

$$R = \{\langle \alpha/\equiv, \beta/\equiv \rangle : [\alpha \, \mathsf{R} \, \beta] \in M\}$$

and the operator O is defined for $X \subseteq A$ so that

$$O(X) = \begin{cases} Ov_i\Phi/\equiv & \text{if } X = \{v_j/\equiv \in A : \Phi(v_i/v_j) \in M\} \\ v_0/\equiv & \text{if there is no such formula } \Phi \end{cases}$$

For the particular assignment s where $s_i = v_i/\equiv$, we wish to show that $\vDash_{\mathfrak{A}} \Phi_0[s]$ does not hold where $\mathfrak{A} = \langle A, A_*, R, O \rangle$ is the structure just defined. This cannot be done quite directly: one must prove by induction that if Φ is a formula, α is a term, and s is an assignment where $s_i = \alpha_i/\equiv$, then

$$\vDash_{\mathfrak{A}} \Phi[s] \text{ iff } \Phi(v_0/\alpha_0, v_1/\alpha_1, \ldots, v_n/\alpha_n, \ldots) \in M$$

and

$$\|\alpha[s]\|_{\mathfrak{A}} = \alpha(v_0/\alpha_0, v_1/\alpha_1, \ldots, v_n/\alpha_n, \ldots)/\equiv$$

where the notation on the right-hand sides indicates simultaneous substitution of terms for free variables. Again this step is just like the corresponding step in the usual proofs, and conditions (i)–(iv) on M were explicitly chosen so that the argument would work out.

In case the additional axioms of Sections 1 or 2 were added, the structure $\mathfrak{A}$ just obtained could be modified directly to obtain the structure in the earlier sense that is required.

4. Eliminability

A *sentence* is a formula without free variables. A *theory* is a set of sentences containing all universally valid sentences and closed under the rule of modus ponens. The operator O is *eliminable* in a theory T, if for each formula Φ there is a formula Ψ not containing O such that

$$\forall v_0 \forall v_1 \cdots \forall v_{m-1}[\Phi \leftrightarrow \Psi]$$

belongs to T, where the free variables of Φ and Ψ are among $v_0, v_1, \ldots, v_{m-1}$.

The theory based on schemata (I1)–(I3) (where O replaces the symbol I) is a theory in which O is eliminable. Similarly for the theory based on (Q1)–(Q3) (where O replaces the abstraction operator). The purpose of this section is to give necessary and sufficient model-theoretic conditions for O to be eliminable in a theory T. The conditions found will be very close to those of Beth's Definability Theorem (cf. Robinson 1963).

If $\mathfrak{A} = \langle A, A_*, R, O \rangle$ and $\mathfrak{A}' = \langle A', A'_*, R', O' \rangle$ are two structures, we say that $\mathfrak{A}$ and $\mathfrak{A}'$ are *weakly isomorphic* if there is a one–one function mapping the set A on to the set A' such that for all $a, b \in A$,

$$\langle a, b \rangle \in R \text{ iff } \langle f(a), f(b) \rangle \in R'$$

When s is an assignment with values in A, we let $f((s))$ denote the assignment with values in A' such that

$$f((s))_i = f(s_i)$$

The condition of f to give a weak isomorphism can be equivalently stated as

$$\vDash_{\mathfrak{A}} \Phi[s] \text{ iff } \vDash_{\mathfrak{A}} \Phi[f((s))]$$

for all assignments s with values in A and all formulas Φ not containing the operator O. We shall say that f gives a *strong isomorphism* if this last biconditional holds for *arbitrary* formulas Φ.

A *model* for a theory is of course a structure for which all sentences of the theory are true. We can now state the theorem on eliminability:

The operator O *is eliminable in a theory* T *if and only if whenever two models of* T *are weakly isomorphic by a certain one–one function, they are also strongly isomorphic by the same function.*

If O is eliminable in T, then it is clear that weak isomorphism implies strong isomorphism. The converse will be proved by applying Beth's theorem to a suitable first-order theory with many predicate symbols but without operators.

Let T be a theory for which weak isomorphism implies strong isomorphism. Introduce new predicate symbols S^{Φ} corresponding to each formula Φ in the original sense. The predicate S^{Φ} will be an m-place predicate, where m is the least integer such that the free variables of Φ are among $v_0, v_1, \ldots, v_{m-1}$. Consider the extension of T obtained by adjoining these sentences as axioms:

$$\forall v_0 \forall v_1 \cdots \forall v_{m-1} [S^{\Phi}(v_0, v_1, \ldots, v_{m-1}) \leftrightarrow \Phi]$$

The theory T_0 is the set of sentences of the extension of T involving the predicates R and S^{Φ} but *not* O. It is obvious that O is eliminable in T if and only if *all* the S^{Φ} are definable in T_0 in terms of the predicate R in the ordinary sense of first-order definability.

According to Beth's theorem, to show that the S^{Φ} are definable in terms of R it is enough to show that two models

$$\mathfrak{A}_0 = \langle A, R_0, \ldots, S^{\Phi}, \ldots \rangle$$

and
$$\mathfrak{A}'_0 = \langle A', R'_0, \ldots, S'^\Phi, \ldots \rangle$$
of T_0, where $\langle A, R_0 \rangle$ and $\langle A', R'_0 \rangle$ are isomorphic by a function f, are also isomorphic by the same function f. To prove this we will construct structures
$$\mathfrak{A} = \langle A, A_*, R, O \rangle$$
$$\mathfrak{A}' = \langle A', A'_*, R', O' \rangle$$
such that for all assignments s with values in A and for all formulas Φ
$$\vDash_{\mathfrak{A}_0} S^\Phi(v_0, v_1, \ldots, v_{m-1})[s] \text{ iff } \vDash_{\mathfrak{A}} \Phi[s]$$
Similarly for $\mathfrak{A}'_0$ and $\mathfrak{A}'$. Now by assumption $\langle A, R_0 \rangle$ and $\langle A', R'_0 \rangle$ are isomorphic. Hence $\mathfrak{A}$ and $\mathfrak{A}'_0$ are weakly isomorphic; therefore strongly isomorphic. But this means that $\mathfrak{A}_0$ and $\mathfrak{A}'_0$ are isomorphic, all by the same function we started with. It will be enough to show how to construct $\mathfrak{A}$ from $\mathfrak{A}_0$; actually it will be easier to construct a structure $\bar{\mathfrak{A}}$ which is strongly isomorphic to the structure we want.

First let U be the set of all pairs $\langle \alpha, s \rangle$ where α is a term of the original language and s is an assignment with values in A. We define an equivalence relation $\equiv$ on the set U:
$$\langle \alpha, s \rangle \equiv \langle \beta, t \rangle \text{ iff } \vDash_{\mathfrak{A}_0} S^{\alpha = \beta'}(v_0, \ldots, v_{m-1}, v_m, \ldots, v_{m+n-1})[u]$$
where m is the least integer such that the free variables of α are among $v_0, \ldots, v_{m-1}$; n is the least integer such that the free variables of β are among $v_0, \ldots, v_{n-1}$; β' is the term $\beta(v_0/v_m, \ldots, v_{n-1}/v_{m+n-1})$; and u is the assignment where
$$u_i = \begin{cases} s_i & \text{if } i < m \\ t_{i-m} & \text{if } i \geq m \end{cases}$$
We let $\langle \alpha, s \rangle / \equiv$ be the equivalence class of $\langle \alpha, s \rangle$ in U and put
$$\bar{A} = \{\langle \alpha, s \rangle / \equiv \, : \, \langle \alpha, s \rangle \in U\}$$
and
$$\bar{A} = \{\langle v_0, s \rangle / \equiv \, : \, \langle v_0, s \rangle \in U\}$$
We note that A and $\bar{A}$ are in a one-one correspondence by the function e such that for $a \in A$,
$$e(a) = \langle v_0, s \rangle / \equiv$$

for all assignments s with $s_0 = a$. The relation $\bar{R}$ on $\bar{A}_*$ is such that

$$\langle\langle\alpha, s\rangle/\equiv, \langle\beta, t\rangle/\equiv\rangle \in \bar{R} \text{ iff}$$

$$\vDash_{\mathfrak{A}_0} S^{\alpha R \beta'}(v_0, \ldots, v_{m-1}, v_m, \ldots, v_{m+n-1})[u]$$

where m, n, β', and u are determined as before. Finally to define $\bar{O}$, we let r be a fixed assignment with values in A and for $X \subseteq \bar{A}$ we set

$$\bar{O}(X) = \begin{cases} \langle Ov_i\Phi, s\rangle/\equiv & \text{if } X = \{e(a) \in \bar{A} : \vDash_{\mathfrak{A}_0} S^{\Phi}(v_0, \ldots, v_{m-1})[s(i/a)]\} \\ \langle v_0, r\rangle/\equiv & \text{if there is no such formula } \Phi \text{ and assignment } s \end{cases}$$

The desired properties of the structure

$$\bar{\mathfrak{A}} = \langle \bar{A}, \bar{A}_*, \bar{R}, \bar{Q}\rangle$$

will be established by proving for all formulas Φ, all terms α, and all assignments s with values in A that

$$\vDash_{\mathfrak{A}} \Phi[e((s))] \text{ iff } \vDash_{\mathfrak{A}_0} S^{\Phi}(v_0, \ldots, v_{m-1})[s]$$

and

$$\|\alpha[e((s))]\|_{\mathfrak{A}} = \langle \alpha, s\rangle/\equiv$$

This result on eliminability is not very satisfactory. The operators of Sections 1 and 2 are eliminable in a much stronger sense: for example, the schema (IE) gives practically a wholesale way of eliminating the descriptive operator. Similar things may be said for Quine's abstraction operator. In other words to eliminate an occurrence of $Ix\Phi$ we need only examine the context in which this term is found; we do not have to make our elimination depend on any peculiarities of the formula Φ within the scope of the operator. The author has no idea what kind of model-theoretic conditions would correspond to this *uniform* eliminability that we always have when operators are introduced by contextual definitions. It seems to be an interesting problem.

References

Bernays, P. and Fraenkel, A. A., *Axiomatic Set Theory*, North-Holland, Amsterdam (1958).
Hailperin, T. and Leblanc, H., "Non-designating singular terms," *Philosophical Review*, vol. 68 (1959) pp. 239–243.
Hilbert, D. and Bernays, P., *Grundlagen der Mathematik*, Bd. I (1934), Bd. II (1939).

Hintikka, J., "Existential presuppositions and existential commitments," *Journal of Philosophy*, vol. 56 (1959a) pp. 125–137.

Hintikka, J., "Towards a theory of definite descriptions," *Analysis*, vol. 19 (no. 4) (1959b) pp. 79–85.

Mostowski, A., "On the rules of proof in the pure functional calculus of the first order," *J. of Symbolic Logic*, vol. 16 (1951) pp. 107–111.

Quine, W. V., *Set Theory and Its Logic*, Harvard University Press (1963).

Rescher, N., "On the logic of existence and denotation," *Philosophical Review*, vol. 69 (1959) pp. 157–180.

Robinson, A., *Introduction to Model Theory and to the Metamathematics of Algebra*, North-Holland, Amsterdam (1963).

Smiley, T., "Sense without denotation," *Analysis*, vol. 20 (no. 4) (1960) pp. 125–135.

3
A Free IPC is a Natural Logic: Strong Completeness for Some Intuitionistic Free Logics

Carl J. Posy

Abstract

IPC, the intuitionistic predicate calculus, has the property

(i) $Vc(\Gamma \vdash Ac/x) \Rightarrow \Gamma \vdash \exists x A$

Furthermore, for certain important Γ, IPC has the converse property

(ii) $\Gamma \vdash \exists x A \Rightarrow Vc(\Gamma \vdash A^c/x)$

(i) may be given up in various ways, corresponding to different philosophic intuitions and yielding different systems of intuitionistic free logic. The present paper proves the strong completeness of several of these with respect to Kripke-style semantics. It also shows that giving up (i) need not force us to abandon the analog of (ii).

0. Introduction

A. Intuitionism

One goal of intuitionistic mathematics is the elimination of nonconstructive inferences from mathematical practice. Yet the usual intuitionistic logic as formulated by Heyting (see Heyting 1930), does validate intuitively nonconstructive inferences. Consider, for instance, the inference from $t = t$ to $\exists x(x = t)$ which is valid in Heyting's logic. Or imagine the situation in which $\exists x Px$ has not yet been decided, though $\exists x Px \to \exists! x Px$ has been established. (This will occur, e.g., when Px is $(f(t) = x)$ and f stands for a function which is not known to be defined at the argument represented by t). In this circumstance, clearly the intuitionist ought to be free to accept the inference from $\neg\neg\exists xPx$ to $P(\iota xPx)$ without falling into the trap (which would be imposed by Heyting's logic) of thereby concluding $\exists xPx$.

Current formalizations of intuitionism disarm such inferences by limiting all function terms to range over provably total function and generally restricting singular terms to those which are guaranteed to denote. (See, e.g., Kleene and Vesley 1965; Myhill 1968; Troelstra 1973, pp. 15, 17; and Stenlund 1975.) However, not only does this technique overly constrain the syntax of any intuitionistic language, it does not conform to Brouwer's actual intuitionistic practice. Thus, for instance, Brouwer (e.g., Brouwer 1924, 1975) explicitly introduces singular terms which are not known to denote any intuitionistic object; and certainly a major thrust of Brouwer's writing is the claim that many expressible real-valued functions are not total. For this reason I suggest that a natural logic for intuitionism is a "free-IPC": a logic derived from a quantification theory that is free of existential assumptions and that is built on an intuitionistic propositional base. In what follows I will discuss the semantics for several different free-IPCs.

B. Free Semantics

In fact free IPCs are doubly natural; for the intuitionistic bias provides a single framework to express and even combine the intuitively prominent elements of some competing approaches to the semantics of ordinary free logic. Specifically there are four main schools of thought concerning semantics for sentences with nondesignating singular terms. Each position has its own strengths and weaknesses, and the intuitionistic style of semantics can simultaneously accommodate the best features of each. In more detail as follows.

1. The *outer domain* approach combines the belief that the property of truth is derived from the basic relation of predication[1] with the view that there may be logically contingent truths containing nondenoting terms. It merges these two beliefs by assigning to such terms a referent which is outside the range of the variables (i.e., a nonexistent object). The domain of the actual world consists of the union of an inner domain (of actually existing things) with this outer domain of nonexistents. (See Leblanc and Thomason 1968.)

All the remaining approaches question the ontological status of such nonexisting objects, and refuse to base any model theory upon them.

2. The second approach, called by van Fraassen the method of *classical valuations* (see van Fraassen 1966) surrenders the universal dependence of truth on predication and simply *assigns* truthvalues to atomic sentences containing nondenoting singular terms. One may limit the arbitrariness of such assignments: for instance, one may insist that '$t = t$' always be made true, or that particular natural statements like 'Pegasus is a horse' come out true. But in these cases the limits are matters of "convention" and are not reflected in the model theory itself. (See van Fraassen 1966.) In any event, though, no matter how arbitrary the assignments, logical theorems (not involving identity) will always come out true.
3. The third approach, developed by R. Schock (see Schock 1968), once again adheres to the primacy of predication, and as a result simply declares any atomic sentence with a nondenoting singular term to be false. On this approach as well, logical theorems, even those containing nondenoting terms, will come out true (see also Burge 1974).
4. The *partial valuation* approach also maintains allegiance to predication and to the view that nothing (atomic) can be true of a nonexistent object. But in this case the consequence of these views is simply to withhold any truthvalue at all from atomic sentences with singular terms lacking denotations. Ordinarily on this approach compound sentences obtain their truthvalues in the standard way from their atomic parts; so those compound sentences containing nondenoting terms will inherit the truthvalueless status of the atomic parts in which those terms occur. However, van Fraassen, in his method of *supervaluations* had provided a means of assuring the truth of the theorems of (free) logic even under a partial valuation approach (see van Fraassen 1966).

The partial valuation approach, even when augmented by supervaluations, does not admit the possible truth of logically contingent sentences with nondenoters. To be sure, one may guarantee the truth of some such sentences by adopting conventions as in (2). (van Fraassen argues on conceptual grounds that the validity of '$t = t$' be preserved in this way.) But once again, in order to make this move, one must renounce the primacy of predication. Moreover, the standard free logics are not strongly complete (argument complete) with respect to the semantics of supervaluations: There are B and A such that $B \Vdash A$ under the semantics of supervaluations but $B \nvdash A$ in the deductive systems of such logics. (For instance, for Pt atomic, $Pt \Vdash \exists x(x = t)$ under this semantics, since under the partial valuation approach 'Pt' could only be true if 't' denotes; but in most of these logics $\vdash Pt \to \exists x(x = t)$ and $Pt \vdash \exists x(x = t)$ do not hold.[2]

The general semantics for free IPCs that I will propose below is designed to preserve an allegiance to the claim that truth must be based on predication, so it is adaptable to approaches (1), (3), or (4). But it also eliminates some of the drawbacks of each. Thus, for instance, since the modeling is intuitionistic we are not forced—even under the outer domain approach—to attribute an "actual world" truthvalue to contingent statements containing nondenoting singular terms. In the intuitionistic semantics, attributing the undesignated value to a formula at a world need not connote falsity, but merely lack of knowledge or evidence. On the other hand, like (1) and (2), the semantic approach sketched below, need not deny truth to all atomic sentences with nondenoting terms. Indeed, the approach of this paper provides a versatile medium for expressing competing intuitions about the behavior of $(t = t)$ and generalizes it (if desired) to allow other semantically necessary atomic truths involving nondenoting terms without sacrificing the dependence of these truths on predication. (I shall call these truths "analytic" in a loose sense.) In this respect the present technique is similar to Bencivenga's approach (Bencivenga 1980). It differs from that, though, in avoiding the need to assign truthvalues to atomic sentences at a world in a manner inconsistent with the naive description of that world.[3] The major device of this paper for dealing with such truths may be called (after Cocchiarella 1975) a "doubly secondary" semantics: Kripke models that make use of inaccessible nodes. This device has the added formal advantage of generating strong completeness theorems for the partial valuation as well as the outer domain versions of the semantics.

C. Free IPC

There are already three free intuitionistic logics in the literature: an (inadvertent) one of Thomason's (Thomason 1965);[4] one devoted to the logic of definite descriptions due to Stenlund (1975); and the logic of Leblanc and Gumb (1981). Each of the first two has only a limited degree of freedom from existential assumptions; and the third, though comparable in strength to the present approach, does not base truth on predication. Specifically, one can grade a free modeling relation according to some of the standard existential inferences that it invalidates:

(a) It may invalidate $At \Vdash \exists x(x = t)$ for all A.
(b) It may invalidate $At \Vdash \exists x(x = t)$ for some class of matrices but still always validate $Pt \Vdash \exists x(x = t)$ (Pt atomic, but other than $(t = t)$).
(c) It may invalidate even $Pt \Vdash \exists x(x = t)$.

Formally these are arranged in order of increasing "freedom." Heuristically a modeling relation of grade (a) holds that true sentences contain only denoting terms. Sentences with nondenoting terms must be truthvalueless. Stenlund's semantics is of this grade.[5] (b) allows that there may be a class of formulas that are free of existential assumptions, but excludes contingent atomic formulas from that class. Thomason's semantics is of this grade. It invalidates, for example, $\neg At \Vdash \exists x(x = t)$ and similar inferences. When supplemented by Notes 3 and 5, Thomason (1965) invalidates $t = t \Vdash \exists x(x = t)$.[6] Semantics of grade (c) allow formally contingent atomic truths, with nondenoting singular terms.

The semantics of Leblanc and Gumb (1981) is of grade (c). However, it achieves this by adopting a substitutional approach. Thus when (atomic) Pt is true, it is true simply by stipulation. To be sure, that will be a stipulation that is independent of the assignments governing $\exists x(x = t)$; nevertheless, the conditions governing Pt have no connection with the intuitive criteria of predication. In particular, to assure the validity of $(t = t)$, Leblanc and Gumb, like van Fraassen, resort to convention: simply building $(t = t)$ into the structure of an acceptable valuation.

In contrast, the general approach sketched below provides nonsubstitutional procedures for achieving a semantics of grade (c). Such a procedure can include letting the denotation function be undefined for some terms. I shall call that a "Russellian" approach and discuss it in Section 2 below. The Russellian treatment, however, commits us to the view that atomic sentences with nonreferring terms are naturally truthvalueless. So Section 1 will look at a more general attitude: The "Meinongian" or outer domain approach.[7] We shall see that the flexibility of the general framework will allow even the Meinongian approach to be adapted to reflect the intuitions of those who are uncomfortable with this approach in the classical setting.

The reader may gain familiarity with the standard treatment of Kripke models for intuitionistic logic from Smorynski (1973). However, for convenience, I have outlined the usual syntax and semantics of IPC in Appendix 1. In the text I will sketch only the heuristics of the longer proofs, providing more details in Appendix 2.

With but few exceptions this paper concentrates on the theory of unanalyzed terms.

1. The Meinongian Approach

A. Basic Semantics

A morphology, L, is a structure consisting of a set of logical symbols: $\rightarrow$,

¬, ∨, &, ∃, ∀, and $E!$; a denumerable set, V_L, of variables; a countable set, C_L, of constants, and for each finite i a set of i-ary predicate letters, P^i. W_L is the set of well-formed formulas of L.

A Meinongian frame (mf) is a structure $\langle K, R, \alpha_0, D, E \rangle$, where K is a nonempty set, R is a reflexive and transitive relation on K, α_0 is a minimal element under R, D is a nonempty set, and E is a function assigning to each α in K a characteristic function $E_\alpha: D \to \{e, n\}$. E has the property:

$$\Lambda\alpha_{\in K}\Lambda\beta_{\in K}\Lambda d_{\in D}[(\alpha \text{ R } \beta \text{ and } E_\alpha(d) = e) \Rightarrow E_\beta(d) = e]$$

('Λ', 'V', and '⇒' are metasymbols standing for the metalanguage notions of universal quantification, existential quantification, and implication, respectively). Heuristically e represents existence, and n nonexistence. $\{d \mid E_\alpha(d) = e\}$ is the "inner domain" at α; $\{d \mid E_\alpha(d) = n\}$ is the "outer domain" at α. Notice that the presence of E allows us to work with a constant domain for the model structure, an unusual convenience for intuitionistic Kripke models. Notice also that α_0 is described as a minimal element; it need not be the only minimal element, though, of course, any additional minimal element is not related to α_0 by R.

An interpretation, I, of L on $\langle K, R, \alpha_0, D, E \rangle$ is a function with the following properties:

(i) $\Lambda x \in V_L[I(x) \in D]$
(ii) $\Lambda c \in C_L[I(c) \in D]$
(iii) For each P^0, $I(P^0)$ is a function from K to $\{1, 0\}$ such that $[(\alpha \text{ R } \beta \text{ and } I_\alpha(P^0) = 1) \Rightarrow I_\beta(P^0) = 1]$.
(iv) For each $P^i (i \geq 1)$, $I(P^i)$ is a function from K to $2^{(D)}$ such that $[\alpha \text{ R } \beta \Rightarrow I_\alpha(P^i) \subseteq I_\beta(P^i)]$.

In addition to clause (iii), I induces an assignment of semantic values to the formulas of W_L at each $\alpha \in K$ as follows [t and t_j will range over $V_L \cup C_L$, and $I_\alpha(A) \in \{1, 0\}$]:

(0a) $I_\alpha(P^i t_1 \cdots t_i) = 1$ iff $\langle I(t_1) \cdots I(t_i) \rangle \in I_\alpha(P^i)$

[Notice that (0a), the basic predication clause, is insensitive to $E_\alpha(I(t_j))$, $1 \leq j \leq i$.]

(0b) $I_\alpha(E!t) = 1$ iff $E_\alpha(I(t)) = e$

Clauses (1)–(4) for the propositional connectives are standard. [See Appendix 1 for details.]

(5) $I_\alpha(\forall x B) = 1$ iff $\Lambda\beta_{\in K}[\alpha \text{ R } \beta \Rightarrow \Lambda d_{\in D}(E_\beta(d) = e \Rightarrow (Id/x)_\beta(B) = 1)]$

(6) $I_\alpha(\exists xB) = 1$ iff $\mathrm{V}d_{\in D}[E_\alpha(d) = e$ and $(Id/x)_\alpha(B) = 1]$

[Id/x is the interpretation that is like I with the possible exception that it assigns $d \in D$ to x.]

Lemma 1. $I_\alpha(At/x) = (II(t)/x)_\alpha(A)$.

Proof. By induction on A. ●

Definitions

Given L, a (naive) *Meinongian model* is a pair $\langle\langle K, R, \alpha_0, D, E\rangle, I\rangle$, where I interprets L on the mf $\langle K, R, \alpha_0, D, E\rangle$.

A is *satisfied* by the model if $I_{\alpha_0}(A) = 1$.

A set $\Gamma \subseteq W_L$ is *simultaneously satisfied* by the model if $I_\alpha(A) = 1$ for all $A \in \Gamma$.

$\Gamma \Vdash A$ if, for every Meinongian model $\langle\langle K, R, \alpha_0, D, E\rangle, I\rangle$, $(I_{\alpha_0}(B) = 1$ for all $B \in \Gamma) \Rightarrow I_{\alpha_0}(A) = 1$.

A is *valid* ($\Vdash A$) if $I_{\alpha_0}(A) = 1$ for every Meinongian model $\langle\langle K, R, \alpha_0, D, E\rangle, I\rangle$. (I.e., $\varnothing \Vdash A$.)

B. Syntax

Deductive Structure. We need not concern ourselves with the details of particular formalizations of any of the free-IPCs. It will suffice to attend to the overall features of the deducibility relation. '⊢' will stand for the deducibility relation in the basic version of free-IPC. In expressing its features '$\Gamma \vdash A$' will mean that A is deducible from $\Gamma (\subseteq W_L)$. A horizontal line indicates that if the deducibility relation described above it holds, so too does the one described below. ⊢, the basic relation, is the standard intuitionistic deducibility relation (see Appendix 1) with the following changes:

($\forall I$) is replaced by $F(\forall I)$:

$$\frac{\Gamma \cup \{E!t\} \vdash At/x}{\Gamma \vdash \forall xA}$$

t not free in A or any element of Γ.

($\forall E$) is replaced by $F(\forall E)$:

$$\frac{\Gamma \vdash \forall xA}{\Gamma \cup \{E!t\} \vdash At/x}$$

($\exists I$) is replaced by $F(\exists I)$:

$$\frac{\Gamma \vdash At/x}{\Gamma \cup \{E!t\} \vdash \exists xA}$$

($\exists E$) is replaced by $F(\exists E)$:

$$\frac{\Gamma \vdash \exists xA \quad \Gamma \cup \{At/x\} \cup \{E!t\} \vdash B}{\Gamma \vdash B}$$

t not free in A, B or any element of Γ.

The following general condition is derivable:

($\forall E!$): $\emptyset \vdash \forall x E!x$

Free-Saturated Sets. Definition: A set, Γ, of formulas is consistent if for some A, not $\Gamma \vdash A$.

Lemma 2. (a) $[\Gamma \text{ consistent} \Rightarrow \Gamma \nvdash A \, \& \, \neg A]$ for all $A \in W_L$.
(b) Γ *consistent* $\Rightarrow$ *at least one of* $\Gamma \cup \{A\}$ *or* $\Gamma \cup \{\neg A\}$ *is consistent.* ●

Definition. A free-L-saturated set is a set $\Gamma \subseteq W_L$, with the following properties:

(1) Γ is consistent.
(2) Γ is deductively closed. [I.e., $\Gamma \vdash A \Rightarrow A \in \Gamma$.]
(3) For A and $B \in W_L$, $A \vee B \in \Gamma \Rightarrow A \in \Gamma$ or $B \in \Gamma$.
(4) For all $x \in V_L$, and $A \in W_L$, [$\exists xA \in \Gamma \Rightarrow$ for some $c \in C_L$, $Ac/x \in \Gamma$, and $E!c \in \Gamma$].

Let underlined Greek capital letters range over free saturated sets.

Lemma 3. *Say* $\Gamma \cup \{A\} \subseteq W_L$ *and* $\Gamma \nvdash A$. *Let* $\{c_1, c_2, \ldots\}$ *be distinct symbols foreign to* L. *Let* L' *be like* L *except that* $C_{L'} = C_L \cup \{c_1, c_2, \ldots\}$. *Then there is a free-$L'$-saturated set* $\underline{\Gamma}$ *such that* $\Gamma \subseteq \underline{\Gamma}$ *and* $A \notin \underline{\Gamma}$.

The proof of this lemma (see Appendix 2) is a variation on the standard Lindenbaum construction. $\underline{\Gamma}$ is built up by step-by-step accretion in infinitely many stages, starting with Γ itself. Each odd stage helps to guarantee that $\underline{\Gamma}$, the infinite sum, will satisfy clause (3) in the definition of saturation by adding either B or B' for disjunctions, $B \vee B'$, which are known at that stage to be ultimate elements of $\underline{\Gamma}$. The even stages add sentences of the form Bc/x and $E!c$ for existentials $\exists xBx$ which will belong to $\underline{\Gamma}$. This will assure that $\underline{\Gamma}$ satisfies clause (4).

The following three corollaries are proved by adapting this same style of construction (see Appendix 2).

Corollary 1. *If Γ and A are as above, $B \in W_L$, and $\Gamma \cup \{B\} \nvdash A$, then there is a $\underline{\Gamma} \subseteq W_L$, such that $B \in \underline{\Gamma}$, $A \notin \underline{\Gamma}$, and $\Gamma \subseteq \underline{\Gamma}$.*

Corollary 2. *If $\Gamma \cup \{\exists x E! x\} \nvdash A$, then we can require that $E! c \in \underline{\Gamma}$ for some $c \in \{c_1, c_2, \ldots\}$. Indeed, we can find $\underline{\Gamma}$ such that $E! c \in \underline{\Gamma}$ for all $c \in \{c_1, \ldots\}$ and $\Gamma \subseteq \underline{\Gamma}$ and $A \notin \underline{\Gamma}$.*

Corollary 3. *If $\Gamma \cup \{\exists x E! x\} \nvdash (B \mathbin{\&} \neg B)$, then we can find $\underline{\Gamma} \subseteq W_L$, such that $\Gamma \subseteq \underline{\Gamma}$ and for all $t \in V_L \cup C_L$, either $E! t \in \underline{\Gamma}$ or $\neg E! t \in \underline{\Gamma}$.*

C. Soundness

Theorem 1. *If $\Gamma \vdash A$ then $\Gamma \Vdash A$.*

Proof sketch. Suffice it for now to show $\varnothing \vdash \forall x E! x$. Suppose $\langle\langle K, R, \alpha_0, D, E\rangle, I\rangle$ is a Meinongian model such that $I_\alpha(\forall x E! x) = 0$. Then $\mathsf{V}\beta_{\in K} \mathsf{V} d_{\in D}(\alpha_0 \mathsf{R} \beta$ and $E_\beta(d) = e$ and $(Id/x)_\beta(E! x) = 0)$. But $(Id/x)(x) = d$; so $(Id/x)_\beta(E! x) = 1$, contradicting the initial assumption. Each of the additional cases is proved similarly. ●

D. Strong Completeness

Lemma 4 (Main Lemma). *If $\underline{\Gamma}$ is a free-L-saturated set, then there is a Meinongian model $\langle\langle K, R, \alpha_0, D, E\rangle, I\rangle$ such that $\underline{\Gamma} = \{A : I_{\alpha_0}(A) = 1\}$.*

Once again the proof is a variation on a standard technique, in this case the construction of a model out of linguistic elements themselves. The construction begins by describing a hierarchy of languages $\{L_i\}_i$ which starts with $L = L_0$ and builds the final language L' by adding new constants to each language L_i. Then K for the desired model is the set of all sets of formulas that contain $\underline{\Gamma}$ as a subset and that are L_i-saturated for some i. R will be set-theoretic inclusion, α_0 will be $\underline{\Gamma}$ itself. D will be the union of V_L with $C_{L'}$ (the sum of all the constants eventually added), and E will be defined so that

$$E_{\underline{\Delta}}(t) = \begin{cases} e & \text{if } E! t \in \underline{\Delta} \\ n & \text{if not} \end{cases} \quad \text{for } \underline{\Delta} \in K$$

This is indeed a legitimate mf. One then constructs an interpretation, I, with the property that $I_{\underline{\Delta}}(A) = 1$ iff $A \in \underline{\Delta}$ for each $\underline{\Delta} \in K$. $\underline{\Gamma}$ itself is one of these $\underline{\Delta}$'s and is by definition α_0 of the mf. That will prove the lemma. For details see Appendix 2.

Theorem 2 (Strong Completeness). $\Gamma \Vdash A \Rightarrow \Gamma \vdash A$.

Proof. Say $\Gamma \Vdash A$ and $\Gamma \nvdash A$. By Lemma 3 there is a $\underline{\Gamma}$, such that $\Gamma \subseteq \underline{\Gamma}$, and $A \notin \underline{\Gamma}$. By Lemma 4 there is a Meinongian model $\langle K, R, \alpha_0, D, E \rangle, I \rangle$ such that $\underline{\Gamma} = \{B : I_{\alpha_0}(B) = 1\}$.

Thus I simultaneously satisfies Γ in α_0 but not A in α_0. This contradicts $\Gamma \Vdash A$. ●

Theorem 3 (Weak Completeness). $\Vdash A \Rightarrow \vdash A$.

Proof. Take $\Gamma = \emptyset$ in Theorem 2. ●

E. Weakly Modified Meinongian Approach (Soundness and Completeness)

According to the intuitive reading of Kripke models suggested in Kripke (1965), the move from one node in K to a later (R-related) node corresponds to an increase in evidence. When one adapts that conception to the modeling described above it becomes clear that a Meinongian model represents epistemic advance along two fronts. First there is a nondecreasing collection of terms that are known to denote, and secondly a similarly growing collection of atomic truths containing both denoting and nondenoting terms.

As noted above, there is at least one point of view that will find this modeling objectionable on the grounds that no atomic sentence can be true of a nonexisting object, and thus that no such sentence containing nondenoting terms can be known. (See the discussion, e.g., in Burge (1974), Cocchiarella (1975), Schock (1968), and Stenlund (1975).) This opinion can be accommodated in the present framework by changing clause (iv) in Section 1.A as follows:

(iv') For each P^i, $I_\alpha(P^i) \in 2^{(D^i)}$ such that
  (a) $I_\alpha(P^i) \subseteq \{d : E_\alpha(d) = e\}$ and
  (b) $\alpha \mathsf{R} \beta \Rightarrow I_\alpha(P^i) \subseteq I_\beta(P^i)$ (see Note 8).

Let us call Meinongian models in which I satisfies (iv') *weakly modified Meinongian models*. Models of this sort will obviously validate $(P^i t_1 \cdots t_i \to (E! t_1 \& \cdots \& E! t_i))$ for all P^i and t of L. Indeed, we can add the following condition to the description or $\vdash$:

$$(E!I): \quad \frac{\Gamma \vdash P^i t_1 \cdots t_i}{\Gamma \vdash (E! t_1 \& \cdots \& E! t_i)}$$

and easily get the following theorem.

Theorem 4. $\vdash'$ *(which is* $\vdash$ *supplemented by* $(E!I)$*) is sound with respect to the weakly modified Meinongian semantics.* ●

Remark. When we add $(E!I)$ to $\vdash$, then condition (2) in the definition of free-L-saturation entails that every free-L-saturated set, $\underline{\Gamma}$, satisfies:

$$\Lambda t_1, \ldots, \Lambda t_i \in V_L \cup C_L(P^i_j(t_1 \cdots t_i) \in \underline{\Gamma} \Rightarrow (E!t_1 \& \cdots \& E!t_i) \in \underline{\Gamma})$$

for all $i, j < \omega$. The proofs of the counterparts to Lemma 3 and its corollaries will proceed straighforwardly.

Lemma 5. *If* $\underline{\Gamma}$ *is a free-L-saturated set (in the new sense) then there is a weakly modified Meinongian model:*

$$\langle\langle K, R, \alpha_0, D, E\rangle, I\rangle \text{ where } \underline{\Gamma} = \{A: I_{\alpha_0}(A) = 1\}$$

The proof is similar to that of Lemma 4. See Appendix 2 for detail.

Theorem 5. $\vdash'$ *is strongly complete with respect to the modified Meinongian semantics.* ●

Theorem 6. $\vdash'$ *is weakly complete as well.* ●

F. Identity and the Strongly Modified Meinongian Approach

Identity

The weakly modified Meinongian semantics makes no allowance for any atomic truths containing nondenoting terms. That includes even atomic formulas of the form $(t = t)$. Were we to adapt this approach to a language with '=' it would correspond to the two-valued view which validates $E!t \equiv (t = t)$. (See the discussions in, e.g., Burge (1974) and Lambert (1962).) A less radical position holds that nothing *factual* can be true of a nonexistent object (and, a fortiori, that one can have no factual knowledge such an object) but nevertheless one can make certain truth-bearing "analytic" claims containing nondenoting terms.

To see how to accommodate this position, let us first single out self-identity claims as a natural family of such linguistically necessary statements. Were we to follow Thomason (1965) we would define an mfi (model frame with identity) as a structure $\langle K, R, \alpha_0, D, E, S\rangle$, where $\langle K, R, \alpha_0, D, E\rangle$ is an mf and S takes $\alpha \in K$ into an equivalence relation $\sim_\alpha$ on D with the provisos: (α R β and $d_1 \sim_\alpha d_2 \Rightarrow d_1 \sim_\beta d_2$), and ($d_1 \sim_\alpha d_2 \Rightarrow E_\alpha(d_1) = E_\alpha(d_2)$).

We can then add '=' to the basic morphology and adapt the satisfaction relation accordingly. In particular we must add the following amendment to clause (iv) in the definition of I:

(iva) For $i \geq 1$ and all $P^i \in L$, if $d_1 \sim_\alpha d_2$ then $I_\alpha(P^i)$ is invariant under replacement of d_1 by d_2.

And we must add the following clause to the inductive definition of satisfaction:

(0c) $I_\alpha(t_1 = t_2) = 1$ iff $I(t_1) \sim_\alpha I(t_2)$

It follows immediately that $(t = t)$ is valid for all terms t. Similarly, the standard axioms for identity all hold.

One can object that (iva) and (0c) clearly violate the spirit of (iv'). The excuse that self-identity is "analytic" and hence rightfully immune to (iv') might then be rejected as further ad hoc-ary. A more satisfying picture emerges if we do indeed adopt (iv') and modify (0c) accordingly:

(0c') $I_\alpha(t_1 = t_2) = 1$ iff

(1) $E_\alpha(I(t_1)) = E_\alpha(I(t_2)) = e$ and $I(t_1) \sim_\alpha I(t_2)$; or

(2a) $\vee \beta_{\in K}[\alpha \mathrel{R} \beta$ and $E_\beta(I(t_1)) = E_\beta(I(t_2)) = e]$ and

(2b) $\wedge \gamma_{\in K}[(E_\gamma(I(t_1)) = E_\gamma(I(t_2)) = e \Rightarrow (I(t_1) \sim_\gamma I(t_2))]$.

Clause (2a) is designed to avoid the anomalous conclusion that $\neg E! t_0$ entails $t = t_0$ for all t. It does this, to be sure, by insuring $\neg E! t_0 \to (t_0 \neq t)$ for all t (including t_0). Perhaps this is not an unreasonable requirement once we recall that intuitionistic negation is implicitly modal—and that $\neg E! t$ means that t *must* not represent an existent thing. (Or that t represents a necessarily nonexistent thing—a situation that is unavoidable at the level of analyzed terms).

Notice that though this semantics validates $\neg E! t \equiv \neg(t = t)$, it does *not* validate (the more controversial) $E! t \equiv (t = t)$. On the other hand it continues to validate "logical theorems" even for nondenoting terms. For example, $\neg E! t \to (Pt \to Pt)$ and even $\neg E! t \to (t = t \to t = t)$. The techniques of Section G will provide a means of invalidating $\neg E! t \to t \neq t$, while still validating $t \neq t \to \neg E! t$.[9]

The Strongly Modified Meinongian Semantics

Let us take this treatment of identity as our cue for developing a Meinongian treatment of linguistic necessity sensitive to (iv') and the possibility of nonexistence.

$\langle\langle K, R, \alpha_0, D, E\rangle, I\rangle$ is a *strongly modified* Meinongian model if $\langle K, R, \alpha_0, D, E\rangle$ is an mf, and I is an interpretation of L on it satisfying (iv′) and having the following condition instead of (0a):

(0a′) $I_\alpha(P^i t_1 \cdots t_i) = 1$ iff either

(1) $\langle I(t_1), \ldots, I(t_i)\rangle \in I_\alpha(P^i)$ or

(2a) $\mathsf{V}\beta_{\in K}(\alpha \mathrel{\mathsf{R}} \beta \text{ and } E_\beta(I(t_1)) = \cdots = E_\beta(I(t_i)) = e)$ and

(2b) $\Lambda\gamma_{\in K}((E_\gamma(I(t_1)) = \cdots = E_\gamma(I(t_i)) = e) \Rightarrow$
$$(\langle I(t_1), \ldots, I(t_i)\rangle \in I_\gamma(P^i))).$$

Let us denote by $\vdash^+$ the deducibility relation obtained by adopting the conditions on $\vdash$ as in Section 1.B and adding the following schema, a weak sister of $(E!I)$:

$$(\neg E!E): \quad \frac{\Gamma \vdash^+ \neg(E!t_1 \,\&\, \cdots \,\&\, E!t_i)}{\Gamma \vdash^+ \neg P^i t_1 \cdots t_i}$$

If $\vdash^+$ represents semantic entailment with respect to the class of strongly modified Meinongian models, then the following is readily checked:

Theorem 7 (Soundness). *If* $\Gamma \vdash^+ A$ *then* $\Gamma \Vdash^+ A$. ●

It is clear that the strongly modified semantics no longer validates $Pt \to E!t$ nor in general $(E!I)$. Indeed, $\vdash^+$ is strongly complete with respect to this semantics. To see this it is necessary first to prove Lemma 3 and its corollaries with $\vdash^+$ replacing $\vdash$ in the definition of saturation and in the lemma. That can be done simply by inspection of the original proof. Then we must prove the following version of the main lemma:

Lemma 6. *If* $\underline{\Gamma}$ *is a free-L-saturated set* (*in the new sense*) *then there is a strongly modified Meinongian model* $\langle\langle K, R, \alpha_0, D, E\rangle, I\rangle$ *such that* $\underline{\Gamma} = \{A : I_{\alpha_0}(A) = 1\}$.

The proof of this lemma (for details see Appendix 2) is similar to that of Lemma 4 with the following major exception: K of the required modified Meinongian model will be set equal to $K_0 \cup K_1$. K_0 has $\underline{\Gamma}$ itself as a minimal node. (Once again $\underline{\Gamma} = \alpha_0$.) The remaining elements of K_0 resemble the elements of the K used in Lemma 4 (with some added provisos required by $\vdash^+$). R once again will be the subset relation. K_1 will consist of all those formulas (of the expanded languages) that are $\vdash^+$-derivable from the set of atomic formulas in $\underline{\Gamma}$ and from the assumption $E!t$ for any t in one of the expanded languages. Ordinarily, K_1 will be a node, unrelated by R to any element of K_0, that has the property that $E_{K_1}(d) = e$ for all d. (In general, K_1 will be totally isolated under R; though that is easily changed).

One needs such a node to handle cases like that in which $(E!t \to Pt) \in \Gamma$, $\neg\neg E!t \in \Gamma$, but $E!t \notin \Gamma$ and $Pt \notin \Gamma$. In this case, if we had only K_0 alone, (0a')(2) would be satisfied. K_1 must contain the $\vdash^+$-consequences of the atomic elements of Γ to assure that $I\Gamma(Pt) = 1$ when $Pt \in \Gamma$ but $E!t \notin \Gamma$. Without that addition, K_1 would violate (0a')(2b).

Theorem 8. $\Gamma \Vdash^+ A \Rightarrow \Gamma \vdash^+ A$. ●

Theorem 9. $\Vdash^+ A \Rightarrow \vdash^+ A$. ●

G. Two Variations on the Strongly Modified Semantics

If in (2b) of (0a') we were to add the condition that $\alpha \, \mathsf{R} \, \gamma$ we would get a condition (call it (0a'')) which does not allow satisfaction at a node to be influenced by inaccessible nodes. This captures a weak (or perhaps relativized) sort of analyticity. Notice for instance that a semantics with (0a'') would validate:

$$(\forall x Px \, \& \, \neg\neg E!t) \to Pt$$

In effect, it identifies analytic truths with universal truths. This is likely to be thought of as a shortcoming. For, even though under the intuitionistic reading universal truths are nomic, that does not ordinarily qualify them as 'analytic'.[10]

The second variation entails a strong notion of analyticity. The move here is to drop the requirement that $\alpha \, \mathsf{R} \, \beta$ in (2a) of (0a'). Call the resulting clause (0a*). That is:

(0a*) $I_\alpha(P^i t_1 \cdots t_i) = $ iff either

(1) $\langle I(t_1), \ldots, I(t_i) \rangle \in I_\alpha(P^i)$ or

(2a) $\mathsf{V}\beta_{\in K}(E_\beta(It_1) = \cdots = E_\beta(It_i) = e)$ and

(2b) $\Lambda\gamma_{\in K}((E_\gamma(It_1) = \cdots = E_\gamma(It_i) = e) \Rightarrow$

$[\langle I(t_1), \ldots, I(t_i) \rangle \in I_\gamma(P^i)])$.

The semantics stemming from (iv') and (0a*) has the advantage that it not only invalidates $Pt \to E!t$ and $(E!I)$, but also $\neg E!t \to \neg Pt$ and $(\neg E!E)$. That is desirable; for one may be quite willing to affirm $\neg E!t$, even admitting the modal character of the claim, and still want to assert Pt as analytic and necessary.

The distinctive feature of (0a*) is that it validates $(Pt \lor (Pt \to E!t))$.

The first disjunct is of course a consequence of the claim that Pt is an "analytic truth"; while the second follows from the denial of that claim. Thus (0a*) plays on a seemingly nonintuitionistic conception of analyticity according to which it is a decidable matter whether or not Pt is analytic. As it happens, though, there is in the literature a sense of analyticity that does satisfy this criterion, and that can even be applied intuitionistically. (See Hintikka 1973, chap. VIII and p. 136, and Posy 1974, sect. 12.) If we let $\vdash^*$ be $\vdash$ augmented by the principle

$$F(v \to I): \varnothing \vdash P^i t_1 \cdots t_i \lor (P^i t_1 \cdots t_i \to (E!t_1 \& \cdots \& E!t_i))$$

(all i, P^i) and let $\Vdash^*$ represent the semantic entailment derived from $\Vdash$ by substituting (iv') for (iv) and (0a*) for (0a), then it is possible to show that $\vdash^*$ is sound and strongly complete with respect to $\vdash^*$.

Lemma 7. *If I' is an interpretation of L on $\langle K, R, \alpha_0, D, E \rangle$ satisfying (iv') and (0a*), there is an unmodified I interpreting L on $\langle K, R, \alpha_0, D, E \rangle$ such that*

$$\Lambda \alpha_{\in K} \Lambda A_{\in W_L}(I'_\alpha(A) = I_\alpha(A))$$

Proof. Let $I(t) = I'(t)$, $I_\alpha(P^0) = I'_\alpha(P^0)$; and for $i > 0$, $I_\alpha(P^i) = I'_\alpha(P^i) \cup \{\langle I(t_1), \ldots, I(t_i) \rangle : I'_\alpha(P^i t_1 \cdots t_i) = 1\}$. ●

Corollary 4. $\Gamma \vdash A \Rightarrow \Gamma \Vdash^* A$.

Proof. $\Gamma \not\Vdash^* A \Rightarrow \Gamma \not\Vdash A$ (by Lemma 7) $\Rightarrow \Gamma \not\vdash A$ (by Theorem 1). ●

Theorem 10. $\Gamma \vdash^* A \Rightarrow \Gamma \Vdash^* A$.

Proof. After Corollary 4 there remains only to show the $\Vdash^*$-validity of $F(v \to I)$. Say $I_\alpha(P^i t_1 \cdots t_i \lor (P^i t_1 \cdots t_i \to (E!t_1 \& \cdots \& E!t_i))) = 0$. So $I_\alpha(P^i t_1 \cdots t_i) = 0$; and there is a β such that $\alpha \, R \, \beta$, $I_\beta(P^i t_1 \cdots t_i) = 1$, and $I_\beta(E!t_1 \& \cdots \& E!t_i) = 0$. Thus (by (iv')) $\langle I(t_1), \ldots, I(t_i) \rangle \notin I_\beta(P^i)$. The only remaining way for $I_\beta(P^i t_1 \cdots t_i) = 1$ to be justified is by (2a) and (2b) of (0a*). This in turn entails $I_\alpha(P^i t_1 \cdots t_i) = 1$. ●

After revising Lemmas 2 and 3 and Corollaries 1–3 for $\vdash^*$, one proves:

Lemma 8. *If $\underline{\Gamma}$ is a free-L-saturated set (in the sense of $\vdash^*$), then there is an mf $\langle K, R, \alpha_0, D, E \rangle$ and an interpretation, I, of L on $\langle K, R, \alpha_0, D, E \rangle$ satisfying (iv') and (0a*) such that $\underline{\Gamma} = \{A : I_{\alpha_0}(A) = 1\}$.*

The proof is quite similar to that of Lemma 6; see Appendix 2.

Theorem 11. $\Gamma \Vdash^* A \Rightarrow \Gamma \vdash^* A$. ●

Theorem 12. $\Vdash^* A \Rightarrow \vdash^* A$. ●

Given these results and the ready interpretation of $F(v \to I)$, the package of $\vdash^*$ and $\Vdash^*$ commends itself to us as a natural intuitionistic rendition of the view that there may be "analytic" truths containing nondenoting terms.

H. Two More Variations

Two other heuristic intuitions are worth pursuing for their philosophical interest. One is the view that actually existing objects are fully determined and thus complete in all atomic properties and relations. We may express this view by the schema:

$$\frac{\Gamma \vdash (E!t_1 \& \cdots \& E!t_i)}{\Gamma \vdash (P^i t_1 \cdots t_i \vee \neg P^i t_1 \cdots t_i)}$$

As it turns out, this schema holds in intuitionistic number theory. It has some additional philosophical applications as well. The schema is validated if we amend clause (iv) of Section A as follows:

(iv)(b) $\{(E_\alpha(d_1) = \cdots = E_\alpha(d_i) = e)$ and $(\langle d_1, \ldots, d_i \rangle \notin I_\alpha(P^i))$ and $(\alpha \mathrel{R} \beta)\}$

$$\Rightarrow \langle d_1, \ldots, d_i \rangle \in I_\beta(P^i).$$

Incorporating (iv)(b) into the definition of an interpretation has the effect of restricting the growth of knowledge to just one front. The state of knowledge progresses by the discovery of the existence of entities (previously only potentially known to exist). However, once an entity of that sort is hit upon, it is fully determinate in all its observable properties (including relations to other existing entities).[11]

One disadvantage of the modified modelings explored so far is that they validate $\neg E!t \to (Pt \vee \neg Pt)$. Thus we may assert that Pegasus does not exist and that Pegasus is a horse. But if we are not prepared to affirm that Pegasus is white, we must then *deny* it (with the full force of intuitionistic negation). This situation may be remedied by adopting (iv′) and the following changes in the semantics:

(0a**) $I_\alpha(P^i t_1 \cdots t_i)$

= 1 if (1) $\langle I(t_1), \ldots, I(t_i) \rangle \in I_\alpha(P^i)$ or

(2) for some $j, 1 \leqslant j \leqslant i, E_\alpha(t_j) = n$; but for some $\beta \in K$, $E_\beta(t_1) = \cdots = E_\beta(t_i) = e$ and for all such β, $\langle I(t_1), \ldots, I(t_i) \rangle \in I_\beta(P^i)$.

$= u$ if for each β such that $\alpha \mathrel{R} \beta$, for some j $(1 \leqslant j \leqslant i)$, $E_\beta(t_j) = n$; but there is a β in K such that $E_\beta(t_1) = \cdots = E_\beta(t_i) = e$, and $\langle I(t_1), \ldots, I(t_i) \rangle \in I_\beta(P^i)$; and there is a γ in K such that $E_\gamma(t_i) = \cdots = E_\gamma(t_i) = e$ and $\langle I(t_1), \ldots, I(t_i) \rangle \notin I_\gamma(P^i)$.

$= 0$ otherwise.

(1**) $I_\alpha(B \to C)$

$= 1$ if for all β such that $\alpha \mathrel{R} \beta$, $(I_\gamma(B) = 0$ or $I_\beta(C) = 1$ or $I_\beta(B) = I_\beta(C) = u)$,

$= 0$ otherwise.

(4**) $I_\alpha(\neg B)$

$= 1$ if for all β such that $\alpha \mathrel{R} \beta$, $I_\beta(B) = 0$,

$= 0$ otherwise.

The remaining clauses are as in Appendix 1. One can readily check that this logic validates the following schema:[12]

$$\frac{\Gamma \vdash \neg(E!t_1 \& \cdots \& E!t_i)}{\Gamma \vdash (P^i t_{t_1} \cdots t_1 \vee \neg P^i t_1 \cdots t_i \vee \neg\neg P^i t_1 \cdots t_i)}$$

In both of these last two cases soundness and completeness are now routine.

2. Russellian Version

Some people object to outer domain semantics on the grounds that no (contingent) sentences containing nondenoting terms can bear truthvalue. These objections can be met by the modified Meinongian systems of Sections 1.E to 1.H. However, some purists may remain dissatisfied with even that watered down approach—since it does allow $I(t)$ to be defined at worlds where $E!t$ fails to hold.[13] In the present section I indicate how to work with partial valuations, interpretations in which that does not happen. The main observation is the availability of some strong completeness results even for these sorts of semantics.

A. Basic Semantics

A Russellian frame (rf) is a structure, $\langle K, R, \alpha_0, D \rangle$, where K, R, and α_0

are as usual, and D is a function taking the elements of K into sets and satisfying $\alpha \mathrel{R} \beta \Rightarrow D_\alpha \subseteq D_\beta$.

For every $t \in V_L \cup C_L$, rather than defining $I(t)$ once and for all, we must define $I_\alpha(t)$ and leave open the possibility that $I_\alpha(t)$ is undefined.

(i)(a) $I_\alpha(t) = d \Rightarrow d \in D_\alpha$;
(ii)(b) $(I_\alpha(t) = d$ and $\alpha \mathrel{R} \beta) \Rightarrow I_\beta(t) = d$;[14]
(iv) $\quad I_\alpha(P^i) \subseteq (D_\alpha)^i$ and $\alpha \mathrel{R} \beta \Rightarrow I_\alpha(P^i) \subseteq I_\beta(P^i)$.

(0a) $\quad I_\alpha(P^i t_1 \cdots t_i) = 1 \quad$ iff $\langle I_\alpha(t_1), \ldots, I_\alpha(t_i)\rangle \in I_\alpha(P^i)$

(0b) $\quad I_\alpha(E!\,t) = 1 \quad$ iff $I_\alpha(t)$ is defined

Remaining clauses are as usual. The pair $\langle\langle K, R, \alpha_0, D\rangle, I\rangle$ is a *Russellian model*.

B. Relation to Meinongian Semantics

Obviously the basic Russellian semantics is most closely related to the weakly modified Meinongian version discussed in Section 1.E. Clearly the Russellian version also validates $E!\,I$. Indeed,

Theorem 13. $\Gamma \vdash' A \Rightarrow \Vdash_R A$. ●

(Recall, $\vdash'$ is $\vdash$ augmented by $(E!\,I)$ and $\Vdash_R$ represents semantic entailment in the basic Russellian semantics). Once again the proof is straightforward.

Let us turn to the strong completeness of $\vdash'$ with respect to this semantics. Fix L and let **M** = the class of weakly modified Meinongian models, and **R** = the class of Russellian models.

Define the map $f: \mathbf{M} \to \mathbf{R}$ as follows: Let $M = \langle\langle K, R, \alpha_0, D, E\rangle, I\rangle \in \mathbf{M}$, $f(M)$ will be $\langle\langle K, R, \alpha_0, D'\rangle, I'\rangle \in \mathbf{R}$. (That is, $f(M)$ will be a Russellian model with the same K, R, α_0.) $D': K \to 2^D$ such that $D'_\alpha = \{d : E_\alpha(d) = e\}$. $I'_\alpha(t) = I(t)$ if $E_\alpha(I(t)) = e$, $I'_\alpha(t)$ is undefined if not.
$I'_\alpha(P^i) = I_\alpha(P^i)$.

Lemma 9. For any $A \in W_L$, for any $M \in \mathbf{M}$, for any $\alpha \in K_M$: $I_\alpha(A) = (I_{f(M)})_\alpha(A)$.

The proof will be by induction on the number of logical symbols in A. See Appendix 2.

Corollary 5. $\Gamma \Vdash_R A \Rightarrow \Gamma \Vdash' A$ (*where $\Vdash'$ is entailment in the weakly modified Meinongian semantics*).

Proof. If $\Gamma \Vdash_R A$ and not $\Gamma \Vdash' A$, then $\forall M \in \mathbf{M}((I_M)_{\alpha_0}(C)) = 1$ for all $C \in \Gamma$, and $(I_M)_{\alpha_0}(A) = 0)$. Then, by Lemma 9,

$$(I_{f(M)})_{\alpha_{0_{f(M)}}}(C) = 1 \text{ for all such } C \text{ and } (I_{f(M)})_{\alpha_{0_{f(M)}}}(A) = 0,$$

which contradicts $\Gamma \Vdash_R A$. •

Theorem 14. $\Gamma \Vdash_R A \Rightarrow \Gamma \vdash' A$.

Proof. By Corollary 5 and Theorem 5. •

Theorem 15. $\Vdash_R A \Rightarrow \Vdash A$. •

Finally, one is free to change (0a) of the basic Russellian semantics to (0a'), (0a''), (0a*), or (0a**), and achieve the corresponding results. The transformation, f, will relate the modified semantics to the appropriate class of strongly modified Meinongian models.

The counterparts to Lemma 9 go through fairly straightforwardly. And so we have strong completeness (and soundness) for these logics as well.[15]

3. Explicit Definability

A. Introduction

IPC is known to have the property that for certain sets $\Gamma \subseteq W_L$

$$\Lambda A_{\in W_L} \forall a_{\in C_L} (\Gamma \vdash_i \exists x A \Rightarrow \Gamma \vdash_i A^a/x) \tag{1}$$

where $\vdash_i$ is standard intuitionistic derivability (as in Appendix 1). The most notable cases are when $\Gamma = \emptyset$ (logic) and $\Gamma = HA$ (intuitionistic arithmetic). The obvious free analog of this explicit definability property is

$$\Lambda A_{\in W_L} \forall a_{\in C_L} [\Gamma \vdash \exists x A \Rightarrow \Gamma \vdash (A^a/x \,\&\, E!\,a)] \tag{2}$$

where $\vdash$ is the basic deducibility relation of Section 1.B. It is tempting to conjecture that one can characterize the Γ's satisfying (2) by appealing directly to the intuitionistic case. (That is, Γ satisfies (1) $\Rightarrow$ Γ satisfies (2).) However, that this cannot be done is seen by taking

$$C_L \neq \emptyset \text{ and } \Gamma = \{\exists x Px, \forall x Px\}$$

In the intuitionistic case, those Γ's known to satisfy (1) are identified by showing them to satisfy certain general constraints that entail explicit definability. These constraints can be relativized to the free intuitionistic case in a straightforward way, and will in their turn yield the explicit definability property (2). Section B applies the methods of Axcel (1968),

Kleene (1962) and others to yield necessary and sufficient conditions for the logical closure of Γ to be saturated. That in turn yields (2) as a consequence.

B. Adaptation of the Axcel–Kleene Slash

In this section suppose that $\Gamma \subseteq W_L$ contains only closed wffs, and that $C_L \neq \emptyset$.

Definition. $\Gamma \mid A$ is defined inductively as follows: $\Gamma \mid A$ iff

(0) A is atomic and $\Gamma \vdash A$. (This does not distinguish among different atomic sentences. A can be P^0, $P^i t_1 \cdots t_i$, or $E!t$.) Or
(1) $A = B \,\&\, C$ and ($\Gamma \mid B$ and $\Gamma \mid C$). Or
(2) $A = B \vee C$ and ($\Gamma \mid B$ or $\Gamma \mid C$). Or
(3) $A = B \to C$ and (($\Gamma \mid B \Rightarrow \Gamma \mid C$) and $\Gamma \vdash B \to C$). Or
(4) $A = \neg B$ and ($\Gamma \nvdash B$ and $\Gamma \vdash \neg B$). Or
(5) $A = \exists x B$ and $\Gamma \mid (B^c/x \,\&\, E!c)$ for some $c \in C_L$. Or
(6) $A = \forall x B$ and $\{(\Gamma \mid B^c/x \text{ for all } c \in C_L \text{ such that } \Gamma \mid E!c) \text{ and } \Gamma \vdash \forall x B\}$.

The following definition and lemma will give intuitive content to the relation $\Gamma \mid A$, by providing a model theoretic interpretation. If Γ is consistent, then by Lemmas 3 and 4 there is a saturated $\underline{\Gamma} \supseteq \Gamma$, and an (unmodified) Meinongian model M (with $D_M = C'_L$) such that

$$\underline{\Gamma} = \{A : (I_M)_{\alpha_0}(A) = 1\}$$

Let M^+ be derived from M as follows:

$$K_M^+ = K_M \cup \{\alpha^+\}(\alpha^+ \notin K_M)$$

$$D_M^+ = D_M = C'_L \text{ (recall } L' = \bigcup_{j<\omega} \{L_j\})$$

$$R_M^+ = R_M \cup \{\langle \alpha^+, \beta \rangle : \beta \in K_M^+ \text{ and } \langle \alpha_0, \beta \rangle \in R_M\} \cup \{\langle \alpha^+, \alpha^+ \rangle\}$$

$$(E_M^+)_{\alpha^+}(c) = e \text{ iff } \Gamma \vdash E!c \text{ and } c \in C_L$$

$$(E_M^+)_\alpha(c) = (E_M)_\alpha \text{ for } \alpha \in K_M$$

$$(I_M^+)_{\alpha^+}(P^0) = 1 \text{ iff } \Gamma \vdash P^0$$

$$(I_M^+)_{\alpha^+}(P^i) = \{\langle c_1, \ldots, c_i \rangle : \Gamma \vdash P^i c_1 \cdots c_i \text{ and } \{c_1, \ldots, c_i\} \subseteq C_L\} \ (i > 0)$$

$$(I_M^+)_\alpha(P^i) = (I_M)_\alpha(P^i) \text{ for all } i \geq 0 \text{ and } \alpha \in K_M$$

Remarks

(1) $\{A : (I_M^+)_{\alpha^+}(A) = 1\} \subseteq \{A : \Gamma \vdash A\}$.

(2) $\{A: (I_M^+)_{\alpha^+}(A) = 1\}$ is a free-L-saturated set.
(3) $\{A: (I_M^+)_{\alpha^+}(A) = 1\}$ is the maximal free-L-saturated subtheory of $\{A: \Gamma \vdash A\}$ (see Smorynski 1973, sect. 5.1.14).

Lemma 10. $\Gamma \mid A$ iff $(I_M^+)_{\alpha^+}(A) = 1$.

The proof will be by induction on the length of A. See Appendix 2.

Theorem 16. *Let Γ be a consistent set of sentences. The following are equivalent*: (i) $\{A: \Gamma \vdash A\}$ *is a free-L-saturated set.* (ii) $\{A: \Gamma \vdash A\} = \{A: \Gamma \mid A\}$.

The proof makes use of Lemma 10 and remarks (1) and (2) above. See Appendix 2.

Corollary 6. *A sufficient condition for Γ to satisfy* (2) *is that* $\{A: \Gamma \vdash A\} = \{A: \Gamma \mid A\}$. ●

Corollary 7. *If Γ is a consistent set of sentences, then the following are equivalent*: (i) $\{A: \Gamma \vdash A\}$ *is a free-L-saturated set.* (ii) $\Gamma \subseteq \{A: \Gamma \mid A\}$.

Proof. (i) $\Rightarrow$ (ii) by Theorem 16. Conversely, $\{A: \Gamma \mid A\}$ is deductively closed (by remark (2) or Lemma 10). Thus (ii) $\Rightarrow \{A: \Gamma \vdash A\} \subseteq \{A: \Gamma \mid A\} \Rightarrow \{A: \Gamma \vdash A\} = \{A: \Gamma \mid A\} \Rightarrow$ (i) (by Theorem 16). ●

Thus a sufficient condition for Γ to have the existential definability property is that $A \in \Gamma \Rightarrow \Gamma \mid A$. Another useful such condition will be provided by Corollary 8. [That will be an adaptation of the condition used by Axcel to demonstrate the applicability of (1) to HA.]

Definition. Let A be a sentence. $RH(A)$ holds if every occurrence of '$\vee$', '$E!$', or '$\exists$' in A is either in the antecedent, B, of a well-formed part of A of the form $B \rightarrow C$, or is a well-formed part of a well-formed part of A of the form $\neg B$. (This version of the RH condition is due to Robinson (1965).)

Corollary 8. *Let Γ be a consistent set of sentences. The following are equivalent*: (i) $\{A: \Gamma \vdash A\}$ *is a free-L-saturated set.* (ii) $\Gamma \mid B$ *for all $B \in \Gamma$ such that not $RH(B)$.*

The proof of Corollary 8 is gotten by combining Corollary 7 with the following lemma.

Lemma 11. $\{A: RH(A) \text{ and } \Gamma \vdash A\} \subseteq \{A: \Gamma \mid A\}$. [*Thus* $RH(A) \Rightarrow (\Gamma \vdash A \Leftrightarrow \Gamma \mid A)$.]

The proof will be induction on A. See Appendix 2.

Finally, the standard applications. First, trivially:

Theorem 17. *Each of the logics described in Part 1 satisfies* (2). ●

Second, simple observation reveals that all of the considerations of the present section apply without change to each of these logics. Consider a language for arithmetic with equality, a relation symbol, S, for successor and the standard arabic numerals as its individual constants. Let FHA denote the closure under any one of those logics of the following set of axioms and schemata:

(1) $E!\bar{0}$

(2)   (a) $S(\bar{n}, \overline{n+1})$

         (b) $E!\,\overline{(n+1)}$

         (c) $\forall x[S(\bar{n}, x) \rightarrow x = \overline{n+1}]$

(3) $\forall x(\neg S(x, \bar{0}))$

(4) $\forall x \forall x' \forall y \forall y'\{[S(x, x') \,\&\, S(y, y') \,\&\, (x' = y')] \rightarrow x = y\}$

(5)   For any wff, A, with free variable x:

$$\{A^{\bar{0}}/x \,\&\, \forall x \forall y[(A \,\&\, S(x, y)) \rightarrow Ay/x]\} \rightarrow \forall x A$$

(where $\bar{n}$ is the numeral for the natural number n).

If desired one may add axioms defining primitive recursive relations. The resulting system is a free version of HA (with numerals; and functions replaced by relations; see, e.g., Troelstra (1973, sect. 1.3.8) and Smorynski (1973, sect. 5.2.3)).

Theorem 18. *Each FHA satisfies* (2).

The proof applies Corollary 8, taking Γ as the axioms of FHA. The only elements of Γ not satisfying RH are (1) and the instances of (2b) and (5). For any such B, $\Gamma \mid B$ can be shown. See Appendix 2 for details.[16]

APPENDIX 1: STANDARD SYNTAX AND SEMANTICS FOR IPC

A. Properties of the Deducibility Relation

R: $A \in \Gamma \Rightarrow \Gamma \vdash A$

A Free IPC is a Natural Logic

T: $$\frac{\Gamma \vdash A}{\Gamma \cup \Delta \vdash A}$$

C: $\Gamma \vdash A \Rightarrow \forall \Gamma'(\Gamma' \vdash A)$ (Γ' finite and $\Gamma' \subseteq \Gamma$)

$\rightarrow I$: $$\frac{\Gamma \cup \{A\} \vdash B}{\Gamma \vdash A \rightarrow B}$$

$\rightarrow E$: $$\frac{\Gamma \vdash A \quad \Gamma \vdash A \rightarrow B}{\Gamma \vdash B}$$

$\vee I$: $$\frac{\Gamma \vdash A}{\Gamma \vdash A \vee B} \quad \frac{\Gamma \vdash B}{\Gamma \vdash A \vee B}$$

$\vee E$: $$\frac{\Gamma \cup \{A\} \vdash C \quad \Gamma \cup \{B\} \vdash C \quad \Gamma \vdash A \vee B}{\Gamma \vdash C}$$

$\& I$: $$\frac{\Gamma \vdash A \quad \Gamma \vdash B}{\Gamma \vdash A \& B}$$

$\& E$: $$\frac{\Gamma \vdash A \& B \quad \Gamma \cup \{A\} \vdash C}{\Gamma \vdash C}$$

$$\frac{\Gamma \vdash A \& B \quad \Gamma \cup \{B\} \vdash C}{\Gamma \vdash C}$$

$\neg I$: $$\frac{\Gamma \cup \{A\} \vdash B \quad \Gamma \cup \{A\} \vdash \neg B}{\Gamma \vdash \neg A}$$

$\neg E$: $$\frac{\Gamma \vdash A \quad \Gamma \vdash \neg A}{\Gamma \vdash B}$$

$\forall I$: $$\frac{\Gamma \vdash A}{\Gamma \vdash \forall x At/x} \quad (t \text{ not free in any member of } \Gamma)$$

$\forall E$: $$\frac{\Gamma \vdash \forall x A}{\Gamma \vdash At/x}$$

$\exists I$: $$\frac{\Gamma \vdash At/x}{\Gamma \vdash \exists x A}$$

$\exists E$: $$\frac{\Gamma \vdash \exists x A \quad \Gamma \cup \{At/x\} \vdash B}{\Gamma \vdash B}$$

(t not free in A, B or any member of Γ)

B. Standard Semantics

A model structure is a triple $\langle K, R, D \rangle$. K and R as usual, D a function assigning a nonempty set D_α to each $\alpha \in K$. $\alpha \mathrel{R} \beta \Rightarrow D_\alpha \subseteq D_\beta$. An interpretation, I, of a morphology, L (without $E!$) on $\langle K, R, D \rangle$ is defined as follows:

(i) $x \in V_L \Rightarrow I(x) \in \bigcup_{\alpha \in K} (D_\alpha)$.

(ii) $c \in C_L \Rightarrow I(c) \in \bigcup_{\alpha \in K} (D_\alpha)$.

(iii) $I_\alpha(P^0) \in \{1, 0\}$ for each P^0.

(iv) For $i > 0$, $I_\alpha(P^i) \subseteq (D_\alpha)^i$, $\alpha \mathrel{R} \beta \Rightarrow I_\alpha(P^i) \subseteq I_\beta(P^i)$.

Interpretation of wffs: In each case $I_\alpha(A) = 0$ iff $I_\alpha(A) \neq 1$.

(0) $I_\alpha(P^i t_1 \cdots t_i) = 1$ iff $\langle I(t_1), \ldots, I(t_i) \rangle \in I_\alpha(P^i)$

(1) $I_\alpha(B \to C) = 1$ iff $\wedge \beta_{\in K}(\alpha \mathrel{R} \beta \Rightarrow (I_\beta(B) = 1 \Rightarrow I_\beta(C) = 1))$

(2) $I_\alpha(B \vee C) = 1$ iff $I_\alpha(B) = 1$ or $I_\alpha(C) = 1$

(3) $I_\alpha(B \mathbin{\&} C) = 1$ iff $I_\alpha(B) = I_\alpha(C) = 1$

(4) $I_\alpha(\neg B) = 1$ iff $\wedge \beta_{\in K}(\alpha \mathrel{R} \beta \Rightarrow I_\beta(B) = 0)$

(5) $I_\alpha(\forall x B) = 1$ iff $\wedge \beta_{\in K}(\alpha \mathrel{R} \beta \Rightarrow \wedge d \in D_\beta((Id/x)_\beta(B) = 1))$

(6) $I_\alpha(\exists x B) = 1$ iff $\vee d_{\in D_\alpha}((Id/x)_\alpha(B) = 1)$

APPENDIX 2: PROOF SKETCHES

Though what follows differs in several details from Thomason (1965), some of my initial moves are taken from there. I will refer the reader to the appropxiate section of Thomason (1965) when the argument requires that we slavishly mimic Thomason's construction.

Lemma 3 (Like Thomason's L1). *Say* $\Gamma \cup \{A\} \subseteq W_L$ *and* $\Gamma \nvdash A$. *Let* $\{c_1, c_2, \ldots\}$ *be distinct symbols foreign to* L. *Let* L' *be like* L *except that* $C_{L'} = C_L \cup \{c_1, c_2, \ldots\}$. *Then there is a free-$L'$-saturated set* $\underline{\Gamma}$ *such that* $\Gamma \subseteq \underline{\Gamma}$ *and* $A \notin \underline{\Gamma}$.

Proof Sketch. To construct $\underline{\Gamma}$ enumerate $W_{L'}$, set $\Gamma = \Gamma_0$, and define Γ_k inductively as follows for $k > 0$, k even. Let $\exists x B$ be the first formula of the enumeration of the form $\exists x C$ which has not yet been considered such that $\Gamma_k \vdash \exists x C$, and let c be the first element of $\{c_1, \ldots\}$ not to occur in any member of Γ_k. Then $\Gamma_{k+1} = \Gamma_k \cup \{Bc/x\} \cup \{E!c\}$.

If there is no such formula $\exists x B$, then $\Gamma_k = \Gamma_{k+1}$, k odd. Let $B \vee B'$ be the first such element of $W_{L'}$ of the form $C \vee C'$ such that $\Gamma_k \vdash C \vee C'$. If $\Gamma_k \cup \{B\} \nvdash A$ set $\Gamma_{k+1} = \Gamma_k \cup \{B\}$; if $\Gamma_k \cup \{B\} \vdash A$, then $\Gamma_{k+1} = \Gamma_k \cup \{B'\}$. Set $\underline{\Gamma} = \Gamma_\omega = \bigcup_{i<\omega} \Gamma_i$. The proof that $\underline{\Gamma}$ satisfies the lemma is straightforward, so long as one is careful to appeal to $F(\exists E)$ in showing $\underline{\Gamma} \nvdash A$. ●

Corollary 1. *If Γ and A are as above, $B \in W_L$, and $\Gamma \cup \{B\} \nvdash A$, then there is a $\underline{\Gamma} \subseteq W_{L'}$, such that $B \in \underline{\Gamma}$, $A \notin \underline{\Gamma}$ and $\Gamma \subseteq \underline{\Gamma}$.*

Proof. As above except that $\Gamma_0 = \Gamma \cup \{B\}$. ●

Corollary 2. *If $\Gamma \cup \{\exists x E! x\} \nvdash A$, then we can require that $E!c \in \underline{\Gamma}$, $c \in \{c_1, c_2, \ldots\}$. Indeed, we can find $\underline{\Gamma}$ such that $E!c \in \underline{\Gamma}$ for all $c \in \{c_1, \ldots\}$ and $\Gamma \subseteq \underline{\Gamma}$ and $A \in \underline{\Gamma}$.*

Proof. Set $\Gamma_0 = \Gamma \cup \{E!c : c \in \{c_1, \ldots\}\}$, and proceed as above, except that when k is even the condition on c is that c occurs in $E!c$ but in no other formula of Γ_k. ●

Corollary 3. *If $\Gamma \cup (\exists x E! x) \nvdash (B \& \neg B)$, then we can find $\underline{\Gamma} \subseteq W_{L'}$ such that $\Gamma \subseteq \underline{\Gamma}$ and for all $t \in V_L \cup C_{L'}$, either $E!t \in \underline{\Gamma}$ or $\neg E!t \in \underline{\Gamma}$.*

Proof. Enumerate $V_{L'} \cup C_{L'}$. Let $\Gamma^0 = \Gamma$ and $\Gamma^{i+1} = \Gamma^i \cup \{E!t_i\}$ if $\Gamma^i \cup \{E!t_i\}$ is consistent; otherwise $\Gamma^{i+1} = \Gamma^i \cup \{\neg E!t_i\}$. (Invoke Lemma 2.) Let $\Gamma_0 = \bigcup_{i<\omega} \Gamma^i$, and proceed as above. ●

Lemma 4. *If $\underline{\Gamma}$ is a free-L-saturated set, then there is a Meinongian model $\langle \langle K, R, \alpha_0, D, E \rangle, I \rangle$ such that $\underline{\Gamma} = \{A : I_{\alpha_0}(A) = 1\}$.*

Proof Sketch. The proof is similar to that of T2 in Thomason (1965). Suffice it for now to sketch the bare outline and note the main points in Thomason's proof requiring change. Let $L_0 = L$, and let L_{i+1} be like L_i except $C_{L_{i+1}} = C_{L_i} \cup \{c_1^{i+1}, c_2^{i+1}, \ldots\}$, where the c_k^{i+1} are distinct symbols foreign to L_i. Let $K = \{\underline{\Delta} : \underline{\Gamma} \subseteq \underline{\Delta}$ and for some $i \geq 0$, $\underline{\Delta}$ is a free L_i-saturated set$\}$. For $\underline{\Delta}$ a free L_i-saturated element of K:

(i) $B \to C \in \underline{\Delta} \Leftrightarrow \Lambda\underline{\Delta}'_{\in K}(\underline{\Delta} \subseteq \underline{\Delta}' \Rightarrow (B \in \underline{\Delta}' \Rightarrow C \in \underline{\Delta}'))$.
(ii) $\neg B \in \underline{\Delta} \Leftrightarrow \Lambda\underline{\Delta}'_{\in K}(\underline{\Delta} \subseteq \underline{\Delta}' \Rightarrow B \notin \underline{\Delta}')$.
(iii) $\forall x B \in \underline{\Delta} \Leftrightarrow$ for all L_{i+j}-saturated $\underline{\Delta}' \in K$ such that $\underline{\Delta} \subseteq \underline{\Delta}'$, $\Lambda t_{\in C_{L_{i+j}} \cup V_L}[E!t \in \underline{\Delta}' \Rightarrow B^t/x \in \underline{\Delta}']$ $(j \geq 0)$.

The proofs of (i) and (ii) are as in Thomason (1965). The ⇒ direction of (iii) is as in Thomason (1965) except that appeal is to $F(\forall E)$. In the ⇐ direction one reasons as follows: If $\neg \exists x E! x \in \underline{\Delta}$, then automatically $\forall x B \in \underline{\Delta}$. $\neg \exists x E! x \notin \underline{\Delta} \Rightarrow \underline{\Delta} \cup \{E!c\}$ is consistent for all $c \in C_{L_{i+1}} - C_{L_i}$ (Corollary 2). Fix such a c, and suppose $\forall x B \notin \underline{\Delta}$. By $F(\forall I)$, $\underline{\Delta} \cup \{E!c\} \vdash B^c/x$. By Corollary 1, there is a L_{i+1}-saturated, $\underline{\Delta}' \supseteq \underline{\Delta}$, in K, such that $E!c \in \underline{\Delta}'$ but $B^c/x \in \underline{\Delta}'$. A similar claim and argument hold for wffs of the form $\exists x B$.

One defines an mf $\langle K, R, \alpha_0, D, E \rangle$ as follows. $\alpha_0 = \underline{\Gamma}$, R is set-theoretic inclusion. $D = V_L \cup C_{L'}$, where L' is like L except $C_{L'} = \bigcup_{i < \omega} C_{L_i}$. And

$$E_{\underline{\Delta}}(t) = \begin{cases} e & \text{if } E!t \in \underline{\Delta} \\ n & \text{if not} \end{cases}$$

The interpretation, I', of L' is gotten by: $I'(t) = t$; $I'_{\underline{\Delta}}(P^0) = 1$ iff $P^0 \in \underline{\Delta}$; for $i \geqslant 1$, $I'_{\underline{\Delta}}(P^i) = \{\langle t_1, \ldots, t_i \rangle : P^i t_1 \cdots t_i \in \underline{\Delta}\}$. As in Thomason (1965), one proves by induction on $A \in W_{L_i}$ that for all free-L_i-saturated $\underline{\Delta} \in K$, $I'_{\underline{\Delta}}(A) = 1$ iff $A \in \underline{\Delta}$. For illustration, take $A = \forall x B$. $A \in \underline{\Delta} \Leftrightarrow \wedge \underline{\Delta}'_{\underline{\Delta} R \underline{\Delta}'} \wedge t(E_{\underline{\Delta}'}(t) = e \to Bt/x \in \underline{\Delta}')$ (by iii) $\Leftrightarrow \wedge \underline{\Delta}'_{\underline{\Delta} R \underline{\Delta}'} \wedge t(E_{\underline{\Delta}'}(t) = e \to (I't/x)_{\underline{\Delta}'}(B) = 1)$ (by Lemma 1 and induction hypothesis) $\Leftrightarrow I'_{\underline{\Delta}}(B) = 1$.

Thus, $\underline{\Gamma} = \{A : I'_{\Gamma}(A) = 1\}$, in particular. Let I be the restriction of I' to L. $\langle \langle K, R, \alpha_0, D \rangle, I \rangle$ is the Meinongian model which meets the requirements of the lemma. ●

Lemma 5. *If $\underline{\Gamma}$ is a free-L-saturated set* (*in the sense of Section 1.E*) *then there is a weakly modified Meinongian model*:

$$\langle \langle K, R, \alpha_0, D, E \rangle, I \rangle \text{ where } \underline{\Gamma} = \{A : I_{\alpha_0}(A) = 1\}$$

Proof. Proceeds as in Lemma 4 with the following additions:
—Add claim (iv): For all $\underline{\Delta} \in K$, if $\underline{\Delta}$ is a free-L_k-saturated set then for all i and all $t_1, \ldots, t_k \in V_L \cup C_{L_k}$, $P^i t_1 \cdots t_i \in \underline{\Delta} \Rightarrow (E!t_1 \& \cdots \& E!t_i) \in \underline{\Delta}$. (Proof by the remark following Theorem 4 in the text).
—Define the mf $\langle K, R, \alpha_0, D, E \rangle$ precisely as in the proof of Lemma 4, and I' as in Thomason (1965, p. 5), except that $I'_{\underline{\Delta}}(P^i) \subseteq (\{t : E_{\underline{\Delta}}(t) = e\})^i$.
—In proving $\underline{\Delta} = \{A : I'_{\underline{\Delta}}(A) = 1\}$ we need to observe in the basis clause that $I'_{\underline{\Delta}}(P^i t_1 \cdots t_i) = 1 \Leftrightarrow \langle t_1, \ldots, t_i \rangle \in I'_{\underline{\Delta}}(P^i) \Leftrightarrow (P^i t_1 \cdots t_i \& E!t_1 \& \cdots \& E!t_i) \in \underline{\Delta} \Leftrightarrow P^i t_1 \cdots t_i \in \underline{\Delta}$. (By $E!I$.) ●

Lemma 6. *If $\underline{\Gamma}$ is a free-L-saturated set* (*in the new sense*) *then there is a strongly modified Meinongian model $\langle \langle K, R, \alpha_0, D, E \rangle, I \rangle$ such that $\underline{\Gamma} = \{A : I_{\alpha_0}(A) = 1\}$.*

Proof Sketch. Let $L = L_0$ and let L_{i+1} be like L_i except that $C_{L_{i+1}} = C_{L_i} \cup \{c_1^{i+1}, \ldots\}$, where each c_1^{i+1} is foreign to L_i and all are distinct. L' is like L except $C_{L'} = \bigcup_{i<\omega} C_{L_i}$. Let $K = K_0 \cup K_1$. $K_0 = \{\underline{\Gamma}\} \cup \{\underline{\Delta}: \underline{\Gamma} \subseteq \underline{\Delta}, \underline{\Delta}$ is L_k-saturated for some $k \geq 0$, and for all $t \in V_L \cup C_{L_k}(E!t \in \underline{\Delta}$ or $\neg E!t \in \underline{\Delta})\}$. (Such $\underline{\Delta}$ can always be found by Corollary 3.)

$$K_1 = \{A: (\Gamma_1 \cup \Gamma_2) \vdash^+ A\}$$

$$\Gamma_1 = \{P^i t_1 \cdots t_i: P^i t_1 \cdots t_i \in \underline{\Gamma}\}$$

$$\Gamma_2 = \{E!t: t \in V_L \cup C_L\}$$

The following claims hold for all $\underline{\Delta} \in K_0$:

(i) $B \to C \in \underline{\Delta}$ iff for all $\underline{\Delta}' \in K_0$ such that $\underline{\Delta} \subseteq \underline{\Delta}'$, $C \in \underline{\Delta}'$ if $B \in \underline{\Delta}'$.
(ii) $\neg B \in \Delta$ iff for all $\underline{\Delta}' \in K_0$ such that $\underline{\Delta} \subseteq \underline{\Delta}'$, $B \notin \underline{\Delta}'$.
(iii) $\forall x B \in \underline{\Delta}$ iff for all free-L_k-saturated $\underline{\Delta}' \in K_0$ such that $\underline{\Delta} \subseteq \underline{\Delta}'$ and for all $t \in C_{L_k} \cup V_L(E!t \in \underline{\Delta}' \Rightarrow Bt/x \in \underline{\Delta}')$.

Proof as in Lemma 4, except that K there is replaced by K_0.

Define R as set-theoretic inclusion and $\alpha_0 = \underline{\Gamma}, D = V_L \cup C_L$, and

$$E_\alpha(t) = \begin{cases} e & \text{if } E!t \in \alpha \\ n & \text{if not} \end{cases}$$

$$I'(t) = t$$

$$I'_\alpha(P^i) = (\{\langle t_1, \ldots, t_i \rangle: P^i t_1 \cdots t_i \in \alpha\} \cap$$
$$\{\langle t_1 \cdots t_i \rangle: [E!t_1 \& \cdots \& E!t_i] \in \alpha\}), i > 0$$

$$I'_\alpha(P^0) = 1 \text{ iff } P^0 \in \alpha$$

The recursion for satisfaction is as usual, with (0a') in place of (0a).

The next claim to be proved is that for all $\underline{\Delta} \in K_0$ and all $A \in W_L$, $I'_{\underline{\Delta}}(A) = 1$ iff $A \in \underline{\Delta}$. The proof is by induction on A. For compound A the cases follow easily or by use of claims (i)–(iii), and the arguments can be adapted from Thomason (1965) with practically no change. When A is $E!t$ or P^0 the claims are trivial. The only remaining case then is the basis case: $A = P^i t_1 \cdots t_i$, $t_1, \ldots, t_i \in C_{L_k} \cup V_L$. Say $A = P^i t_1 \cdots t_i$ and $\underline{\Delta} \in K_0$. Subcases as follows:

(i) $A \in \underline{\Delta}$. (a) $(E!t_1 \& \cdots \& E!t_i) \in \underline{\Delta}$. Then clause (1) of (0a') is satisfied, and $I'_{\underline{\Delta}}(A) = 1$. (b) Neither $(E!t_1 \& \cdots \& E!t_i)$ nor $\neg(E!t_1 \& \cdots \& E!t_i)$ is in $\underline{\Delta}$. By the conditions on K_0, this can only occur when $\underline{\Delta} = \underline{\Gamma}$. Then $A \in \underline{\Delta}'$ for all $\underline{\Delta}' \in K_0$, and $A \in \Gamma'$ as well. Thus (2b) of (0a') is satisfied.

(2a) is satisfied in virtue of Corollary 1. So $I'_{\underline{\Delta}}(A) = 1$. (Notice: the possibility of $(E!t_1 \& \cdots \& E!t_i) \in \underline{\Delta}$ is ruled out by $(\neg E!E)$.)

(ii) $A \notin \underline{\Delta}$. Thus (2b) is violated by Γ', and (1) is violated by $\underline{\Delta}$. So $I'_{\underline{\Delta}}(A) = 0$.

Finally, to get the desired model, let I be the restriction of I' to L. ●

Lemma 8. *If $\underline{\Gamma}$ is a free-L-saturated set (in the sense of $\vdash^*$, then there is an mf $\langle K, R, \alpha_0, D, E \rangle$ and an interpretation, I, of L on $\langle K, R, \alpha_0, D, E \rangle$ satisfying* (iv') *and* (0a*) *such that* $\underline{\Gamma} = \{A : I_{\alpha_0}(A) = 1\}$.

Proof. The proof is identical to that of Lemma 6 except that in showing $P^i(t_1 \cdots t_i) \in \underline{\Delta}(\in K_0) \Rightarrow I'_{\underline{\Delta}}(P^i t_1 \cdots t_i) = 1$ one parses the cases as follows.

(i) $A \in \underline{\Delta}$. (a) $(E!t_1 \& \cdots \& E!t_i) \in \underline{\Delta}$: Then (1) of (0a*) is satisfied, so $I'_{\underline{\Delta}}(A) = 1$. (b) $(E!t_1 \& \cdots \& E!t_i) \notin \underline{\Delta}$: Then $(A \to (E!t_1 \& \cdots \& E!t_i)) \notin \underline{\Delta}$ and thus $(A \to (E!t_1 \& \cdots \& E!t_i)) \notin \underline{\Gamma}$, (by construction of K_0). But then (by $F(v \to I)$ and the saturation of $\underline{\Gamma}$) $A \in \underline{\Gamma}$. Hence $A \in \underline{\Delta}$ for all $\underline{\Delta} \in K_0$ and $A \in \Gamma'$ and $(E!t_1 \& \cdots \& E!t_i) \in \Gamma'$. (2b) and the new (2a) are satisfied, and $I'_{\underline{\Delta}}(A) = 1$.

(ii) $A \notin \underline{\Delta}$. Again, (1) is violated by $\underline{\Delta}$ and (2b) by Γ', so $I'_{\underline{\Delta}}(A) = 0$.[17] ●

Lemma 9. *For any $A \in W_L$, for any $M \in \mathbf{M}$, for any $\alpha \in K_M$: $I_\alpha(A) = (I_{f(M)})_\alpha(A)$.*

Proof. By induction on the number of logical symbols in A. Say $M = \langle \langle K_M, R_M, \alpha_{0_M}, D_M, E_M \rangle, I_M \rangle$ (which I will write without subscripts). Let $\alpha \in K$ be given.

Case (0a): $A = P^i t_1 \cdots t_i$. If $I_\alpha(A) = 1$ then trivially $(I_{f(M)})_\alpha(A) = 1$. $I_\alpha(A) = 0$ reduces to two possibilities. (i) $E_\alpha(I(t_1)) = \cdots = E_\alpha(I(t_i)) = e$. Then $\{d_1, \ldots, d_i\} \subseteq D_\alpha$. $I(t_i) = d_j$ $(1 \leq j \leq i)$ and $\langle d_1, \ldots, d_i \rangle \notin I_\alpha(P^i) \Rightarrow \langle d_1, \ldots, d_i \rangle \notin (I_{f(M)})_\alpha(P^i) \Rightarrow (I_{f(M)})_\alpha(A) = 0$. (ii) $E_\alpha(d_j) = n$ for some j $(1 \leq j \leq i)$. But then $(I_{f(M)})_\alpha(t_j)$ is undefined, and $(I_{f(M)})_\alpha(A) = 0$.

Case (0b): $A = E!t$. If $I_\alpha(A) = 1$ then $E_\alpha(I(t)) = e$. Let $d = I(t)$, $d \in (D_{f(M)})_\alpha$, and thus $(I_{f(M)})_\alpha(t)$ is defined. So $(I_{f(M)})_\alpha(A) = 1$. If $I_\alpha(A) = 0$, then $E_\alpha(I(t)) = n \Rightarrow (I_{f(M)})_\alpha(t)$ is undefined $\Rightarrow (I_{f(M)})_\alpha(A) = 0$.

The remaining propositional cases are straightforward.

Case 5: $A = \forall xB$. $I_\alpha(A) = 0 \Rightarrow \vee \beta_{\in K_M} \vee d_{\in D}(\alpha \mathrel{R} \beta$ and $E_\beta(d) = e$ and $(Id/x)_\beta = 0)$. Fixing β and d, note that $M' = \langle \langle K, R, \alpha_0, D, E \rangle, Id/x \rangle \in \mathbf{M}$, and thus $\langle \langle K_{f(M)}, R_{f(M)}, \alpha_{0_{f(M)}}, D_{f(M)} d/x \rangle$ is a well defined element of $\mathbf{R}$, which in fact bears the same relation to $f(M)$ as M' bears to $\mathbf{M}$. By the induction hypothesis, $(I_{f(M)} d/x)_\beta(B) = 0$. But then $(I_{f(M)})_\alpha(A) = 1$. And similarly Case 6, when $A = \exists xB$. ●

Lemma 10. $\Gamma \mid A$ *iff* $(I_M^+)_{\alpha^+}(A) = 1$.

Proof. (Basically as in Smorynski (1973), sect. 5.1.18.) By induction on the length of A. For any atomic A, $\Gamma \mid A$ iff $(I_M^+)_\alpha^+(A) = 1$ (by definition in each case). As a representative inductive case take $A = \neg B$. Say $(I_M^+)_{\alpha^+}(A) = 1$ then $(I_M^+)_{\alpha^+}(B) = 0$, and by the induction hypothesis, $\Gamma \nvdash B$. Also by remark (1), $\Gamma \vdash \neg B$, so $\Gamma \mid \neg B$. Conversely, say $(I_M^+)_{\alpha^+}(\neg B) = 0$. If $(I_M^+)_\alpha B = 1$, then, by the hypothesis of induction, $\Gamma \mid B$ and $\Gamma \nvdash \neg B$. If $(I_M^+)_\alpha B = 0$ as well, then for some β such that α_0 R β, $(I_M^+)_\beta(B) = 1$; and thus $\Gamma \nvdash \neg B$, again entailing $\Gamma \nvdash \neg B$. Remaining cases similarly. •

Theorem 16. *Let Γ be a consistent set of sentences. The following are equivalent*: (i) $\{A : \Gamma \vdash A\}$ *is a free-L-saturated set.* (ii) $\{A : \Gamma \vdash A\} = \{A : \Gamma \mid A\}$.

Proof. (i) ⇒ (ii). (a) $\{A : \Gamma \mid A\} \subseteq \{A : \Gamma \vdash A\}$ by Lemma 10 and remark (1) following the definition of M^+. (b) Say $\{A : \Gamma \vdash A\}$ is a free-L-saturated set. And suppose $\Gamma \vdash A$. $\Gamma \mid A$ follows by induction on A: A atomic and $A = B \& C$ by definition. $A = B \vee C$ and $A = \exists x B$, by saturation. For $A = B \to C$ we must show $\Gamma \mid B \Rightarrow \Gamma \mid C$. By (a), $\Gamma \mid B \Rightarrow \Gamma \vdash B$; and since $\Gamma \vdash B \to C$, $\Gamma \vdash C$. So, by the induction hypothesis, $\Gamma \mid C$. For $A = \neg B$ the result follows by (a) and consistency. Finally, for $A = \forall x B$ we need to show $\Gamma \mid E!c \Rightarrow \Gamma \mid Bc/x$. But $\Gamma \mid E!c \Rightarrow \Gamma \vdash E!c \Rightarrow$ (by $F\forall E$) $\Gamma \vdash Bc/x \Rightarrow$ (by induction hypothesis) $\Gamma \mid Bc/x$.

(ii) ⇒ (i) follows from Lemma 10 and remark (2). •

Lemma 11. $\{A : RH(A) \text{ and } \Gamma \vdash A\} \subseteq \{A : \Gamma \mid A\}$.

Proof. By induction on A. For instance, for A atomic it is true by definition. Now say for example that $A = (B \to C)$.

Then $RH(A) \Rightarrow RH(C)$. Say $\Gamma \vdash A$. To prove $\Gamma \mid A$, we must show $\Gamma \mid B \Rightarrow \Gamma \mid C$. Now $\Gamma \mid B \Rightarrow \Gamma \vdash B$, and thus $\Gamma \vdash C$. Then $\Gamma \mid C$ by the hypothesis of induction.

Other cases are treated similarly. The cases of $A = (B \vee C)$, $A = E!c$, and $A = \exists x B$ do not arise, since for these $RH(A)$ fails. •

Theorem 18. *Each FHA satisfies* (2).

Proof. Apply Corollary 8, taking Γ as the axioms of *FHA*. The only ones not satisfying RH are (1) and instances of (2)(b) and (5). $\Gamma \mid B$ immediately when $B = E!0$ or $B = E!\overline{n+1}$. To show $\Gamma \mid B$ when B is an instance of (5) apply Lemma 10. (Without loss of generality assume x is the only free variable of A.) Say $I_\alpha^+(B) = 0$. Since B is an axiom $I_{\alpha_0}(B) = 1$, and thus $I_\alpha(B) = 1$ for all $\alpha \in K_M$. (Or for all $\alpha \in (K_0)_M$, if we are dealing with one of the modified logics.) Thus $I_\alpha^+(A^0/x \& \forall x \forall y[(A \& S(x, y)) \to Ay/x]) = 1$, but $I_\alpha^+(\forall x A) = 0$. Hence there is an $\alpha \in K_M^+$ and a least natural number,

n, such that $I_\alpha(A^{\bar{n}}/x) = 0$. This is quickly seen to be impossible, since $I_\alpha(A^{\bar{0}}/x) = 1$ and $I_\alpha(\forall x \forall y[(A \& S(x, y)) \to Ay/x]) = 1$. ●

References

Axcel, P., "Saturated intuitionistic theories," in *Contributions to Mathematical Logic* (ed. Schmidt, Schütte, and Thiele), North Holland, Amsterdam (1968) pp. 1–11.
Bencivenga, E., "Free semantics," *Boston Studies in the Philosophy of Science*, vol. 47 (1980) pp. 31–48.
Brouwer, L. E. J., "Intuitionistische Zerlegung mathematischen Grundbegriffe," *Jber. Deutsch. Math. Verein.*, vol. 33 (1924) pp. 251–256 [pp. 275–80 in Brouwer 1975].
Brouwer, L. E. J., "Über Definitionsbereiche von Funktionen," *Math. Annalen*, vol. 97 (1927) pp. 60–75 [pp. 390–405 in Brouwer 1975].
Brouwer, L. E. J., "Willen, weten, spreken," *Euclides*, Groningen, vol. 9 (1933) pp. 177–93 [pp. 443–46 in Brouwer 1975].
Brouwer, L. E. J., *Collected Works*, vol. 1 (ed. A. Heyting), North Holland, Amsterdam (1975).
Burge, T., "Truth and singular terms," *Nous*, vol. VIII (1974) pp. 309–25.
Cocchiarella, N., "On the primary and secondary semantics of logical necessity," *Journal of Philosophical Logic*, vol. 4 (1975) pp. 13–28.
Grandy, R., "Predication and singular terms," *Nous*, vol. XI (1977) pp. 163–67.
Heyting, A., "Die formalen Regeln der intuitionistischen Logik," *Sitzungsber. preuss. Akad. Wiss.* (1930) pp. 42–56.
Hintikka, J., *Logic, Language Games and Information*, Oxford University Press (1973).
Kleene, S. C., "Disjunction and existence under implication in elementary intuitionistic formalisms," *J. of Symbolic Logic*, vol. 27 (1962) pp. 11–18.
Kleene, S. C. and Vesley, R., *Foundations of Intuitionistic Mathematics*, North Holland, Amsterdam (1965).
Kripke, S., "Semantical analysis of intuitionistic logic, I," in *Formal Systems and Recursive Functions* (ed. Crossley and Dummett), North Holland, Amsterdam (1965) pp. 92–130.
Lambert, K., "Notes on E!, III: A theory of descriptions," *Philosophical Studies*, vol. 13 (1962) pp. 51–59.
Leblanc, H. and Gumb, R., "Soundness and completeness proof for three brands of intuitionistic logic," in *Essays in Epistemology and Semantics* (ed. Leblanc, Gumb and Stern), Haven (1981).
Leblanc, H. and Thomason, R., "Completeness theorems for some presupposition-free logics," *Fundamenta Mathematicae*, vol. 62 (1968) pp. 125–64.
Myhill, J., "Formal systems of intuitionistic analysis, I," in *Logic Methodology and Philosophy of Science, III* (ed. van Rootselaar and Staal), North Holland, Amsterdam (1968) pp. 161–78.
Posy, C., "Brouwer's constructivism," *Synthese*, vol. 27 (1974) pp. 125–59.
Robinson, T., "Interpretations of Kleene's metamathematical predicate in intuitionistic arithmetic," *J. of Symbolic Logic*, vol. 3 (1965) pp. 140–54.

Schock, R., *Logics Without Existence Assumptions*, Almqvist & Wiksell, Stockholm (1968).
Smorynski, C., "Applications of Kripke models," in *Metamathematical Investigations of Intuitionistic Arithmetic and Analysis* (ed. A. S. Troelstra), Springer (L.N.M. 344) (1973) pp. 324–91.
Stenlund, S., "Descriptions in intuitionistic logic," in *Proc. Third Scandinavian Logic Symp.* (ed. S. Kanger), North Holland, Amsterdam (1975) pp. 197–212.
Thomason, R., "On the strong semantical completeness of the intuitionistic predicate calculus," *J. of Symbolic Logic*, vol. 33 (1965) pp. 1–7.
Troelstra, A., "Intuitionistic formal systems," in *Metamathematical Investigations of Intuitionistic Arithmetic and Analysis* (ed. A. Troelstra), Springer (L.N.M. 344) (1973) pp. 1–96.
Van Fraassen, B., "The completeness of free logic," *Zeitschrift für math. Logik und Grundlagen der Math.*, vol. 12 (1966) pp. 219–34.
Van Fraassen, B., "Singular terms, truth-value gaps, and free logic," *Journal of Philosophy*, vol. 67 (1966) pp. 481–95.

Notes

1. A semantics respects predication when it holds that predicates and singular terms are representational and bases the truth of an atomic sentence on the manner in which the entities represented by the predicate and singular terms occurring in that sentence are related. I wish to make no claims about the ontological nature of these represented entities. In particular I do not require that the entities represented by singular terms be "objects" and that the representation relation be standard reference. (See Note 8.)
2. There are other semantic techniques for free logic (e.g., the method of intermediate truthvalues) but these are philosophical variations of the four considered above.
3. I am referring here to Bencivenga's "secondary auxiliary valuations." These are valuations based on a structure M', which is a "completion" of a structure M. If, for example, $\neg E!t$ is true in M, then under these valuations $\neg E!t$ is true at M', even though t might denote in M'. In note 20 to Bencivenga (1980) the author speaks of M and M' as "possible worlds."
4. Thomason's system was accidental. He simply neglected to guarantee that each term denote at every node in a model structure. When the omission was pointed out to him, he decided not to change the semantics on the grounds that a free logic was more appropriate for intuitionism than the logic originally proposed in Thomason (1965).
5. As a consequence, Stenlund's semantics cannot capture certain of Brouwer's claims. For instance, in Brouwer (1975, pp. 390–405), Theorem 1, Brouwer clearly wants to establish the truth of $A(f, \xi_0) \to \neg \exists y[f(\xi_\omega) = y]$ where A states that f is discontinuous at ξ_0. On Stenlund's semantics the consequent and thus the entire conditional must be truthvalueless. (By contrast, one can construct Thomason-style model structures in which $\neg \exists x(x = t)$ is true at some nodes. Though it turns out that $\neg \exists x(x = t)$ will never be true at a base node (representing the actual world) whenever there is such a node).

6. In Thomason's semantics the inference $(t = t) \Vdash \exists x(x = t)$ is not valid. We can also invalidate $At \Vdash \exists x(x = t)$ for the case in which A is a logical theorem by allowing the domain D_α to be empty for some α in K. Similarly for Kripke's original semantics in Kripke (1965).
7. Strictly speaking an outer domain approach should qualify as "Meinongian" only if the outer domain contains impossible objects like the round square. Once the language admits definite descriptions, the corresponding semantics cannot avoid taking a stand on the status of such objects. For the present, even though this paper does not deal directly with analyzed terms, I have chosen to call the outer domain version a "Meinongian" semantics on the weaker excuse that the intuitionistic negation in $\neg \exists x(x = t)$ has a quasimodal character.
8. Ontological purists might still object to the fact that $I(t)$ is defined at nodes in which $E_\alpha(t) = n$. Often these purists wonder what can be the referent of a nondenoting term. However, the advocate of (iv') might respond that in general $I(t)$ is the *sense* rather than the traditional *reference* of t.
9. To guarantee the unqualified validity of $t = t$ one can adopt a quasi-Fregean approach, requiring $I(t) \in U_{\alpha \in K}\{d: E_\alpha(d) = e\}$. This of course is most natural on the view that $I(t)$ is the *referent* of t and not the sense as suggested in Note 8. The quasi-Fregean device is not transferable to semantic systems for analyzed terms. For in the theory of analyzed terms there are t such that $\neg E!t$ is a logical theorem. Of course it is precisely such terms that call the general validity of $t = t$ into question.
10. The validity of $(\forall xPx \,\&\, \neg\neg E!t \to Pt)$ together with the validity of $(\forall E!)$ (Section 1.B) does not, however, entail $\neg\neg E!t \to E!t$. For $E!$ is not treated as an ordinary unary property. *Even in free logic, existence need not be a predicate.*
11. Assigning $I_\alpha(A) = 0$ in this case has the effect of full-blooded irreversible falsity for atomic A of the form $P^i t_1 \cdots t_i$ where all the terms denote. It does not have this effect for other wffs.
12. The double negation here is of course not the usual intuitionistic one. It expresses possibility in a broader context. Thus $\neg\neg$(Pegasus is white) does not entail that we must encounter a white Pegasus somewhere along every branch stemming from α_0, as it would in ordinary intuitionistic semantics. It entails only that a white Pegasus is "conceptually possible."
13. These purists might be viewed as insisting that $I(t)$ be the reference of t, and not its sense. That insistence, together with a refusal to countenance nonexistent objects, would recommend the Russellian semantics.
14. Even though the Russellian purist is likely to take $I_\alpha(t)$ as the *referent* of t, the relation R is still read epistemically, as a growth of knowledge about some objects, their properties and their relations. The nodes accessible to α under R represent the growth of knowledge about the temporally stable state described by R. That is why once $I_\alpha(t)$ is defined, it may not change from α to accessible β.
15. We can add identity in the same way as in Section 1.F. $\sim_\alpha$ will partition D_α. We could then treat definite descriptions as follows: Define I for terms and formulas simultaneously, with the final clause (modulo (iib)) as: $I_\alpha(\iota xA)$ is defined if there is a $d \in D_\alpha$ such that $I_\alpha(\exists x(\forall y(A^y/x \to y = x))) = 1$, in which case $I_\alpha(\iota xA)$ is set equal to any such d. $I_\alpha(\iota xA)$ is undefined otherwise.

This semantics will validate $\exists!xA \to A(\iota xA)$, but will invalidate $(B \to \exists!xA) \to \exists x(B \to A)$. That is the combination that motivated Stenlund's (1975) approach.

Once again, the seeming impropriety of letting ιxA have a single designation in all the nodes accessible to α is mitigated when we think of R in epistemic rather than ontological terms. Thus, having established $t = \iota xA$ at α means that no future evidence will refute that claim. It does not mean that the world may not evolve in such a way as to contradict a similar claim made about some other time or situation.

16. I am indebted to E. Bencivenga who guided my first steps in free logics and to K. Lambert and R. Thomason for many helpful comments.

17. The proof of Lemma 8 can be adapted along the following lines for use in a generalization of the semantics to include analyzed terms. Let $\Gamma^* = \{A : \Gamma' \vdash A\}$, $\Gamma' = \Gamma_1 \cup \Gamma_2 \cup \Gamma_3$. $\Gamma_1 = \{P^i t_1 \cdots t_i : P^i t_1 \cdots t_i \in \Gamma\}$. $\Gamma_2 = \{E!t : P^i t_1, \ldots, t, \ldots, t_i \in \Gamma \text{ for some } i, P^i, \text{ and } t_1, \ldots, t_i\}$, $\Gamma_3 = \{E!t : \neg E!t \in \Gamma\}$.

The cases in the above argument are handled as follows:

(i) $A \in \Delta$: as in Lemma 8.

(ii) $A \notin \Delta$. Then $A \notin \Gamma$ and $A \notin \Gamma^*$. (a) $\neg A \in \Gamma$. Then $\neg A \in \underline{\Delta}$ and $\neg A \in \underline{\Delta}'$, for all $\underline{\Delta}' \in K_0$. So either (2a) or (2b) of (0a*) is violated, and $I'\underline{\Delta}'(A) = 0$ for all $\underline{\Delta}'$ including $\underline{\Delta}$. (b) $\neg A \notin \Gamma$. Then $\neg (E!t_1 \& \cdots \& E!t_i) \notin \underline{\Gamma}$ (by $F(v \to I)$), since by saturation $A \to (E!t_1 \& \cdots \& E!t_i) \in \underline{\Gamma}$) and $(E!t_1 \& \cdots \& E!t_i) \in \underline{\Gamma}^*$. But then (2b) is violated. So since (1) is violated by $\underline{\Delta}$, $I'_{\underline{\Delta}}(A) = 0$.

4
Singular Terms, Truthvalue Gaps, and Free Logic

Bas C. van Fraassen

In Strawson's paper "On Referring"[1] the idea was advanced that a simple, syntactically well-formed statement may in certain circumstances be neither true nor false. The circumstances in question are those in which some singular term occurring in the statement does not have a referent. The present paper is not concerned with Strawson's work on this subject (nor does it use his terminology). Rather, we wish to explore the consequences of this view, as we have formulated it above, for logic and formal semantics.

In the above formulation, 'statement' is used to refer to formulas, in simple artificial languages (such as are studied in elementary logic), which do not contain free variables. 'Singular term' is used to mean "name or definite description." Strawson's own example of a statement that is neither true nor false by the above criterion is

1. The king of France is wise.

Another such example, in which the singular term is a name, is

2. Pegasus has a white hind leg.

We assume that whatever reasons incline one not to assign a truthvalue to 1 are operative also in the case of 2. Thus, the fact that Pegasus does not exist may be taken to mean that the question whether he has a white hind leg does not arise, and so forth. (While I do not intend to defend the position, I must admit that, although I can think of many artificial ways to bestow a certain truthvalue on (2), I cannot find a single plausible reason to call it true or to call it false.) Since definite descriptions introduce certain complexities, we shall from here on assume that all our singular terms are names.

1

Among logicians, there have been two basic reactions to the idea. One is that the logic of a language for which this is so can also be explored by the usual methods, though we may expect the relevant logical system to be quite unordinary. Thus Prior, in *Time and Modality*, explores many-valued logics for this reason. The other reaction is that, for ordinary purposes, the logician can restrict his attention to languages for which this is not so. Thus, if in a given language

3.   The king of France is bald.

means what Russell said it meant, then it has a truthvalue. And even if Strawson is correct concerning ordinary discourse, sentences that in his view are neither true nor false are "don't cares" for all ordinary purposes, and there is therefore no reason why we should not arbitrarily assign them some truthvalue. This will be convenient, since it will make standard logical techniques applicable.

Both reactions appear to start from the tacit supposition that a language in which some statements are neither true nor false must have a very unordinary logical structure. (This supposition may have been motivated by the custom of logicians to treat 'is neither true nor false' on a par with 'has a third value which is neither *True* nor *False*'.) My thesis will be that this supposition is not quite correct. Before going on, I want to note that my conviction on this point is a matter of hindsight on the basis of the development of so-called *free logic* and that I have no quarrel with the reactions outlined above. I think they are correct (certainly not mutually exclusive) and well-supported; only to accept them now would be to overlook something which has come to light since they were formulated.

2

The view we wish to consider is that a simple statement (an *n*-adic predicate followed by *n* proper names) has a truthvalue if and only if all the names it contains have referents. As long as we deal with the logic of unanalyzed propositions, however, the only relevant distinction we are able to draw is that between those sentences that are true, those that are false, and those that are neither. (For example, we cannot distinguish there between those statements that contain names and those that do not.) This suggests that for our purposes it will be best to regard propositional logic as but a truncated part of the first-order predicate calculus. However, we shall begin with an intuitive discussion of the logic of propositions, with the purpose of introducing some relevant concepts.

3

For the sake of perspicuity, let us consider an argument with English sentences:

4. a. Mortimer is a man.
    b. If Mortimer is a man, then Mortimer is mortal.
    c. Mortimer is mortal.

Should our present view, that (4a) is neither true nor false if 'Mortimer' does not refer, cause us to qualify our precritical reaction that (4) is a valid argument? I think not. This naive reaction is not based simply on the conviction inculcated by elementary logic courses that questions of validity can be decided on the basis of syntactic form. It can also be based quite soundly on the semantic characterization of validity found in many logic texts:

5. An argument is valid if and only if, were its premisses true, its conclusion would be true also.[2]

The fact that Mortimer has to exist for the premisses of (4) to be true is just as irrelevant to the validity of that argument as any other factual precondition for the truth of those premisses. Were Mortimer a man and were it the case that if he is a man then he is mortal, then it would be the case that he is mortal—this is exactly why (4) is valid. Hence, acceptance of (5) leads to the very welcome conclusion that all the same arguments are still valid, as far as propositional logic is concerned.

The reader may have thought my conclusion about all arguments rather audacious, since my reasoning pertained to a single example. Let us therefore repeat this reasoning in a more abstract form. If a statement is neither true nor false, then it is not true. Hence the view we are presently considering has as consequence for truth that, under certain stated conditions, a given statement is not true. In the simple language of unanalyzed propositions that we are presently considering, these conditions cannot be made explicit in the premisses of an argument. An argument is valid provided, for any conditions under which the premises are true, the conclusion is true also. On the view above, which entails that under certain conditions (failure of reference) a statement cannot be true, this is *redundantly equivalent* to: for any conditions under which the premises are true and the relevant names all refer, the conclusion is also true. But that means that this view has no consequences vis-à-vis validity (in the present context; the situation would be different if failure of reference could be made explicit in the premises or conclusion).

From this we conclude that, as long as this view is not accompanied by other aberrant views on the propositional connectives, classical propositional logic validates just the right arguments. This property of propositional logic we may call its "argument-completeness" with respect to the present view.

However, validity of argument is not the only subject of interest in the logic of propositions. A second, related, subject is *logical truth* of statements. If we knew how to characterize logical truth, in accordance with the view we are presently investigating, then we could ask whether classical propositional logic is "statement-complete" with respect to this view—that is, whether all the logical truths are theorems. We could then also ask whether all the theorems of this logic are logical truths on this view (*soundness*).

It is not easy to see how we can characterize logical truth in the present context. The intuitive guide is the idea that logical truth is truth in all possible situations. But as long as we deal only with unanalyzed propositions, how can we reconstruct the relevant notion of situation? There is another possible approach: it seems plausible that the question of whether the theorems are exactly the logical truths cannot be independent of the question whether the arguments justified by the rules and theorems are exactly the valid arguments. It would seem that there is a certain connection between validity and logical truth, and if this connection were spelled out, we might be able to use the former notion to explicate the latter. We shall not look into this possibility further (though it suggests that the set of logical truths will turn out to be not too unfamiliar). Rather, we shall turn to the full first-order logic, with respect to which we *have* a plausible reconstruction of the notion of truth in all possible situations.

4

Let us consider, then, a simple first-order language with predicates, names, variables, connectives, and quantifiers. We wish to investigate to what extent standard logic remains applicable to this language if we do not assume all its names to refer. It will be convenient to use the logical system—which we shall call *ML*—of Quine's *Mathematical Logic*.[3] ML was formulated for a language like ours, except that no names belong to its vocabulary (we disregard the part that concerns set theory). The logical system is essentially a specification to the effect that all statements in that language that have *a certain form* are theorems of ML. We shall analogously designate all the statements of our language that have that form as "theorems of ML." For example, Quine specifies that all statements of the form

$(x)(A \supset B) \supset \cdot (x)A \supset (x)B$

are theorems. This means that, if *F* and *G* are monadic predicates, *R* a dyadic predicate, and *b* a name, of our language, then the following two statements are examples of theorems of ML (in our language):

$(x)(Fx \supset Gx) \supset \cdot (x)Fx \supset (x)Gx$

$(x)(Rxb \supset Gx) \supset \cdot (x)Rxb \supset (x)Gx$

Here our first example is a theorem of ML (in our language) that does not contain names, and the second is one that does.

Now truthvalue gaps occur in the present context only in connection with names. It follows that the standard logic will apply to all those statements in which no names occur. So we conclude:

6. If a statement is a theorem of ML and if no names occur in it, then it is logically true.

This much we know before we have even answered the question of how we can characterize logical truth here—provided, again, that the view regarding failure of reference that is presently under consideration is not accompanied by further exotic views on logic. I shall now try to show that furthermore we can drop the qualification about names from (6): *any* statement that is a theorem of ML in our language—even if it does contain names—can be regarded as logically true even on the present view.

To show this, we must first explicate what is meant by 'logically true', or 'true in all possible situations'. The explication runs as follows. To present a specific interpretation of a language, I must specify a *domain of discourse* and the *extensions* the predicates have in that domain. Suppose L is the language, *D*—which may be any nonempty set of things—the domain of a discourse, and *f* a function such that if *F* is a predicate of

L then $f(F)$ is the extension of F in D. Then the couple $(f; D)$ is called "a model of L." It is not difficult to see what counts as *truth* in such a model. For example

$(x)Fx$

will be true in this model exactly if the extension $f(F)$ of F is the whole of the domain D. These models are the reconstructed counterparts of Leibniz's possible worlds. Thus, a sentence is *logically true* if and only if it is true in any model $(f; D)$, no matter how D and f are chosen.

In this account, no mention is made of singular terms. Let us suppose that the language L has, in addition to predicates, variables, quantifiers, and the usual propositional connectives; also a set of names. Let us suppose that some of these names have referents and that some do not. In terms of the model, this means that, for some names t, $f(t)$ is defined and is a member of the domain D, and that for some other names that function is not defined. For example, let e be a name, and let its referent $f(e)$ be a member of the extension $f(F)$ of the predicate F. Then

Fe

is true in this model. If, on the other hand, the name e does not have a referent, then, according to the position we are presently examining, the sentence

Fe

is neither true nor false in this model. We must now ask which sentences of *this* language are logically true. In answer, I shall characterize the set of logically true sentences of this language, proceeding in two steps.

First, recall the proposal that the troublesome truthvalue gaps be eliminated by simply assigning truthvalues to the offending statements in some arbitrary manner. We begin by taking up this suggestion.

7. A classical valuation over a model is a function v that assigns T or F to each statement, subject to:
   a. if A is an atomic statement containing no nonreferring names, then $v(A)$ is determined by the model, in the indicated manner, and
   b. if A is a complex statement, then $v(A)$ is determined by what v assigns to the simpler statements, in the usual manner.

We have not put any conditions on $v(A)$ when A is an atomic statement containing some nonreferring name—except that $v(A)$ is either T or F. This means that, if there is any name e that has no referent in the domain of a given model and if $A(e)$ is an atomic statement in which e occurs, then there are at least two classical valuations over this model: one which assigns T to $A(e)$ and one which assigns F to $A(e)$.

8. A statement is *CL-true* (*false*) if and only if it is assigned T (F) by all classical valuations over all models.

It will not surprise anyone that all theorems of ML are CL-true by the above definition. Some familiar theorems of other systems, however, are not considered:

9. $Fa \supset (\exists x)Fx$

This is not CL-true by the above definition. Nor can it be a theorem of ML, since the latter allows us to deduce only:

$(y)(Fy \supset (\exists x)Fx)$

If we added to ML some rule to replace variables by names, then (9) would become a theorem. But since (9) is not logically true once we take into account the possibility that the name a does not refer, we shall of course not add such a rule.

The second point we wish to bring to the reader's attention is that the classical valuations go beyond the model to which they belong, just with respect to those terms that have no referent in the model. But there, they go beyond the model *in all possible ways*. The nature of the model determines in a given valuation not what is peculiar to it, but what it has in common with the other valuations over this model. So what the classical valuations over a model have in common we can take as correctly reflecting truth and falsity in the model.

10. A *supervaluation over a model* is a function that assigns T (F) exactly to those statements assigned T (F) by all the classical valuations over that model.

Supervaluations have truthvalue gaps. The following diagram shows how the supervaluation compares with the classical valuations over a model $(f; D)$ when the language is supposed to contain exactly one predicate F (monadic) and exactly two names, a and b, such that $f(a)$ is defined and $f(b)$ is not defined. There are here exactly two classical valuations, v_1 and v_2; the supervaluation we call s. Let us assume that $f(a) \in f(F)$ and that all of D is included in $f(F)$.

Singular Terms, Truthvalue Gaps, and Free Logic

	v_1	v_2	s
Fa	T	T	T
$\sim Fa$	F	F	F
Fb	T	F	—
$\sim Fb$	F	T	—
$Fb \vee \sim Fb$	T	T	T
$(x)Fx$	T	T	T
$(x)Fx \supset Fb$	T	F	—

The dashes indicate truthvalue gaps.

Now we can characterize logical truth in accordance with the view that is the subject of this paper.

11. A statement is *SL-true* if and only if it is assigned T by all supervaluations.

And the third point to make is that the set of CL-truths and the set of SL-truths are exactly the same! For if a supervaluation assigns T to *A* only if all corresponding classical valuations do, then *all* supervaluations assign T to *A* only if all classical valuations do. And conversely, since if all classical valuations over a model assign T to *A*, then so does the corresponding supervaluation, it follows that if all classical valuations over all models assign T to *A*, then so do all supervaluations.

Hence, since all theorems of ML are CL-true, they are also all SL-true. We have now shown what we set out to show in this section, namely that the qualification about names may be dropped from principle (6) above. Along the way we have also answered a question left open in the preceding section: all tautologies may be regarded as logically true, in complete accordance with the present view. (Because, of course, all theorems of the classical propositional calculus are also theorems of ML.)

We said above that the domain of discourse of a model must be a *nonempty* set. This was necessary because ML is not valid for the empty domain. However, Hailperin[4] and Quine[5] have shown how a slight revision in the axioms of ML—to yield what we shall call *revised ML*—makes it valid for empty domains also. The above argument comes through in its entirety when we allow domains to be empty, and substitute 'revised ML' everywhere for 'ML' (see Note 11).

5

Besides propositional inference and quantification, elementary logic comprises one further subject: *identity theory*. If we add the identity symbol to our language, we must ensure that at least the following hold:

12. $\vdash (x)(x = x)$

13. $\vdash (x)(y)(x = y \supset \cdot A(x) \supset A(y))$, no names occurring in $A(x)$.

(where '$\vdash$' is used as Quine uses it in ML: all instances of the schema given become theorems provided we supply initial universal quantifiers to bind any free variables in them.) This is simply to say that we must retain the usual identity theory at least for contexts in which truthvalue gaps cannot occur. (From now on, the distinction between names and variables is very important, but we also want to be able to talk about both at once. Let us use x, y, z for variables; a, b, c for names; and t, t' for either.)

In addition, it seems to me that we cannot plausibly reject that

14. $t = t'$

is false when t has a referent and t' does not.[6] (We should point out that (14) may not be a statement, since either t or t' may be a free variable. A free variable is always to be regarded as having been assigned some referent.) For example, that Santa Claus does not exist is sufficient reason to conclude that the president of the United States is not Santa Claus; that the Scarlet Pimpernel is a fictional character, sufficient reason to conclude that the man in the iron mask was not the Scarlet Pimpernel.

Given only this, it already follows that

   t exists

is appropriately expressed as

   $(\exists y)(y = t)$

and, furthermore, that the logical truth

15. $((x)A(x) \,\&\, (\exists y)(y = t)) \supset A(t)$

is the valid counterpart of "universal instantiation." In addition, however, I propose that we accept entirely without restriction:

16. $\vdash t = t$

17. $\vdash t = t' \supset \cdot A(t) \supset A(t')$

This extension appears to me to be reasonable because, first, it seems to me reasonable to adopt the principle that, if a sentence A is logically true, then any sentence obtained from A through a consistent substitution of singular terms for singular terms must also be logically true. Thus, if

Cicero = Cicero

is to be taken as *logically* true, then so must

Pegasus = Pegasus[7]

Furthermore, the system obtained through this extension of revised ML is not only correct, but also *statement-complete* under the present interpretation (disregarding for the moment the complications of description theory).

6

The logical system obtained through the extension discussed in the preceding paragraph is called "free logic" (the term is due to Karel Lambert). This variety of logic was developed by Leonard,[8] Hailperin and Leblanc,[9] Hintikka,[10] and Lambert.[11] We should mention here that the interpretation just given is not the only one under which free logic is sound and statement-complete. That is, its theorems are exactly the logical truths of the language by definitions (8) and (11), but also by an infinity of other definitions. These definitions yield a spectrum of interpretations. On the one extreme, we find logical truth identified with CL-truth; on the other extreme of the spectrum we find logical truth identified with SL-truth. Equivalently, on the one extreme, all sentences in the language are assigned a truthvalue (classical valuations), and this set of sentences that are either true or false diminishes as we move to the other extreme, where it is at a minimum. It is not difficult to see why some philosophers might prefer some intermediate interpretation on this spectrum. For there certainly are sentences in which there occur nonreferring singular terms and which we do assign a truthvalue. Examples are:

> The ancient Greeks worshipped Zeus.
> Pegasus is to be conceived of as a horse.
> The wind prevented the greatest air disaster in history.

Of these examples, the first two are due to Leonard, and the third to Chisholm. The proof of statement-completeness for the system, for this whole spectrum of interpretations, is given in a forthcoming article by this author.[12] The system has been extended to deal with definite descriptions by Lambert[13] and by van Fraassen and Lambert.[14]

7

We may recall that in the section on propositional logic we distinguished between *argument-completeness* and *statement-completeness*. Free logic is statement-complete under each of the interpretations mentioned: the logical truths are exactly the theorems of the system. But does it validate all the correct arguments? Its only rule is a rule for deriving theorems, not a rule for deriving true statements from other true statements. This rule, which is part of ML, reads:

18. If A and $A \supset B$ are both theorems, then B is also a theorem.

But the system can be used to show that arguments are valid, as follows:

19. Free logic validates an argument '$P_1, \ldots, P_n$; hence Q' if and only if '$P_1 \& \cdots \& P_n . \supset Q$' is a theorem of free logic.[15]

What it means to say that an argument is valid depends of course on the interpretation chosen. If we consider only the two extreme interpretations, then we have the following two characterizations of validity.

20. An argument is C-valid if and only if every classical valuation that assigns T to all its premisses also assigns T to its conclusion.

21. An argument is S-valid if and only if every supervaluation that assigns T to all its premisses also assigns T to its conclusion.

It is not difficult to see that every argument that is C-valid is also S-valid. Furthermore, every argument validated by free logic is C-valid and, hence, S-valid. (If all classical valuations assign T to $A \supset B$, then there cannot be a classical valuation that assigns T to A but not to B.) This means that (19) is a correct definition.

But this does not yet answer the question whether free logic is argument-complete. Let us first ask whether all the C-valid arguments are validated by free logic. Suppose that A and B are such that every classical valuation that assigns T to A also assigns T to B. There are then two possibilities. Those classical valuations that assign T to A also assign T to $A \supset B$. There is only one other sort of classical valuation: those that assign F to A. These also assign T to $A \supset B$. Therefore, if 'A, hence B' is C-valid, then $A \supset B$ is CL-true, and a theorem of free logic. This establishes that free logic is *argument-complete* under the interpretation that uses classical valuations.

Unfortunately, this is not true for the interpretation that uses supervaluations.[16] The difference does not show up in arguments constructed from unanalyzed propositions. But it does show up in arguments involving some essential use of the quantifiers. This is because, using the quantifiers, *we can state in the language itself that a given name does not refer.* Hence,

22. If $A(c)$ is atomic, then both '$A(c)$, hence $(\exists x)(x = c)$' and '$\sim A(c)$, hence $(\exists x)(x = c)$' are S-valid.

Of course, the corresponding conditionals '$A(c) \supset (\exists x)(x = c)$' and '$\sim A(c) \supset (\exists x)(x = c)$' are not SL-true. This is because those supervaluations that do not assign T to $A(c)$—respectively, $\sim A(c)$—comprise some that do not assign F to that statement either; and these assign neither T nor F to the corresponding conditional.

So we have found that, although free logic is statement-complete for the whole spectrum of interpretations considered, we can regard it as argument-complete only for the interpretation that uses classical valuations. This is not too surprising from a historical point of view, because several of the logicians who developed free logics have indicated repeatedly that they adhere firmly to the principle of bivalence (namely, Lambert and Leonard).

That not all the S-valid arguments are specifiable by means of free logic does not mean that we do not know which arguments are S-valid. Given any argument, we can in principle tell whether it is S-valid or not, because we have a complete semantic account of what this amounts to. (There is of course the interesting, but purely *technical* question of what sort of logical system validates exactly the S-valid arguments. That is exactly the question: What are the peculiar *syntactic* features of these arguments? From a philosophical point of view, this is not quite so important.)

This is also the situation for the interpretations intermediate between the two we have just discussed. All these interpretations take failure of reference seriously. For all of them, free logic gives us an exact syntactic description of what statements are logically true. All of them agree that all traditional *propositional* arguments are valid. All agree further that, if a given conditional statement is logically true, then the argument from its antecedent to its conclusion is valid. The differences are these. They differ on when failure of reference results in lack of truthvalue; and they differ on which arguments, beyond the ones mentioned, are valid. But the "upper limit" of these disagreements is reached in the interpretation using supervaluations: it withholds truthvalues from the largest class of statements and recognizes as valid the largest class of arguments.[17]

8

We have seen so far that the interpretations involving truthvalue gaps do not lead to very unusual logical systems. Their unorthodoxy is mainly one of semantics. To throw a clearer light on this, we may distinguish between the logical law of the excluded middle and the semantic law of bivalence. The first says that any proposition of the form $P \vee \sim P$ is logically true. The second says that every proposition is either true or false, or, equivalently, that one of P and $\sim P$ is true, the other false.

Clearly the law of bivalence fails for supervaluations and, indeed, for all the interpretations other than that based on classical valuations. But all our interpretations agree that $P \vee \sim P$ is logically true. This shows that (contrary to usage) the two laws must be strictly distinguished. This distinction, just like the distinction between statement-completeness and argument-completeness, is a distinction without a difference in classical contexts. But the admission of truthvalue gaps gives it content.

One interpretation of Aristotle's remarks on future contingencies is that he wished to deny the law of bivalence while retaining the law of the excluded middle. Whether this is historically accurate or not is not now to the point; what is important is that William and Martha Kneale[18] deny that this is a tenable position. Mrs. Kneale, who wrote the relevant chapter in their book, says on p. 48:

> In other words Aristotle is trying to assert the Law of Excluded Middle while denying the Principle of Bivalence. We have already seen that this is a mistake.

The last sentence is a reference to her argument of pp. 46–67:

> In chapter 9 of *De Interpretatione* Aristotle questions the assumption that every declarative sentence is true or false. It might seem that he is clearly committed to this thesis already, but this is not so; for when he says that to be true or false belongs to declarative sentences alone, this may be taken to means that only these are capable of being true or false not that they necessarily are ... Given the definitions of truth which we have quoted, the principles [of Bivalence and of Excluded Middle] are, however, obviously equivalent; for if 'It is true that P' is equivalent to 'P', 'P or not-P' is plainly equivalent to 'It is true that P or it is false that P'.

It is clear that she is talking about a language in which, when P is a sentence, 'It is true that P' is also a sentence. Our language is not thus. But more important is the following observation. Her argument is: 'Aristotle may question the law of bivalence; but if he does reject this law, then his acceptance of the principle

23.   *P* if and only if it is true that *P*

forces him also to reject the law of the excluded middle. Whether this is so depends on how (23) is understood. Mrs. Kneale apparently understands it in such a way that it justifies any argument of the form

24.   —*P*—; hence: —It is true that *P*— (and conversely)

Thus understood, (23) validates her argument. But, since the language we constructed is one for which the law of excluded middle does and the law of bivalence does not hold, it follows that, thus understood, (23) is false. We may formulate this conclusion more strongly: this language provides a counterexample to her conclusion; hence any interpretation of (23) must be such that *either* it makes (23) false *or* it does not justify her reasoning.

Since (23) is such a plausible and widely accepted principle, there remains of course the question of how it is to be understood. It seems to me that the answer to this is that (23) must be construed to mean simply that both the following kinds of argument are valid:

25.   *P*; hence: It is true that *P*
      It is true that *P*; hence: *P*

To say that these are valid simply means that they preserve truth: when the premiss is true, so is the conclusion. This says nothing whatsoever about the truthvalue of the conclusion when the premiss is not true (that is, when the premiss is false *or* when the premiss is neither true nor false[19]).

When (23) is understood in this manner, it does not lead to the incorrect conclusion that bivalence follows from the law of excluded middle. However, it may appear to do so, just because the use of 'if...then' to signify the validity of a certain argument may be misleading. For example, the following reasoning might at first sight appear valid. From (23) follow both:

   a. If *P* then it is true that *P*
   b. If not-*P* then it is true that not-*P*

but the excluded middle yields

   c. *P* or not-*P*

hence, by the rule of Constructive Dilemma:

   d. It is true that *P* or it is true that not-*P*

It is not difficult to spot the fallacy if we keep in mind that (23) is to be construed as (25). For then (a) and (b) amount to

  a*. P; hence: It is true that P
  b*. Not-P; hence: It is true that not-P

From the validity of these two arguments we can deduce the truth of d only if we are given not c, but

  c*. Either P is true or not-P is true

which is not excluded middle but bivalence. ●

A similar argument is the following. From (a) we infer:

  e. If not-(it is true that P) then not-P

by contraposition, and then from (e) and (b) by transitivity:

  f. If not-(it is true that P) then it is true that not-P

But if (e) is really to follow from (a) in the sense of (a*), then it can be understood to mean only:

  e*. P is false; hence, not-P

in which case (f) amounts to the innocuous assertion that

  f*. P is false; hence: not-P is true

is a valid argument.

The plausibility of these two fallacious arguments derives, it seems to me, from the fact that formulation (23) looks like a material biconditional. This suggests that certain familiar patterns of reasoning, which are in fact not applicable, do apply. For this reason, I should like to suggest that formulation (23) not be used. Instead, one may state explicitly that the arguments of form (25) are valid; or, if a biconditional formulation is more convenient, one might use

26.   It is true that P if and only if it is true that (it is true that P)

which is less misleading.

Notes

1. Strawson, P. F., "On referring," *Mind*, vol. 59 (July 1950) pp. 320–44.
2. Dr. Paul Benacerraf, of Princeton University, has pointed out to me that the above characterization of a valid argument, as a syntactic transformation preserving truth, has in the present context as alternative a characterization

as "transformation preserving nonfalsehood." It would indeed be interesting to investigate the properties of those transformations, as well as the properties of those transformations that may take a statement that is neither true nor false into a falsehood but at least do not take a *truth* into a falsehood.
3. Quine, W. V., *Mathematical Logic*, revised edition, Harvard University Press (1951).
4. Hailperin, Theodore, "Quantification and empty individual domains," *J. of Symbolic Logic*, vol. 18 (no. 3) (Sept. 1953) pp. 197–200.
5. Quine, W. V., "Quantification and the empty domain," *J. of Symbolic Logic*, vol. 19 (no. 3) (Sept. 1954) pp. 177–79.
6. I was convinced of this by Dr. Karel Lambert, of West Virginia University (as the reader will see in Section 6, this is only the least of my debts to him).
7. Professor John Lemmon (Claremont Graduate College) does not agree to this. It seems plausible to him to say that such a statement is neither true nor false. He does seem to agree, however, when one of the terms is a definite description.
8. Leonard, Henry S., "The logic of existence," *Philosophical Studies*, vol. 4 (no. 4) (June 1956) pp. 49–64.
9. Leblanc, Hugues and Hailperin, Theodore, "Nondesignating singular terms," *Philosophical Review*, vol. 68 (no. 2) (Apr. 1959) pp. 239–44.
10. Hintikka, Jaakko, "Existential presuppositions and existential commitments,' *Journal of Philosophy*, vol. 56 (no. 3) (Jan. 29 1959) pp. 125–37.
11. Lambert, Karel, "Existential import revisited," *Notre Dame J. of Formal Logic*, vol. 4 (no. 4) (Oct. 1963) pp. 288–92.
12. van Fraassen, Bas, "The completeness of free logic," *Zeitschrift für Mathematische Logik und Grundlagen der Mathematik*, vol. 12 (1966) pp. 219–34.
13. Lambert, Karel, "Notes on E! III: A theory of descriptions," *Philosophical Studies*, vol. 13 (no. 4) (June 1952) pp. 51–59; and "Notes on 'E!' IV: A reduction in free quantification theory with identity and descriptions," *Philosophical Studies*, vol. 15 (no. 6) (Dec. 1964) pp. 85–88.
14. van Fraassen, Bas and Lambert, Karel, "On free description theory," *Zeitschrift für Mathematische Logik und Grundlagen der Mathematik*, vol. 13 (1967) pp. 225–40.
15. We do not observe the use-mention distinction unless the context would not prevent confusion.
16. I realized this after a stimulating discussion at Wayne State University, and especially because of the comments of Mr. Lawrence Powers.
17. By "largest" I do not mean "of greatest cardinality"; I use "*K* is larger than *L*" here to mean "*L* is included in *K*, but *K* also has some members that are not in *L*."
18. Kneale, William and Kneale, Martha, *The Development of Logic*, Oxford University Press, New York (1962).
19. When *P* lacks a truthvalue, we have a choice with respect to 'It is true that *P*': we may regard it as false or as neither true nor false. The above remarks apply regardless of which of these alternatives we adopt.

5
Free Semantics

Ermanno Bencivenga

The fundamental problem to be solved when constructing a semantics for free logic (or free semantics) is clearly the evaluation of sentences containing nondenoting singular terms. This is also, however, a difficult problem, since the notion of *truth* ordinarily applied in contemporary logic (i.e., that of truth as correspondence with *reality*) has no immediate application to sentences referring to "nonexistent objects,"[1] which, as such, cannot be constituents of reality. Therefore, a free semantics should provide, first of all, for a new conception of truth, or at least for a suitable generalization of the correspondence theory, but the efforts made so far cannot be regarded as philosophically satisfactory. Such efforts resulted in accepting one or the other of the following three theses:

(a) An atomic sentence is true if it corresponds to a fact, and false otherwise. But an atomic sentence containing nondenoting singular terms cannot correspond to any fact; hence, it is false.[2]

(b) Nondenoting singular terms are not truly nondenoting: they refer to objects of some special sort (i.e., purely possible objects, null entities, or what have you).[3] (In an interesting variant of this position, nondenoting singular terms "denote" themselves.[4])

(c) A sentence containing nondenoting singular terms receives a truthvalue if and only if, were we to assign arbitrary truthvalues to its atomic constituents containing nondenoting singular terms, and regardless of which truthvalues we assigned, we would always obtain the same truthvalue for the total sentence.[5]

To show the intuitive implausibility of these three theses, consider the sentence

(1)   Pegasus = Pegasus

By (a), we would have to consider (1) false, and this conclusion directly contradicts the intuitions of many free logicians,[6] hence can hardly be the basis for a semantical interpretation of their views. By (b), (1) would indeed be true, but at the cost of admitting creatures that many perceive as a metaphysical burden. By (c), finally, we would have to leave (1) undecided, whereas we could decide on the truth of

(2)   Either Pegasus = Pegasus or Pegasus ≠ Pegasus

and on the falsity of

(3)   Pegasus = Pegasus and Pegasus ≠ Pegasus.[7]

On the other hand, (c) is in my view the most promising starting point for the construction of that generalization of correspondence theory we are looking for. By choosing it, we can say that a sentence is true not only when it corresponds with reality (i.e., when it is *factually* true) but also when every "mental experiment" of a certain sort makes it true (which we can call its being *formally* true). The problem is that such mental experiments are defined according to (c) at the level of unanalyzed sentences, and hence can do no justice to schemata such as the one instantiated by (1), whose validity would depend essentially on the analysis of sentences into terms. My aim here is to develop the suggestions offered by (c) at the level of *analyzed* sentences, and to solve a serious difficulty connected with the fulfillment of this task.

1. The Language and Its Interpretation

The primitive symbols of the language L are:
(a) A denumerably infinite set of *individual variables*
(b) For every n, a denumerably infinite set of n-ary *predicates*
(c) The *identity symbol* $=$ [8]
(d) The *existence symbol* $E!$
(e) The two *connectives* ¬ and &
(f) The *universal quantifier* ∀
(g) The two *parentheses* (and)[9]

All these symbols are autonymous. I will use x, y, z (possibly with subscripts) as metavariables on the set (a), and P^n (or more simply P

when no confusion is possible) as a metavariable on the sets (b). I will assume that an alphabetical order is defined on (a).

Every string of primitive symbols (of L) is a *formula* (of L).[10] The set of wffs (well-formed formulas) is the subset of the set of formulas identified by the following recursive definition:
(i) Every formula of one of the three forms $P^n x_1 \cdots x_n$, $x = y$, $E!x$ is a(n atomic) wff.
(ii) If A and B are wffs, then $\neg A$, $(A \& B)$ and $\forall x A$ are wffs (I will omit outermost parentheses).
(iii) Nothing else is a wff.

I will use A, B, C (possibly with superscripts or subscripts) as metavariables for wffs. I will presuppose ordinary definitions of the connectives $\vee$, $\supset$, and $\equiv$ and of the existential quantifier $\exists$, of the notions of a *free* and a *bound* occurrence of an individual variable in a wff, and of a *wf* (well-formed) *part* of a wff. By $A(y/x)$ and $A(y//x)$ I will understand the result obtained by substituting occurrences of y for every free occurrence of x in A, and any result obtained by substituting occurrences of y for zero or more free occurrences of x in A, respectively (possibly after some relettering to avoid scope confusions).

And now for the semantics. The two fundamental notions of standard semantics are those of a model-structure and of the truthvalue of a wff in such a structure. The first one can easily be adapted to the case in which some singular terms of the language (in our case, some individual variables) are nondenoting: it is sufficient to establish that a *model-structure for* L is an ordered pair $M = \langle D, f \rangle$, where D is a set, *possibly empty*, to be called the *domain*, and f is a unary function (to be called the *function of interpretation*), total on the set of predicates of L and *partial* on the set of individual variables of L, that assigns to every n-ary predicate of L a set of ordered n-tuples of members of D and to every individual variable of L for which it is defined a member of D. (Clearly, the function of interpretation will be defined for no individual variable when the domain is empty.)

It is not so easy to adapt the second fundamental semantical notion: that of the truthvalue of a wff in a model-structure. To start with, only the following is clear: within the approach I have chosen, I must distinguish between factual truth in M and formal truth in M, the latter being defined in a sense (I will have to explain in what sense) in terms of some "mental experiments" made on the basis of M.

Of the two kinds of truth, the factual one is less problematic, but still its definition is not a matter of course, since it requires taking a stand on the following three issues:

Free Semantics

(a) Is the factual truth of a wff A a sufficient condition for the *factual* truth of all the wffs that are tautological consequences of A? For instance, is the factual truth of a wff of the form Px a sufficient condition for the *factual* truth of all wffs of the form $Px \vee Py$, even when y is nondenoting?
(b) Is every wff of the form $E!x$ *factually false* (rather than, e.g., factually truthvalueless) when x is nondenoting?
(c) Is every wff of the form $x = y$ *factually false* (rather than, again, factually truthvalueless) when exactly one of x and y is nondenoting?

My own answers to (a)–(c) are No, Yes, and Yes, respectively. But I am not going to defend such answers here, since my main concern is with *formal* truth and even those who do not share my intuitions on (a)–(c) can make some use of the analysis I carry out.[11]

I now define, for any model-structure M, the *factual valuation* V_M^*.

(4) V_M^* is the partial unary function W from the set of wffs of L into $\{T, F\}$ such that
   (a)   if A is of the form $Px_1 \cdots x_n$ and $f(x_i)$ is defined for every i such that $1 \leq i \leq n$, then $W(A) = T$ if $\langle f(x_1), \ldots, f(x_n)\rangle \in f(P)$, and otherwise $W(A) = F$;
   (b)(1) if A is of the form $x = y$ and both $f(x)$ and $f(y)$ are defined, then $W(A) = T$ if $f(x) = f(y)$, and otherwise $W(A) = F$;
   (b)(2) if A is of the form $x = y$ and exactly one of $f(x)$ and $f(y)$ is defined, then $W(A) = F$;
   (c)   if A is of the form $E!x$, then $W(A) = T$ if $f(x)$ is defined, and otherwise $W(A) = F$;
   (d)   if A is of the form $\neg B$, then $W(A) = T$ if $W(B) = F$ and $W(A) = F$ if $W(B) = T$;
   (e)   if A is of the form $B \& C$ and both $W(B)$ and $W(C)$ are defined, then $W(A) = F$ if $W(B) = W(C) = F$, and otherwise $W(A) = T$;
   (f)   if A is of the form $\forall x B$ and $W(B(y/x))$ is defined for every individual variable y such that $W(E!y) = T$, then $W(A) = T$ if $W(B(y/x)) = T$ for every individual variable such that $W(E!y) = T$, and otherwise $W(A) = F$;
   (g) $W(A)$ is not defined if not by virtue of (a)–(f).

Turning now to formal truth in M, we must begin by making precise the notion of a mental experiment. Since we want to work at the level of analyzed sentences, such an experiment must be defined as the assignment of denotations to nondenoting singular terms (rather than as the

assignment of truthvalues to truthvalueless atomic sentences), and in this regard we have two major options. That is, we can either
  (i) assign denotations only to the nondenoting singular terms occurring in the wff we intend to evaluate, or
  (ii) assign denotations to all nondenoting singular terms at once.

However, in the present context these two alternatives come to the same thing,[12] and hence to simplify matters I choose (ii). Consequently, the notion of a mental experiment is given the following rigorous formulation.

(5) A model-structure $M' = \langle D', f' \rangle$ is a *completion* of a model-structure $M = \langle D, f \rangle$ if and only if (i) D' is a nonempty superset of D; (ii) for every predicate P of L, $f'(P)$ is a superset of $f(P)$; (iii) $f'(x)$ is defined for every individual variable x, and is identical with $f(x)$ whenever $f(x)$ is defined.

At this point, one might think that formal truth in M is to be defined as truth in all completions of M, just as in the semantics of supervaluations formal truth in a valuation V is defined as truth in all the "classical valuations" over V. But this "natural" course would give rise to the difficulty mentioned at the end of the introductory section. For consider the following schema:

(6) $Py \supset \exists x Px$

(6) is provable in standard logic, but not in free logic; so a free semantics would have to invalidate it. But by the course above, invalidating (6) would amount to finding a model-structure M such that either an instance of (6) is factually false in M or an instance of (6) is factually truthvalueless in M but factually false in a completion of M, and it is easy to see that both cases are impossible: no instance of (6) is *ever* factually false. So the outcome of this strategy would be that of collapsing into standard semantics![13]

I cannot develop my approach without resolving this problem. For this purpose, it will be useful to make the following remark. When a particular instance of (6)—call it A—is factually truthvalueless in a model-structure M, not *all* of A is factually truthvalueless. There is always a wf part of A—its consequent—that has a factual truthvalue, and when that truthvalue is T this fact is confirmed in all completions of M. When the factual truthvalue of the consequent is F, on the other hand, there arises a *conflict* between this truthvalue and the one the consequent has in some completion M' of M—or, if you will, between the information provided by M and M' in this regard. The situation is summarized in the following

Free Semantics 103

diagram:

	Py	$\exists xPx$
M	—	F
M'	T	T

When put in this light, the "natural" course discussed earlier can be seen as making a clear choice between the two parties in conflict, that is, as establishing the priority of one source of information over the other. The principle guiding this choice might be formulated as follows: in the course of a mental experiment, all information already available about reality must be forgotten. Then the suggestion might surface of taking the opposite route, and *retaining* instead all information about reality while carrying out a mental experiment. This strategy would make one combine the truthvalue T the antecedent of (6) has in M' (not with the truthvalue T the consequent has in M' but) with the truthvalue F the consequent has *in M*, and hence would allow one to invalidate (6)—and in general to avoid the feared collapse into standard semantics. This suggests that free logic might be committed to the following *Principle of Prevalence of Reality*: in the course of a mental experiment, all information already available about reality prevails over anything offered by the experiment itself. The following five definitions will make all of this precise.

(7) The *valuation* $V_{M'(M)}$ *for M' from the point of view of M* (where M' is a completion of M) is the (total) unary function W from the set of wffs of L into $\{T, F\}$ such that
 (a)(1) if A is an atomic wff and $V_M^*(A)$ is defined, then $W(A) = V_{M'}^*(A)$;
 (a)(2) if A is an atomic wff and $V_M^*(A)$ is not defined, then $W(A) = V_{M'}^*(A)$;
 (b) if A is of the form $\neg B$, then $W(A) = T$ if and only if $W(B) = F$;
 (c) if A is of the form $B \& C$, then $W(A) = F$ if and only if $W(B) = W(C) = F$;
 (d) if A is of the form $\forall xB$, then $W(A) = T$ if and only if $W(B(y/x)) = T$ for every individual variable y such that $W(E!y) = T$.

(8) The *valuation V_M* for a model-structure M is the supervaluation constructed over all valuations $V_{M'(M)}$ for completions M' of M; that is to say, it is the partial unary function W from the set of wffs of L into $\{T, F\}$ such that
 (a) if $V_{M'(M)}(A) = T$ for every completion M' of M, then $W(A) = T$;

(b) if $V_{M'(M)}(A) = F$ for every completion M' of M, then $W(A) = F$;
(c) $W(A)$ is not defined if not by virtue of (a)–(b).

(9) A wff A is *verifiable* (or *falsifiable*, or *not completely determinable*) if and only if $V_M(A) = T$ (or $V_M(A) = F$, or $V_M(A)$ is not defined) for some model-structure M.

(10) A wff A is *invalid* if and only if A is either falsifiable or not completely determinable.

(11) A wff is *valid* if it is not invalid.

(As a consequence, a wff A is valid if and only if $V_M(A) = T$ for every model-structure M.)

2. The Formal Systems

The axiomatic system *FLI* is characterized by the following schemata and rule of inference:

(A0) A, where A is a tautology
(A1) $(\forall x A \,\&\, E!y) \supset A(y/x)$
(A2) $\forall x E! x$
(A3) $\forall x(A \supset B) \supset (\forall x A \supset \forall x B)$
(A4) $A \supset \forall x A$, where x does not occur free in A
(A5) $x = x$
(A6) $x = y \supset (A \supset A(y//x))$
(A7) $\forall x A$, where A is an axiom
(R1) B can be inferred from A and $A \supset B$

The notions of a *derivation* from a set of premisses, a *proof*, a *deductive consequence* of a set of premisses and a *theorem* will be defined as usual for *FLI*.

I intend to prove that *FLI* is weakly adequate with respect to the semantics of Section 1, that is, that its theorems are all and only the valid wffs of that semantics.[14] For this purpose, I will introduce a system of semantic tableaux *STI* and prove that, if *Val* is the set of valid wffs, T_{STI} the set of theorems of *STI* and T_{FLI} the set of theorems of *FLI*,

(a) $T_{STI} \subseteq T_{FLI}$
(b) $T_{FLI} \subseteq Val$
(c) $Val \subseteq T_{STI}$

I begin with the definition of *STI*. I presuppose ordinary definitions of a (*finitely generated*) *tree*, a *point* and a *branch* in a tree, and a point in a tree being the *origin*, being a *successor* of another point, or *preceding* another point in a branch. A point is a *last point* if it has no successor, a branch *closes* if it has a last point, and a tree *terminates* if all its branches close. A semantic tableau for L and *STI* is a tree whose points are wffs of L and in which the property of being a last point and the relation of being a successor are determined by the following rules (S1)–(S7). (To understand these rules, note that an expression E of the form $A_1 \& \cdots \& A_n$ (where $n \geq 0$) stands for any wff of L obtained by adding parentheses to E, that such an expression will be called the *conjunction* of $A_1, \ldots, A_n$ and that $A_1, \ldots, A_n$ will be called *conjuncts* in it, and finally that F and G stand for any conjunctions (possibly empty) such that no conjunct in F is of one of the five forms $x = y$ (where x and y are distinct individual variables), $\neg\neg A$, $\neg(A \& B)$, $\forall x A$, $\neg\forall x A$.)

(S1) A point is a last point if and only if (a) it is a conjunction $A_1 \& \cdots \& A_n$ (where $n > 0$) and (b) either, for some i such that $1 \leq i \leq n$ and for some individual variable x, $A_i = \neg(x = x)$ or, for some i, j such that $1 \leq i \leq n$ and $1 \leq j \leq n$, $A_i = \neg A_j$.

(S2) Every nonlast point of the form $F \& \neg\neg A \& G$ has as its only successor $F \& A \& G$.

(S3) Every nonlast point of the form $F \& \neg(A \& B) \& G$ has exactly the two successors $F \& \neg A \& G$ and $F \& \neg B \& G$.

(S4) Every nonlast point of the form $F \& \forall x A \& G$ has as its only successor $F \& A(y_1/x) \& \cdots \& A(y_n/x) \& G \& \forall x A$, where $y_1, \ldots, y_n$ are all and only the individual variables z such that $E!z$ is a conjunct in F or in G.

(S5) Every nonlast point of the form $F \& \neg\forall x A \& G$ has as its only successor $F \& E!y \& \neg A(y/x) \& G$, where y is alphabetically the first individual variable not occurring either in F or in $\neg\forall x A$ or in G.

(S6) Every nonlast point of the form $F \& x = y \& G$ (where x and y are distinct individual variables) has as its only successor $(F \& x = y \& G)(y/x) \& F \& G \& y = x \& x = y$.

(S7) Every nonlast point of the form F has as its only successor F.

The semantic tableau *for* a wff A is the tableau whose origin is $\neg A$. A wff A is a *theorem* of *STI* if the semantic tableau for A terminates.

And now for the results mentioned above.

Theorem 1. $T_{STI} \subseteq T_{FLI}$.

Proof. Call *degree* of a point C in a semantic tableau the number of points that (a) have at least one branch in common with C and (b) do not precede C in any branch. I will show by induction on n that every point of every degree n of a terminating semantic tableau has a contradiction as a deductive consequence in *FLI*. This will allow me to assert that the origin of any terminating semantic tableau is refutable in *FLI*, and hence that the wff for which the tableau is constructed is a theorem of *FLI*.

So suppose that the desired result holds whenever $n < k$, and prove it for $n = k$. Since the tableau terminates, there is no point in it that results from an application of (S7): therefore I need only consider four cases.

Case 1. C is a last point. Trivial.

Case 2. C has one successor (call it C') constructed by one of the rules (S2), (S4), and (S6). Then C' is an easy deductive consequence (in *FLI*) of C. But by the induction hypothesis C' has a contradiction as a deductive consequence; then by transitivity the same contradiction is a deductive consequence of C.

Case 3. C has two successors (call them C_1 and C_2) constructed by (S3). Then by the induction hypothesis both C_1 and C_2 have a contradiction as a deductive consequence (in *FLI*), and hence the thesis follows by propositional logic.

Case 4. C has one successor (call it C') constructed by (S5). Then C is of the form $F \mathbin{\&} \neg \forall x A \mathbin{\&} G$, C' is of the form $F \mathbin{\&} E!\, y \mathbin{\&} \neg A(y/x) \mathbin{\&} G$ (where y is as specified in (S5)), and by the induction hypothesis C' has a contradiction as a deductive consequence (in *FLI*). Then by the Deduction Theorem and other easy steps the following is a deductive consequence of $F \mathbin{\&} G$: $(E!\, y \mathbin{\&} \neg A(y/x)) \supset (B \mathbin{\&} \neg B)$, where without loss of generality we may assume that y does not occur free in B. Then $\exists y (E!\, y \mathbin{\&} \neg A(y/x)) \supset (B \mathbin{\&} \neg B)$ is also a deductive consequence of $F \mathbin{\&} G$, and since $\neg \forall x A \supset \exists y(E!\, y \mathbin{\&} \neg A(y/x))$ is a theorem of *FLI*, $B \mathbin{\&} \neg B$ is a deductive consequence of C. ●

Theorem 2. $T_{FLI} \subseteq Val$.

Proof. I need to prove that (a) all axioms of *FLI* are valid, and (b) (R1) preserves validity. Thus there are nine cases, but here I will prove only one (Case 2).

Case 2. C is of the form (A1). Then, for some $A, x, y, C = (\forall x A \mathbin{\&} E!\, y) \supset A(y/x)$. Suppose that C is invalid. Then for some model-structure M and some completion M' of M, $V_{M'(M)}(C) = F$. Then by clauses (b) and (c) of (7), $V_{M'(M)}(\forall x A) = V_{M'(M)}(E!\, y) = T$ and $V_{M'(M)}(A(y/x)) = F$. Then by clause (d) of (7), $V_{M'(M)}(A(z/x)) = T$ for every individual variable z such that

$V_{M'(M)}(E!z) = T$, and hence (since $V_{M'(M)}(E!y) = T$) $V_{M'(M)}(A(y/x)) = T$. Thus the assumption that C be invalid entails a contradiction. ●

Theorem 3. $Val \subseteq T_{STI}$.

Proof. Consider a wff C of L. I will prove that C is invalid if it is not a theorem of *STI*. More precisely, I will prove that if the semantic tableau for C contains at least one nonclosing branch, there are two model-structures M and M' such that (a) M' is a completion of M and (b) $V_{M'(M)}(A) = T$ for every wff A occurring as a conjunct in that branch. Since the origin $\neg C$ belongs to all branches, this will entail that $V_{M'(M)}(C) = F$, and hence that C is invalid.

So suppose that X is a nonclosing branch of the semantic tableau for C, and define in the set of individual variables the following binary relation I: xIy if and only if either x and y are the same individual variable or $x = y$ occurs as a conjunct in some point of X. It is easy to see that I is an equivalence relation; call *identity-class* of an individual variable x its equivalence class relative to I. Let $M = \langle D, f \rangle$ be such that (i) D is the set of all and only the individual variables z such that (a) z is the first member of its identity-class that occurs in some point of X and (b) $E!z$ occurs as a conjunct in some point of X; (ii) f assigns to every individual variable z the member of the identity-class of z that belongs to D if such a member exists, and is not defined for any other individual variable; (iii) f assigns to every n-ary predicate P the set of all ordered n-tuples $\langle x_1, \ldots, x_n \rangle$ of members of D such that $Px_1 \cdots x_n$ occurs as a conjunct in some point of X.

Let $M' = \langle D', f' \rangle$ be such that (i) D' is the set of all and only the individual variables z such that either z is the first member of its identity-class that occurs in some point of X or z is the only member of its identity-class; (ii) f' assigns to every individual variable z the member of the identity-class of z that belongs to D'; (iii) f' assigns to every n-ary predicate P the set of all ordered n-tuples $\langle x_1, \ldots, x_n \rangle$ of members of D' such that $Px_1 \cdots x_n$ occurs as a conjunct in some point of X.

M' is clearly a completion of M, hence it remains to show that $V_{M'(M)}(A) = T$ for every wff A occurring as a conjunct in some point of X. This part of the proof will be carried out by induction on the number n of connectives and quantifiers occurring in A.

So suppose that the desired result holds whenever $n < k$, and prove it for $n = k$. I distinguish two cases.

Case 1. A is atomic. I distinguish three subcases.

Subcase 1a. A is of the form $Px_1 \cdots x_n$. Then, by the definitions of M

and M', either $V^*_M(A) = T$ or $V^*_M(A)$ is not defined and $V^*_{M'}(A) = T$. In both cases $V_{M'(M)}(A) = T$.

Subcase 1b. A is of the form $x = y$. Then, by the definitions of M and M', either $f(x) = f(y)$ or neither $f(x)$ nor $f(y)$ is defined and $f'(x) = f'(y)$. In both cases $V_{M'(M)}(A) = T$.

Subcase 1c. A is of the form $E!x$. Then, by the definition of M, $V^*_M(A) = T$, which entails $V_{M'(M)}(A) = T$.

Case 2. A is not atomic. I distinguish three subcases.

Subcase 2a. A is the negation of an atomic wff B. Then B does not occur as a conjunct in any point of X (otherwise, X would close), and the thesis follows from the definitions of M and M'.

Subcase 2b. A is a conjunction $B_1 \& B_2$. Then B_1 and B_2 are also conjuncts in X, and the thesis follows from the induction hypothesis.

Subcase 2c. A is neither a conjunction nor the negation of an atomic wff. Then, since A occurs in some point of X, in some (other) point(s) of X there must occur the result(s) of applying to A one of the rules (S2)–(S5). Simple applications of the induction hypothesis will suffice to prove the thesis in every (subsub)case.

A simple consequence of Theorems 1–3 is

Theorem 4. $T_{STI} = T_{FLI} = Val$,

which establishes the weak adequacy of *FLI* (and *STI*) to the semantics of Section 1.

References

Bencivenga, Ermanno, "Free semantics for indefinite descriptions," *J. Philosophical Logic*, vol. 7 (1978) pp. 389–405.

Bencivenga, Ermanno, "Truth, correspondence, and non-denoting singular terms," *Philosophia*, vol. 9 (1980a) pp. 219–29.

Bencivenga, Ermanno, "Again on existence as a predicate," *Philosophical Studies*, vol. 37 (1980b) pp. 125–38.

Bencivenga, Ermanno, "Free semantics for definite descriptions," *Logique et Analyse*, vol. 92 (1980c) pp. 393–405.

Hintikka, Jaakko, "Existential presuppositions and existential commitments," *Journal of Philosophy*, vol. 56 (1959) pp. 125–37.

Lambert, Karel, "Existential import revisited," *Notre Dame J. of Formal Logic*, vol. 4 (1963) pp. 288–92.

Lambert, Karel, "Free logic and the concept of existence," *Notre Dame J. of Formal Logic*, vol. 8 (1967) pp. 133–44.

Leblanc, Hugues and Hailperin, Theodore, "Nondesignating singular terms," *Philosophical Review*, vol. 68 (1959) pp. 239–43.

Leblanc, Hugues and Thomason, Richmond H., "Completeness theorems for some presupposition-free logics," *Fundamenta Mathematicae*, vol. 62 (1968) pp. 125–64.

Leonard, Henry S., "The logic of existence," *Philosophical Studies*, vol. 7 (1956) pp. 49–64.

Meyer, Robert K. and Lambert, Karel, "Universally free logic and standard quantification theory," *J. of Symbolic Logic*, vol. 33 (1968) pp. 8–26.

Schock, Rolf, "Contributions to syntax, semantics, and the philosophy of science," *Notre Dame J. of Formal Logic*, vol. 5 (1964) pp. 241–89.

Schock, Rolf, *Logics Without Existence Assumptions*, Almqvist & Wiksell, Stockholm (1968).

Scott, Dana, "Existence and description in formal logic," in *Bertrand Russell, Philosopher of the Century* (ed. R. Schoenman), Allen and Unwin, London (1967) pp. 181–200.

van Fraassen, Bas C., "The completeness of free logic," *Zeitschrift für mathematische Logik und Grundlagen der Mathematik*, vol. 12 (1966a) pp. 219–34.

van Fraassen, Bas C., "Singular terms, truth-value gaps, and free logic," *Journal of Philosophy*, vol. 67 (1966b) pp. 481–95.

Woodruff, Peter W., "On supervaluations in free logic," *J. of Symbolic Logic*, vol. 49 (1984) pp. 943–50

Notes

I thank Bas van Fraassen and Hans Herzberger for their useful comments on an earlier draft of this paper. The reader may find it useful to compare the present piece with my (Bencivenga 1980a) paper, which I conceive as a (more philosophically oriented) companion to it.

1. It will be apparent later that one of my goals in the construction of a free semantics is that of eliminating any need for nonexistent objects, which many regard as metaphysically burdensome. But this does not prevent me from using the *phrase* "nonexistent object" (in quotation marks, to emphasize its impropriety) in a colloquial way.
2. See Schock (1964, 1968).
3. See Scott (1967) and Leblanc and Thomason (1968).
4. See Meyer and Lambert (1968).
5. See van Fraassen (1966a, 1966b). The resulting semantics is called "of supervaluations."
6. See Leonard (1956), Leblanc and Hailperin (1959), Hintikka (1959), and Lambert (1963).
7. As a matter of fact, the semantics of supervaluations does validate (1), but only by virtue of an ad hoc device.

8. I will also use = as a metatheoretical symbol, but context will always prevent any confusion.
9. Note that L contains no individual constants. The reason is that there are many free languages with this feature: for example, those in Leonard (1956), Lambert (1963, 1967), and Meyer and Lambert (1968). On the other hand, the addition of constants would create no theoretical complication.
10. I will often omit this qualification.
11. A Yes answer to (a) would only require an obvious modification of (4)(e). A No answer to (c) would require the following modification of (5): $D' - D$ is nonempty when f is nontotal and $f'(x) \in D' - D$ whenever $f(x)$ is not defined. A No answer to (b) would be more troublesome, and hence my Yes answer must not be given up too easily. Bencivenga (1980b) gives an idea of the general line of defense I would put up.
12. They do not when descriptions enter the picture. See Bencivenga (1978) and (1980c).
13. The collapse might not be a total one, since some theorems of standard logic (such as those of the forms $\forall x Px \supset \exists x Px$, $\exists x(x = x)$, $\exists x(x = y)$) could be invalidated in the semantics above on purely factual grounds. However, such standard theorems would be formally true in every model-structure (given the definition suggested here for "formally true"), and hence to invalidate them one would have to accept a criterion of priority among sorts of truth (that is, ultimately, among sources of information) similar to (though less complicated than) the one I will propose later.
14. That a stronger sense of adequacy is out of the question here was established by Woodruff (1984).

6
Applications of Free Logic to Quantified Intensional Logic

James W. Garson

We will argue here that free logic is especially well suited for providing a theory of quantification in intensional languages, and that classical quantification and modality are not a particularly good match. Although classical quantified modal logics can be constructed, they suffer from formal as well as philosophical failings. Free logic provides a much more satisfactory alternative.

1. The Motivation for Using Free Logic in Intensional Languages

A. Fixed and World-Relative Domains

Kripke's historic paper (1963) serves as an excellent starting point for a discussion of the application of free logic in quantified modal logic. Kripke did not adopt free logic, but revealing the difficulties he faced in preserving classical principles will help motivate the adoption of free logic. Kripke's paper lays out two basic options concerning the quantifier domains. The simplest of the two, the *fixed domain approach*, assumes a single domain of quantification that contains all the possible objects. The *world-relative interpretation*, on the other hand, assumes that the domain of quantification changes from world to world, and contains only the objects that exist in a given world. Although the fixed domain approach is less general, it is attractive from the technical point of view because it requires no major adjustments to the classical machinery for the quantifiers. An intensional logic for fixed domains can usually[1] be axiomatized by adding principles of a propositional modal logic to classical quantifier rules plus the Barcan Formula (BF):

(BF) $\forall x \Box A \to \Box \forall x A$

B. The Motivation for World-Relative Domains

The fixed domain interpretation has advantages of simplicity and familiarity, but it does not provide an accurate account of the semantics of quantifier expressions of natural language. We do not think that 'Some man exists who signed the Declaration of Independence' is true, at least not if we read 'exists' in the present tense. Nevertheless, this sentence was true in 1777, and this shows that the domain of the present tense existential quantifier changes to reflect which objects exist at different times. Quantifier domains vary along other dimensions as well. For example, when I announce to my class that everyone did well on the midterm, I am not praising the performance of the whole human race. Time, place, speaker, and even topic of discussion play a role in determining the domain in ordinary communication.

There are also reasons for rejecting fixed domains in modal languages. On the fixed domain interpretation, the sentence $\forall x \Box \exists y (y = x)$ (which reads 'everything exists necessarily') is valid; however, this does not appear to be true, and it would be even harder to argue that it is a logical truth. It seems to be a fundamental feature of common intuitions about modality that the existence of many things is contingent, and that different objects exist in different possible worlds.

The defender of the fixed domain interpretation may respond to these objections by insisting that the domain of quantification for $\forall x$ contains all possible objects, not the objects that exist in a given world or context. Quantifier expressions whose domain is world-dependent can then be defined using $\forall x$ and predicate letters. For example, the present tense quantifier can be defined using $\forall x$ and a predicate letter that reads 'presently exists'.

One difficulty with this proposal is that it requires the invention of predicates for all the different subdomains that we may ever intend for quantifier expressions. It also forces us to represent simple expressions of natural language differently in different contexts of their use. A second problem is that it strips the quantifier domain of the role for which it was presumably intended, namely, to record ontological commitment. When it comes to being true to our ontological intuitions, a domain of existing objects beats a domain of possible objects hands down.

Applications of Free Logic to Quantified Intensional Logic 113

C. Problems Using Classical Quantification with World-Relative Domains

The fixed domain interpretation suffers from serious failings in providing a theory of quantification for natural language. We believe that the best strategy is to acknowledge the context dependence of quantification by adopting world-relative domains. Once this decision is made, major difficulties arise for classical quantification theory. There *are* systems that use classical rules with the world-relative interpretation; however, an examination of their limitations reveals the advantages to be gained from free logic.

To appreciate the problems to be faced in maintaining the classical rules with world-relative domains, notice that $\exists x(x = t)$ is true at a world just in case the extension of t is in the domain for that world. However, $\exists x(x = t)$ is a theorem of classical logic, and so it follows that in each world, t must refer to an object that exists in that world. Let us assume for a moment that t is a rigid designator (i.e., that the referent of t is the same in all worlds). This leads to a difficulty. If t refers to an existent object that is the same in all worlds, its referent must exist in every possible world. However, there need not be any one object that exists in all the worlds. The whole motivation for the world-relative approach was to reflect the idea that objects in one world may fail to exist in another. If standard quantifier rulers are used, however, there are no rigid designators that refer to such objects. For example, if it is assumed that proper names are rigid, it follows that everything that has a name exists necessarily.

D. Kripke's Classical Quantified Modal Logic

One response to this difficulty is simply to eliminate terms. Kripke (1963) gives an example of a system that uses the world-relative interpretation and preserves the classical rules. The system has no terms other than variables, and even the variables are not full-blooded referring expressions. If the semantics for this system were to give variables extensions in the domain, the validity of $\exists x(x = y)$ would demand that the extension of y be a member of every possible world. As a result, all the variables would denote in the same domain, which conflicts with a desire to have world-relative domains. Kripke avoids this difficulty by giving sentences with free variables the closure interpretation. So $\exists x(x = y)$ has the semantical effect of $\forall y \exists x(x = y)$, which is accepted in free logic. From the semantical point of view, Kripke's system has no referring expressions at all, because the variables are really disguised universal quantifiers.

Although Kripke has shown that model logic with classical quantifier rules is possible with the world-relative interpretation, his system underscores a theme that is familiar in the literature on free logic, namely, that adoption of the classical rules forces us into an inadequate account of terms. As Lambert and van Fraassen (1972) show, removal of terms from free logic may result in collapse to the classical rules. So Kripke's preservation of classical principles may reflect more the artificial impoverishment of his language than an endorsement of classical quantification theory.

Another oddity of Kripke's system is that he is forced to weaken the rules of propositional modal logic. The necessitation rule ($\Box$) must be restricted to closed sentences (i.e., sentences that lack free variables).

($\Box$)   If A is a theorem, then so is $\Box A$

If the restriction to closed sentences were not enforced, $\Box \exists x(x = y)$ would be a theorem. Since it is given the closure interpretation, it says that any object of one domain exists in all the others. This, of course, would be fatal to the view that domains are world-relative.

E. Nested Domains

Although Kripke has defined a classical system with world-relative domains, the cost is weakness in both its rules and expressive power. There is still the possibility, nevertheless, that some other classical system may do better. However, any system that uses the principles of classical logic along with the (unrestricted) rule of necessitation ($\Box$), faces a fundamental problem. In such a system, it is possible to prove the converse of the Barcan Formula (CBF).

(CBF) $\Box \forall x A \rightarrow \forall x \Box A$

This fact has serious consequences for the system's semantics. It is not difficult to show that every world-relative model of (CBF) must meet condition (ND) (for "nested domains").

(ND)   If world w' is accessible from world w then the domain of w is a subset of the domain of w'

To see this, notice that $\forall x \exists y(y = x)$ is classically valid, and so true in every possible world. By the validity of ($\Box$), it follows that $\Box \forall x \exists y(y = x)$ is also valid. This entails the validity of $\forall x \Box \exists y(y = x)$ by (CBF). However, $\forall x \Box \exists y(y = x)$ claims that any object that exists in our world must also exist in all worlds that are accessible from ours, and this amounts to

(ND). It follows that any classical logic that adopts ($\Box$) and world relative domains must accept nested domains (ND).

The reason this is a problem is that (ND) conflicts with the point of introducing world-relative domains. We certainly want to allow there to be accessible possible worlds where one of the things in our world fails to exist. This difficulty becomes even more acute in the stronger modal logics. If the accessibility relation is symmetric, then it follows from (ND) that all worlds accessible from ours have exactly the same domains.[2] Even worse, in models of S5 where all worlds are accessible from each other, it follows from (ND) that all domains be the same, in direct conflict with our intention to distinguish the domains.

F. Local Terms

Even if problems with the acceptability of (ND) could be resolved, further adjustments must be made to preserve classical logic. The classically valid $Pt \to \exists x Px$, for example, is not valid on a model where the extension of the predicate letter P is inside the domain $D(w)$ for world w, and the extension of t at w is outside $D(w)$. One way to restore validity to $Pt \to \exists x Px$ is to stipulate that the terms are *local*; that is, the extension of a term at a world must always be in the domain of that world. However, there are serious problems with this requirement. First, nondenoting expressions such as 'Pegasus' and 'the present king of France' cannot count as terms since their extensions are not in the real world. This objection will not impress the classicist, who must deny the termhood of nonreferring expressions in any case. A more serious problem arises for rigid terms, for then the demand that terms be local entails that the referent of any term must exist in all the worlds not just the real world. As we suggested before, there is no reason to believe that there are any objects that exist in all possible worlds. So there may be no targets for term reference. Even if there are, the local terms condition has the odd consequence that objects that are in some domains and not others cannot be rigidly named.

These difficulties are symptomatic of an unpalatable consequence of any standard modal logic that adopts classical rules. Once Existential Instantiation is accepted, $\exists x\, x = t$ is provable. Given unrestricted necessitation ($\Box$), it follows that $\Box \exists x\, x = t$ is a theorem. But $\Box \exists x\, x = t$ claims that t necessarily exists. This undercuts the whole point of introducing world-relative domains, namely to accommodate terms that refer to things that may not exist in other possible worlds.

G. Local Predicates

The requirement that terms be local faces serious difficulties. There is a related idea, however, that may work better. Another way to validate $Pt \to \exists xPx$ is to assume that *predicate letters* are local (i.e., that their extensions at a world must contain only objects that exist at that world). The reason this works is that from the truth of Pt it follows that t refers to an existing object, and from this it follows that $\exists xPx$ is true.

Unfortunately, the local predicate condition does not validate all instances of the classical rules. For example, consider $\sim Pt \to \exists x \sim Pt$. From the truth of $\sim Pt$, it does not follow that the extension of t is an existing object, and so it does not follow that $\exists x \sim Pt$ is true. Not only do we fail to validate the rule of Existential Generalization, but the valid principles cannot be expressed as axiom schemata.[3]

H. Local Predicates with Truthvalue Gaps

These problems can be mitigated somewhat by using a semantics with truthvalue gaps. Strawson argued that sentences that contain nonreferring terms do not make statements, and so lack truthvalues. On an intensional version of this idea, terms are allowed to refer to objects outside of the domain of a given world, and sentences that contain such terms lack truthvalues. Valid sentences are then defined as ones that are never false. As a result, $At \to \exists xAx$ is valid, since any assignment that gives t an extension outside the domain of a world leaves the whole conditional without a value, and assignments that give t an extension inside the domain will make $\exists xAx$ true if At is true.

The introduction of truthvalue gaps, however, results in new problems. We are faced with several choices concerning the truth clause for the modal operator $\Box$. An initial attraction of truthvalue gaps is that they allow us to drop the nesting condition (ND) if we wish (Gabbay 1976, pp. 75 ff.). However, there are technical reasons for keeping (ND). Suppose $\Box At$ is evaluated at world w and the referent of t is in the domain $D(w)$ of w. Then $\Box At$ should receive a truthvalue that depends on the values At has in the worlds accessible from w. But there is no guarantee that t refers to an existing object in all accessible worlds, and so At may be undefined in some of them. Adopting (ND), however, insures that whenever the referent of t is in the domain of w, it is also in the domain of all accessible worlds. So the values of At in all the accessible worlds would be defined, and the determination of the value of $\Box At$ can be carried out in the standard way.

If we drop (ND), however, we must revise the truth clause for □. There are two ways to determine the value of □*At* at *w* depending on whether the failure of *At* to be defined in an accessible world should make □*At* false or undefined. On the first option (Gabbay's system GKc), the necessitation rule must be restricted so that (CBF) is no longer derivable. On the second option (Gabbay's GKs), (CBF) is derivable, but the truth of (CBF) in a model no longer entails (ND). Either way, the principles of the underlying modal logic are nonstandard.

For these reasons, the more popular choice for truthvalue gap theories with local predicates has been to assume (ND) (Hughes and Cresswell 1968). This approach is attractive from a purely formal point of view. We have relatively simple completeness proofs for classical systems based on the major propositional modal logics, provided that language omits identity (=). Proofs are available, for example, for M and S4. In case the modality is as strong as B, the domains become rigid, and completeness can be established using methods developed for systems that validate the Barcan Formula.[4]

Easy technical results should not cause us to overlook the problems with the nested domains condition. (ND) conflicts with the same intuitions that prompt the use of world-relative domains, and in stronger modal systems such as S5 it is flatly incompatible with world-relative domains. Although there are semantics for classical logics without (ND), they require truthvalue gaps and awkward formal principles. Our conclusion is that there is little reason to preserve the classical rules in formulating systems with world-relative domains. As we will see in later sections, the semantics for systems based on free logic is both general and natural, and the formal results are as satisfying as any to be found in classical quantified modal logic.

2. Semantics for Free Quantified Modal Logic

The intensional logicians faces a bewildering variety of choices in defining the semantics of the quantifiers. One complication is that decisions about the quantifiers interact with the treatment of terms and identity. We will argue that the best strategy is to select the more general accounts. To illustrate this point, we will explore some of the difficulties and complications that arise in providing semantics for quantified intensional logics. We will then define a general semantics and explain how to axiomatize it using principles of free logic.

A. Some Semantical Preliminaries

Before we begin a discussion of semantics for quantified modal logic, we must dispose of some basic notational details. We assume that a quantified modal language is constructed from predicate letters, the primitive predicate constant E, terms (which include infinitely many variables) the logical constants $\sim$, $\rightarrow$, $\Box$, $=$, and a quantifier $\forall x$ for each of the variables x. $\exists$ and $\Diamond$ are defined from $\forall$ and $\Box$ in the usual way. The predicate letters come equipped with integers indicating their arity. The propositional variables are taken to be 0-ary predicate letters, and well-formed formulas are defined in the usual way. Given a set D, which serves as a domain of possible objects, the *extensions* of terms and predicate letters are defined just as they are in first-order logic. The extension of a term is some member of D, and the extension of an n-ary predicate letter P is a set of n-length sequences of members of D. Given a set W of situations (typically, possible worlds), an *appropriate intension* for an expression is any function that takes each member of W into an appropriate extension for that expression. For example, an appropriate intension of a term is a function from W into D, which amounts to what Carnap (1947) called an *individual concept*.

Throughout this article, a Q-model $\langle W, R, D, Q, a \rangle$ consists of a set W of possible worlds, a binary relation R on W, a nonempty set D of possible objects, some item Q, which determined the domain of quantification, and an assignment function a, which interprets the terms (including variables) and predicate letters by assigning them appropriate intensions with respect to W and D.

In some semantics, the terms are rigid designators (i.e., their extensions are the same in all possible worlds). Typically such terms are assigned no intensions, but given extensions directly. However, in order to keep the description of a model as general as possible, we will assume that terms always have intensions, and that rigid designators simply meet the added condition that their intensions are constant functions.

The systems discussed in this paper all have semantics with common features. We intend the predicate constant E to pick out the quantifier domain. So to insure this interpretation we assume that a Q-model meets condition (E).

   (E) $a(E)$ is Q

The symbol $=$ will be interpreted as contingent identity. This means that $t = t'$ is ruled true in a world just in case t and t' have the same *extension* in that world. So we will assume that the assignment function satisfies $(=)$

(=) $a_w(t = t')$ is T iff $a_w(t)$ is $a_w(t')$

For a given model $\langle W, R, D, Q, a \rangle$, the truthvalue of a sentence A on a at world w in W (written $a_w(A)$) is defined by induction on the shape of A using the following familiar truth conditions for standard clauses for atomic sentences, $\sim$, $\rightarrow$, and $\square$.

(Atm) $a_w(Pt_1 \cdots t_n)$ is T iff $\langle a_w(t_1), \ldots, a_w(t_n) \rangle \in a_w(P)$

($\sim$) $a_w(\sim A)$ is T iff $a_w(A)$ is F

($\rightarrow$) $a_w(A \rightarrow B)$ is T iff $a_w(A)$ is F or $a_w(B)$ is T

($\square$) $a_w(\square A)$ is T iff for all w' in W, if $w\, R\, w'$ then $a_{w'}(A)$ is T

All that remains to define a given semantics for a quantified modal logic will be to say what Q is, to present the truth clause for the quantifier, and to give truth conditions for sentences of the shape Et.

The quantified modal logics we are going to discuss are all extensions of propositional modal logics S that are adequate with respect to some class R(S) of Kripke frames. For example, S4 is adequate (semantically consistent and complete) with respect to the class R(S4) of Kripke frames $\langle W, R \rangle$ that are reflexive and transitive. When defining quantified modal logics, we will assume that some propositional modal logic S has already been chosen as its foundation, and that the frame $\langle W, R \rangle$ of any Q-model is in R(S). When we need to be explicit, we will talk of S-models, and mean models whose Kripke frames are in R(S). The notions of Q-satisfiability and Q-validity are determined by the concept of a Q-model exactly as in propositional modal logic.

B. The Objectual Interpretation: Q3 Semantics

The most familiar approach to semantics for quantifiers is the objectual interpretation. To begin our discussion, let us formulate a semantics, called Q3,[5] that uses the objectual interpretation with world-relative domains. Let a *Q3-model* be a sequence $\langle W, R, D, Q3, a \rangle$, where Q3 is a function that assigns a subset Q3(w) of D to each possible world w in W. The truth clauses for Et and the quantifier read as follows:

(E3) $a_w(Et)$ is T iff $a_w(t) \in Q3(w)$

(Q3) $a_w(\forall xA)$ is T iff for every $d \in Q3(w)$, $a^d/x_w(A)$ is T

(Here a^d/x is the assignment like a save that $a_w(x)$ is d for each w in W.)

Although Q3 semantics would appear at first to be quite natural, further reflection will show its serious defects. First of all, Q3 is very difficult to axiomatize. Even the rules of free logic, for example, are not Q3-valid. In order to see why, notice that the sentence $(\Box t = t \ \& \ Et) \rightarrow \exists x \Box(x = t)$ is a special case of (F∃G), the free logic principle of Existential Generalization.

(F∃G) $(At \ \& \ Et) \rightarrow \exists x Ax$

Since $\Box(t = t)$ is also provable using principles of identity and the necessitation rule ($\Box$), it follows that (E$\Box$) is a theorem of free modal logic.

(E$\Box$) $Et \rightarrow \exists x \Box(x = t)$

If t reads 'the author of "Counterpart Theory"', then (E$\Box$) says that if the author of 'Counterpart Theory' exists, then there is someone who is necessarily the author of 'Counterpart Theory'. Intuitively, (E$\Box$) is unacceptable, and it is not difficult to back up this insight with a formal counterexample. Let us image a model with two worlds, r (real) and u (unreal) whose domains both contain two objects, namely David Lewis and Saul Kripke. Assume that both worlds are accessible from themselves and each other. Imagine that the extension of t at the real world r is Lewis, but that it is Kripke in the unreal world u. On this model, $\exists x \Box(x = t)$ is false in r because neither Lewis nor Kripke is the extension of t in both worlds. Nevertheless, Et is true in r since the extension of t in the real world, namely David Lewis, is in the domain of r.

This counterexample helps us appreciate the reason why (F∃G) has broken down. There is no question that David Lewis exists, and there is no question that the author of 'Counterpart Theory' is identical to the author of 'Counterpart Theory' in any world we choose. However, the claim that any one person counts as the author of 'Counterpart Theory' in all worlds seems false. One way to help diagnose this situation is to reformulate Q3 semantics in an equivalent, but more complex way. Replace each object in the domain D with the constant function that takes all worlds to that object. Seen this way, the items in our domains are all intensions of rigid terms. (F∃G) is no longer valid because the domains of quantification include only constant term intensions, whereas terms may have nonrigid intensions.

The rules of free logic would be Q3-valid if we were to interpret the primitive predicate E so that Et is true in world w iff the extension $a_w(t)$ of t is a member of Q3(w) and $a(t)$ is a constant function. Notice, however, that the extension of E must then contain term intensions, and not objects. As a result, E is an *intensional* predicate, which means that substitution of identity does not hold for its term position. To see why, let d be the name 'David Lewis', and t the expression 'the author of "Counterpart Theory"'. Although d and t refer to the same thing, Ed is true and Et is false because t is not rigid.

Aldo Bressan (1973) has championed the view that even scientific language requires intensional predicates. His more general semantics defines the extension of a one-place predicate at a possible world as a set of individual concepts (i.e., term intensions) not a set of objects. As a result, he has no difficulty accommodating a primitive predicate that expresses rigidity.

Hintikka (1970) chose more modest methods. He showed how to formulate a correct rule of instantiation for Q3 that does not require an intensional existence predicate. Notice that the sentence $\exists x \Box (x = t)$ is true in a model at world w iff the intension of t has the same value in all worlds accessible from w. Similarly, $\exists x \Box\Box (x = t)$ is true at w just in case the intension of t is constant in all worlds accessible from those worlds. While there is no one sentence that expresses that a term is rigid, a sentence of the shape $\exists x \Box_i x = t$, where $\Box_i$ is a string of i boxes, guarantees that the intension of t is constant across enough worlds so that $\Box_i Pt$ follows from $\forall x \Box_i Px$ when Pt is atomic. This idea is generalized in Hintikka's formulation (HUI) of a valid rule of universal instantiation for nonrigid terms.

(HUI) $\quad \dfrac{\forall x A x}{(\exists x \Box_i (x = t) \,\&\, \cdots \,\&\, \exists x \Box_k (x = t)) \to At}$

where $i, \ldots, k$ is a list of integers that records for each occurrence of x in Ax, the number of boxes whose scope includes that occurrence.

In modal logic as strong as S4, this rule can be simplified considerably because there $\exists x \Box_i (x = t)$ is equivalent to $\exists x \Box (x = t)$. Thomason (1970) demonstrates the adequacy of Q3-S4, using (TUI) as the instantiation rule:

(TUI) $\quad \dfrac{\forall x A x}{\exists x \Box (x = t) \to At}$

It is interesting to note that even in the context of S4, Thomason was forced to adopt other complex rules for identity and the quantifier. We will discuss his system further in Section 3.B(3).

C. The Objectual Interpretation with Rigid Terms

There is a simplifying assumption that can be used to try to rescue the objectual interpretation from these complications. We explained that the domain for the objectual interpretation can be seen as a set of rigid term intensions. To validate (F∃G), we may insure that the treatment of terms and quantifier domains match by demanding that all terms are rigid designators. To insure rigidity of the terms, we stipulate the following:

(aRT) $a_w(t)$ is $a_{w'}(t)$ for all w, and w' in W

To obtain an objectual semantics Q1R with rigid terms simply define a Q1R-model as a Q3-model that satisfies (aRT).

Q1R semantics is fairly well behaved. In Section 3.B, we will explain how to axiomatize it using a general system based on free logic. Nevertheless, Q1R requires that all terms be rigid designators. There are several important reasons why this assumption should be rejected. First, expressions like 'the tallest man' clearly refer to different objects in different worlds. If we want to count descriptions among our terms, as we do, for example, on a Strawsonian account, we cannot accept the rigidity condition. Second, Lewis (1968) contends that it makes no sense to talk of identity of objects across possible worlds. Objects from two different worlds are never identical, although it may make sense to talk of the counterpart of an object in another world. On counterpart theory, it is impossible for the intension of any term to be a constant function. Third, in temporal logic, where objects are time slices, we do not want a thing to consist of the same time slice across different times. The slices of a thing picked out at different times may be quite different, but the world line composed of the slices still represents one unified thing. Since it is important that a logical theory does not rule out reasonable positions, we should relax the restriction that terms are rigid.

D. The Conceptual Interpretation

The systems we have discussed so far are not very satisfying. We have good reasons for wanting to allow nonrigid terms in our language, and yet the rules we need for Q3 are complex unless we use a language with a primitive existence predicate that expresses rigidity. One account of our difficulties, as we explained earlier, is that our terms can be assigned any intension, while the quantifiers range over constant intensions (objects). Perhaps quantifying over nonrigid intensions (or individual concepts) might result in a better match between the quantifiers and the terms and so yield simpler results.

Let us follow this suggestion by defining what we will call the *conceptual interpretation* of the quantifier. Let a Q2 model be a sequence $\langle W, R, Q2, a \rangle$ where Q2 is a function that assigns a domain Q2(w) of members of D to each world w. The clauses for Et and the quantifier now read as follows.

(E2) $a_w(Et)$ is T   iff $a_w(t) \in Q2(w)$

(Q2) $a_w(\forall x A)$ is T   iff for every function $f: W \to D$,

$$\text{if } f(w) \in Q2(w), \text{ then } a^f/x_w(A) \text{ is T.}$$

(Here 'a^f/x' represents the assignment function identical to a except that the intension of x on a^f/x is function f.)

There are several problems with Q2 semantics. One formal difficulty is that no formal system is adequate for it. Whenever we interpret the domain of quantifier as a set of all functions, we run the risk that the language will have the expressive power of second-order arithmetic, with the result that Godel's Theorem applies. This is exactly what happens with Q2 semantics.[6]

There are also philosophical difficulties. The sentence $\exists x \Box \exists y (y = x)$ says that there is something that necessarily exists. Intuitively, this is a strong claim, for it appears to bolster an argument for the existence of God. However, on the conceptual interpretation $\exists x \Box \exists y (y = x)$ will turn out to be true provided only that no domain for any world is empty.[7] So $\exists x \Box \exists y (y = x)$ does not receive its intuitive reading on the conceptual interpretation.

E. The Substantial Interpretation

The conceptual interpretation of the quantifiers does not match the interpretation that we give to quantifier expressions in ordinary language. The sentence $\exists x \Box \exists y (y = x)$, which we interpret as making the very strong claim that some one *thing* exists in every possible world, is valid on the conceptual interpretation as long as no possible world has an empty domain. The difference between our intuitive understanding of $\exists x \Box \exists y (y = x)$, and the conceptual interpretation is that the existence of an individual concept that (say) picks out Saul Kripke in this world, a rock in another, a blade of grass in another, and so on, counts to verify $\exists x \Box \exists y (y = x)$. On the other hand, our intuitions demand that any thing that verifies $\exists x \Box \exists y (y = x)$ must be coherent in some sense; our notion of a thing brings with it some idea of what it would be like in other worlds. Only certain collections of objects (and certainly not a collection consisting of Saul Kripe, a rock, a blade of grass, etc.) could count as the manifestations of a thing, and so only these collections should count to verify $\exists x \Box \exists y(y = x)$.

In order to do justice to these intuitions, we must restrict the domain of quantification to the term intensions that reflect 'the way things are' across possible worlds. Thomason (1969) suggests that the domain should contain only constant functions. The idea is that for $\exists x \Box \exists y(y = x)$ to be true there must be one thing, identical across possible worlds, that exists in each one. This proposal amounts to Q3, the objectual interpretation with nonrigid terms. We have already discussed some of the difficulties with this option.

If we are to accommodate a full variety of conceptions about what things are like, we should not assume that they are represented by the constant term intensions (Q3), nor by all the term intensions (Q2). To be completely general, we introduce a set of individual concepts for each world, to serve as the domain of quantification, and we make no stipulations about what these sets contain. Let us now give a formal account of this approach.

A QS-model is a sequence $\langle W, R, D, QS, a \rangle$, where QS is a function that assigns to each world w in W a set $QS(w)$ of functions from W into D. (We call $QS(w)$ the set of substances for world w.) The truth clauses for Et and the quantifier read as follows:

(ES) $a_w(Et)$ is T   iff $a(t) \in QS(w)$

(QS) $a_w(\forall x A)$ is T   iff for all $f \in QS(w)$, $a^f/x_w(A)$ is T

It is not difficult to show that $\exists x \Box \exists y(y = x)$ is invalid on this semantics.[8]

In Section 3.B, we will show how to axiomatize QS using principles of free logic. We should note here an important restriction on the rule of substitution of identities in QS. The constant E is an intensional predicate, and this means that substitution of term identities does not hold in its term position. When we formulate the rule of substitution for identities, we must make it clear that we do not consider Et to be an atomic sentence, for otherwise we would be able to deduce Et' from $t = t'$ and Et.

3. Completeness in Free Intensional Logics

A. Why Completeness is Hard to Prove in Quantified Modal Logic

Completeness proofs in quantified modal logic are harder than those for propositional modal logic or free logic. One cause of the difficulty may

Applications of Free Logic to Quantified Intensional Logic

be that completeness fails, as in the case of the conceptual interpretation Q2. Even when a system *is* complete, the proof may be elusive, and difficult to formulate in a simple way. Another problem is lack of generality; a proof strategy may only work when the underlying modal logic is fairly strong (e.g., as strong as S4), or when ad hoc conditions are placed on the models.

One of the best ways to understand the methods used in completeness proofs for quantified modal logic is to locate the main difficulty that arises if we simply try to "paste together" proofs for free logic and propositional modal logic. In order to uncover the problem, let us review the crucial steps in the completeness proofs in each kind of logic.

1. Completeness for Propositional Modal Logics

The traditional method for proving completeness of a propositional modal logic S is to use maximally consistent sets. Completeness follows if we can show that any S-consistent[9] set is S-satisfiable. We begin by extending a given S-consistent set H to a maximally consistent set M by Lindenbaum's Lemma. Then we build what we will call the canonical model $\langle W, R, a \rangle$ for S. The set W of possible worlds on the canonical model is the set of all maximally consistent sets. The accessibility relation R is defined by (Def R).

(Def R) $w R w'$ iff for all wffs A, if $\Box A \in w$, then $A \in w'$

Finally, the assignment function a is defined for propositional variables p so that $a_w(p)$ is T iff $p \in w$.

The next project is to prove the Truth Lemma (TL), which shows that membership in w and truth in w on the canonical model amount to the same thing:

(TL) $a_w(A)$ is T iff $A \in w$

Once (TL) is shown, it follows that all members of H are true at M on the canonical model. We may also prove that $\langle W, R \rangle \in R(S)$ (the set of Kripke frames that corresponds to S), and so the canonical model satisfies H.

The proof of (TL) is an induction on the construction of A, and the only really interesting case is when A has the shape $\Box B$. The case for propositional variables is trivial given the definition of the canonical model, and cases for $\sim$ and $\rightarrow$ simply depend on corresponding properties of maximally consistent sets w:

($\sim \varepsilon$) $\sim B \in w$ iff $B \notin w$

($\to \varepsilon$) $B \to C$ iff either $B \notin w$ or $C \in w$

The proof of the case for $\square$ takes the following form:

$a_w(\square A)$ is T   iff for all $w' \in W$, if $w \, R \, w'$, then $a_{w'}(A)$ is T

(1)   iff for all $w' \in W$, if $w \, R \, w'$, then $A \in w'$

(2)   iff $\square A \in w$

The only difficult part is to show the equivalence of (1) and (2). The inference from (2) to (1) is a simple consequence of the way we defined R. In order to show that (1) implies (2), we show ($\sim \square$) instead.

($\sim \square$) if $\square B \notin w$, then there is a maximally consistent set w' such that $w \, R \, w'$ and $B \notin w'$

The proof of ($\sim \square$) makes a second use of the Lindenbaum Lemma. Given that $B \notin w$, we show the consistency of the set $w^* = \{A : \square A \in w\} \cup \{\sim B\}$. Then we use the Lindenbaum Lemma to extend w^* to a maximally consistent set w'. The set w' is such that $w \, R \, w'$ because for each sentence $\square A \in w$, $A \in w'$; it does not contain B since it is consistent and contains $\sim B$.

2. Completeness of Free Logic

In this section we will give a quick review of a completeness proof for FL: extensional free logic with identity. This will prepare the ground for the presentation of technical results for quantified free intensional logics. A complete system FL of free logic with identity can be constructed by adding the following rules to propositional logic:

(FUI) $$\frac{\forall x Ax}{Et \to At}$$

(FUG) $$\frac{B \to (Et \to At)}{B \to \forall x Ax}$$

   where t is a term that does not appear in $B \to \forall x Ax$

($=$ In) $t = t$

($=$ Out) $$\frac{t = t'}{Pt \to Pt'}$$

   where Pt' is atomic and results from replacing an occurrence of t in Pt with t'

Applications of Free Logic to Quantified Intensional Logic

In these rules, and throughout this article, B and Ax are wffs, x is any variable, and At is the result of substituting the term t properly for all occurrences of x in Ax.

It is an easy exercise to show that Et is equivalent in FL to $\exists x\, x = t$ (where x is not t). So we could have defined Et as $\exists x\, x = t$, and avoided the introduction of a special predicate letter E. However, in some intensional logics, there is no way to define Et in terms of the rest of the primitive vocabulary, and so we have prepared for this by assuming that E is primitive.

Although the restriction to atomic sentences in (=Out) may seem strong, it has no effect whatsoever in extensional free logic. A simple induction will show (=Out) insures the substitution of identities in all extensional sentences. However, in intensional logics (=Out) does not guarantee substitution of identical terms that lie in the scope of intensional operators. The restriction is important because $=$ is interpreted as contingent identity, with the result that substitution of identities in intensional contexts is invalid.

FL can be shown to be both consistent and complete for a standard free logic semantics defined as follows: An FL-model $\langle D, Q, a \rangle$ consists of an outer domain D (of possible objects), an inner domain Q of real objects, and an assignment function a, which satisfies the standard truth conditions for atomic sentences, $=$, $\sim$, and $\to$, plus the following for E and the universal quantifier:

(E) $a(E) = Q$

($\forall$) $a(\forall x Ax)$ is T iff for all $d \in Q$, $a^d/x(Ax)$ is T

Notions of FL-validity and FL-satisfiability are defined in the usual way. For use in later discussion, we note now two properties of FL-assignments.

(EQ) $a(Et)$ is T iff $a(t) \in Q$

(SL) $a^{a(t)}/x(Ax)$ is $a(At)$

Condition (EQ) holds because $a(E)$ is Q. (SL) can be established by induction on the form of Ax.

To prove completeness of FL, we show that any FL-consistent set H is FL-satisfiable by first extending H to a maximally consistent set M, written in language L. We then construct a model $\langle D, Q, a \rangle$ from M as follows. The assignment function a is defined so that the extension $a(t)$ of t is $\{t' : t = t' \in M\}$, the equivalence class of terms ruled identical in M. The outer domain D contains $a(t)$ for each term t, and the inner domain Q contains $a(t)$ for terms t such that $Et \in M$. The assignment function a

is defined for n-ary predicate letters P so that $\langle d_1, \ldots, d_n \rangle$ is a member of $a(P)$ just in case each of the $a(t_i)$ is d_i, for $1 \leqslant i \leqslant n$, and $Pt_1, \ldots, t_n \in M$.

Given the principles of identity, it is not difficult to show that (TL) holds for atomic sentences on this model. In order to establish (TL) for all sentences, we must be sure that the set M meets one further condition ($\forall\varepsilon$) concerning the quantifier.

($\forall\varepsilon$) $a(\forall xAx)$ is T iff $\forall xAx \in M$

The proof of ($\forall\varepsilon$) will be insured if we can show that M is *free omega complete* (FOC):

(FOC) If $M \vdash Et \to At$, for every term t of L, then $M \vdash \forall xAx$

Notice that (FOC) is equivalent to (FOC′):

(FOC′) If $M \cup \{\sim\forall xAx\}$ is consistent, then for some term t of L, $M \cup \{\sim(Et \to At)\}$ is consistent

There are maximally consistent sets that disobey (FOC), so when we extend H to M using the Lindenbaum procedure, we must take special steps to guarantee (FOC). Remember that the Lindenbaum method for extending a consistent set to a maximally consistent one begins by ordering the wffs. A series of sets $M_0, M_1, \ldots$ is then formed by letting M_0 be H, and letting M_{i+1} be the result of adding the $(i+1)$st wff to M_i iff doing so would leave M_{i+1} consistent. (Otherwise M_{i+1} is M_i.) The maximally consistent set M desired is the union of all the M_i.

To insure a set is free omega complete during this construction, we do the following. If M_i is the ith set formed in that construction, and $\sim\forall xAx$ is the $(i+1)$st sentence of our ordering of all the wffs, and if adding $\sim\forall xAx$ to M_i would yield a consistent set, then we form M_{i+1} from M_i by adding both $\sim\forall xAx$, *and* a sentence of the form $\sim(Et \to At)$, where t is a term that is new to both $\sim\forall xAx$ and M_i. Adding this second sentence to M_{i+1} cannot cause M_{i+1} to become inconsistent, as long as $M_i \cup \{\sim\forall xAx\}$ was already consistent as we have assumed.[10]

Let us say that a set is *free saturated* iff it is maximally consistent and free omega complete. We can also see from the second formulation (FOC′) of free omega completeness that the result of the construction we have just outlined is free omega complete, and so it is free saturated. It follows that *every FL consistent set has a free saturated extension*. Now suppose we use this fact to produce a free saturated extension M of H. As a result, we can show that ($\forall\varepsilon$) holds in the model constructed from M by the following reasoning.

$a(\forall x Ax)$ is T
iff for all $d \in Q$, $a^d/x(Ax)$ is T (F∀)
iff for all terms t, if $a(t) \in Q$, $a^{a(t)}/x(Ax)$ is T Definition of D
iff for all terms t, if $a(t) \in Q$, $a(At)$ is T (SL)
iff for all terms t, if $a(Et)$ is T, $a(At)$ is T (EQ)
iff for all terms t, $a(Et \to At)$ is T (→ε)
iff for all terms t, $Et \to At \in M$ Hypothesis of the induction
iff $\forall x Ax \in M$ w is (FOC) and closed under (FUI)

Now that we have finished the case for ∀, we have a proof of (TL). It follows that the FL-model we have defined satisfies all the sentences of M and, hence, all sentences of our original set H. We conclude that any FL-consistent set is FL-satisfiable and FL is complete.

3. The Difficulty in Quantified Modal Logics Completeness Proofs

Notice that the method we described for constructing a free saturated set requires that we have an infinite set of terms of L that are foreign to H. Since we may have infinitely many sentences $\sim\forall x Ax$ to add, we need infinitely many 'instances' $\sim(Et \to Pt)$ where t is new to the construction. As a result, the set M that we constructed using this method, contains an infinite set of terms of L that did not appear in H.

Now let us imagine that we hope to prove completeness of a modal logic Q, which adds principles of free logic to the propositional modal logic S. We begin with a Q-consistent set H that we hope to show is Q-satisfiable by extending H to a free saturated set M written in language L. We then hope to construct the canonical model, which will make all sentences of H true at M. Difficulties arise when we try to prove (TL). There is a conflict between what we need to insure (∀) and what we need to insure (□). Condition (∀) demands that the set W of possible worlds be the set of saturated sets in language L, for the terms of L (actually their equivalence classes) determine the domain of the quantification of our model. On the other hand, the proof of condition (□) requires the following. From a given possible world w that contains $\sim\Box B$, we must be able to construct a saturated set w' in language L that is an extension of $w^* = \{A : \Box A \in w\} \cup \{\sim B\}$. The problem is that in order to extend w^* to a saturated set in L, we must find an infinite set of terms of L that do not appear in w^*. However, the world w contains $(Pt \to Pt)$ for each term t of L, with the result that all formulas $Pt \to Pt$ appear in w^*. So there are no terms of L foreign to w^*. If we attempt to remedy the problem at this point by constructing world w' in a larger language L', then we find ourselves in a vicious circle. Now we must prove property ($\forall$) for L' instead of L. This forces us to define W as the set of all saturated sets in language L', so that when we want to extend w^* to a saturated set, we must find infinitely many terms of L' foreign to w^*. However, w is now a saturated set in language L', and contains $(Pt \to Pt)$ for all terms t of L'. Again, we have no guarantee that there are any terms of L' that do not appear in w^*.

B. Strategies for Proving Completeness of Free Modal Logic

In this section, we will illustrate two different strategies for overcoming the difficulty we have just described. The first method works for systems based on a relatively narrow range of propositional modal logics. The second method applies to a much wider range of systems, but it requires a more general formulation of the quantifier rules. Ideally, we would like to find a completely general completeness proof. The proof would demonstrate completeness of the most general semantics we have considered (QS), regardless of the choice of the propositional modal logic S. The proofs for all less general systems would then fall out of the general proof just as proofs for the stronger propositional modal logics result from the completeness proof for K. This would help clarify and unify quantified modal logic. Although strategy 2 goes part of the way toward this goal, it still has limitations as we will see.

1. Strategy 1: Completeness of FS

Let FS be the logic that results from adding the principles FL of free logic to a given propositional modal logics S. We will demonstrate the completeness of FS with respect to QS-semantics using the strategy of Garson (1978). As we will explain in the next subsection, the proof only works for certain choices of S.

Remember that the difficulty described earlier in proving completeness for quantified modal logic was that we needed a way to extend a consistent set w^* to a free saturated one, but we did not have an infinite set of terms missing from w^* in order to carry out the construction. Notice, however, that when the domain of quantification varies from one possible world to the next, we are free to select a different language for each of the free saturated sets that are in W in the canonical model. When it comes time to construct a saturated set from w^*, we may build a saturated set in a language larger than the one in which w is written.

As usual we begin the completeness proof by assuming that set H is consistent in FS, and we extend H to a free saturated set M written in language L. At this point, however, we consider a larger language L^+, which contains infinitely many terms that are not in L. We then define the set W of possible worlds for our canonical FS-model $\langle W, R, D, \text{QS}, a\rangle$ as the set of all free saturated sets written in some language L' such that there are infinitely many terms of L^+ that do not appear in L'. The idea behind this is to guarantee that whenever $w \in W$, there will be infinitely many terms foreign to $w^* = \{A: \Box A \in w\} \cup \{\sim B\}$ so that w^* can be extended to a saturated set in language L^+. The other parts of the definition of the canonical FS-model are straightforward. R is defined by (Def R): $w\, R\, w'$ iff if $\Box A \in w$, then $A \in w'$. The intension $a(t)$ of a term t given by a is defined so that $a_w(t)$ is $\{t: t = t' \in w\}$. QS is defined so that $s \in \text{QS}(w)$ iff s is $a(t)$ for some term t such that $Et \in w$. The domain of possible objects D is simply the set of all term extensions in all the possible worlds. The intension $a(P)$ of an n-ary predicate letter P is given as one would expect: $\langle d_1, \ldots, d_n \rangle \in a_w(P)$ iff each of the $a_w(t_i)$ is d_i, for $1 \leq i \leq n$, and $Pt_1, \ldots, t_n \in w$.

Because the members of w are free saturated sets written in different languages, we cannot prove the Truth Lemma (TL) for this canonical model. If t does not appear in L_w, the language in which the saturated set w is written, then $a_w(\sim Pt)$ is T, but $\sim Pt \notin w$. However, there is a weaker formulation (wTL) that will still serve our purposes:

(wTL) If A is a sentence of L_w, then $a_w(A)$ is T iff $A \in w$

The proof of (wTL) for cases other than $\Box$ and $\forall$ is straightforward. The crucial step in the case for $\Box$ is to demonstrate $(\sim \Box)$.

$(\sim \Box)$ If $\Box B$ is a sentence of L_w, then if $\Box B \notin w$ then there is a w' in W such that $w\, R\, w'$ and $B \notin w'$

We begin the proof by assuming that $\Box B$ is a sentence of L_w, and that $\Box B \notin w$. We construct $w^* = \{A: \Box A \in w\} \cup \{\sim B\}$, which we show to be consistent in the usual way. Since w is a member of W, there must be an infinite set N of terms of L^+ that do not appear in w. By the definition of w^*, it is clear that none of these terms appear in w^* either. We could construct a free logic saturated set w' from w^* using these terms. However, if w' is to be a member of W, there must be an infinite set of terms of L^+ foreign to w'. In order to insure that we do not "use up" all the terms in our construction of w', we divide N into two infinite sets N_1 and N_2. We use N_1 to extend w^* to a free logic saturated set w', and we leave N_2 in reserve to insure that $w' \in W$. When w' is constructed in this way, we can easily prove that $w R w'$, and that $B \notin w'$, and so we have finished the proof of ($\sim\Box$).

The case for $\forall$ involves two ticklish points. First, we must establish the substitution lemma (ISL) for term *intensions*:

(ISL) $a^{a(t)}/x_w(Ax)$ is $a_w(t)$

This can be carried out with an induction on the structure of Ax. Notice that the corresponding lemma (ESL) for term *extensions* $a_w(t)$ does not hold:

(ESL) $a^{a_w(t)}/x_w(Ax)$ is $a_w(At)$

Second, we need to show (EQS):

(EQS) $a(t) \in QS(w)$ iff $Et \in w$

((EQS) is also need to show the case of formulas with the shape Et.)

The proof of (EQS) would seem to be trivial given our definition of $QS(w)$, but it is not. The trouble comes in showing (EQS) from left to right. Suppose that $a(t) \in QS(w)$. Then by the definition of $QS(w)$, there is a term t' such that $a(t)$ is $a(t')$ and $Et' \in w$. For ordinary predicates, this would be enough to insure that $Et \in w$, for when $a_w(t')$ is $a_w(t)$, we have that $t' = t \in w$, and so can substitute t for t'. Remember, however, that E is an intensional predicate for which the rule of substitution of identities does not hold, so this reasoning will not work. We must find some other way to insure that $Et \in w$. Things look bad when we realize that t may not even be in the language L_w, in which case $Et \notin w$. Luckily, our definition of the canonical model insures that whenever $a(t)$ is $a(t')$ then t and t' are the same term. The reason is that when $t \notin L_w$, $a_w(t) = \{t': t = t' \in w\}$ is empty. For any pair of distinct terms t, t' we choose, we can always find a language L_w such that t is in L_w and t' is not. It follows that the only way that $a(t)$ and $a(t')$ can be identical is if t is identical to t'. We have that $Et' \in w$, so we conclude that $Et \in w$ and our proof of (EQS) is finished.

Once (ISL) and (EQS) are available, the case for $\forall$ is demonstrated as follows: Suppose $\forall xAx$ is in L_w. Then

$a_w(\forall xAx)$ is T
iff for all $f \in QS(w)$, $a^f/x_w(Ax)$ is T   (QS)
iff for all t if $a(t) \in QS(w)$, $a^{a(t)}/x_w(Ax)$ Definition of $QS(w)$
   is T
iff for all t, if $a(t) \in QS(w)$, $a_w(At)$ is T   (ISL)
iff for all t, if $Et \in w$, $a_w(At)$ is T   (EQS)
iff for all t, if $Et \in w$, $At \in w$ Hypothesis of the induction[11]

iff for all t, $Et \to At \in w$ $(\to\varepsilon)$
iff $\forall x Ax \in w$ w is (FOC) and closed under (FUI)

Once Lemma (wTL) is established in this way, the completeness of FS is shown fairly easily. We have already extended the FS-consistent set H to a free logic saturated set M, and since there were infinitely many terms foreign to M in L^+, it turns out that $M \in W$. By (wTL), it follows that all members of M (and so all members of H) are true at M on the canonical model, and so H is FS-satisfiable.

2. Limitations of Strategy 1

Although the last result is satisfying because it shows completeness for a general semantical treatment of the quantifiers, it does not work to establish completeness of systems that are less general. For example, we might hope to show the completeness of the objectual interpretation with world-relative domains and *rigid* terms by adding (RT) to FS:

(RT) $\dfrac{t = t'}{\Box t = t'}$ $\dfrac{\sim t = t'}{\Box \sim t = t'}$

We would hope that (RT) would insure that terms are rigid on our canonical model, with the result that all members of QS(w) are constant functions.

However, these hopes cannot be realized using the present definition of the canonical model. In order to insure that (wTL) holds for sentences $t = t'$, we are virtually forced into defining $a_w(t)$ as $\{t : t = t' \in w\}$. If all terms are rigid on this model, then all worlds must be written in the same language.[12] However, the strategy of this completeness proof depends on allowing our languages to shift from one saturated set to the next. Using similar reasoning, we can see that it is pointless to hope for a completeness proof for systems with fixed domains using the canonical model of this section.

There is another respect in which the variable language strategy lacks generality. The method does not work for many propositional modal logics S. The reason is that when possible worlds are written in different languages, we lose an important property ($\Diamond$) that is needed in showing that $\langle W, R \rangle$ on the canonical model is in R(S):

($\Diamond$)   If $w R w'$ and $A \in w'$, then $\Diamond A \in w$

This property fails if term t is in $L_{w'}$ but not in L_w, and A contains t. The sentence $\Diamond A$ cannot be in w because it is not even in the language of w.

For many modal logics (e.g., D, M, and S4), we do not need ($\Diamond$) in order to show that $\langle W, R \rangle \in R(S)$. However, for systems like B, the property is indispensable. There are tricks one can use to overcome the difficulty for individual systems, but the changing language strategy does not provide a proof that is general with respect to the underlying modal logic.

3. Completeness of General Systems: Strategy 2

Thomason's (1970) proof of the completeness of Q3 is the inspiration for the second stragegy we are going to present. We will give a second completeness proof for QS semantics, in this case for systems with a wider range of modalities, and with more powerful quantifier principles. Once we have presented the details, we will show how to modify the proof to obtain completeness results for the objectual interpretation (Q3).

Formulation of GS. To help simplify our presentation, we will adopt a few abbreviations. We use '$\dashv$' for strict implication, so that '$(A \dashv B)$' abbreviates '$\Box(A \to B)$'. We will be working with formulas that have the shape (GF), where parentheses are to be restored from right to left:

(GF) $A_1 \dashv A_2 \dashv \cdots \dashv A_i \dashv B$

(For example, $A \to B \dashv C \dashv D$ amounts to $A \to (B \dashv (C \dashv D))$, or $A \to \Box(B \to \Box(C \to D))$.) We will use '$G(B)$' to represent any sentence with shape (GF), with the understanding that $G(C)$ will be the sentence that results from replacing C for B in $G(B)$. Using this notation, we may now present two general rules for the quantifiers:

(GUI) $\dfrac{G(\forall x Px)}{G(Et \to Pt)}$

(GUG) $\dfrac{G(Et \to Pt)}{G(\forall x Px)}$ where t does not appear in $G(\forall x Px)$

We should make clear that $G(A)$ may represent a sentence where any of arrows (whether $\to$ or $\dashv$) is missing in the pattern (GF). So all of the following, for example, are instances of the rule (GUI):

$\dfrac{\forall x Px}{Et \to Pt} \qquad \dfrac{A \to \forall x Px}{A \to Et \to Pt} \qquad \dfrac{A \dashv \forall x Px}{A \dashv Et \to Pt} \qquad \dfrac{A \dashv B \dashv \forall x Px}{A \dashv B \dashv Et \to Pt}$

The system GS consists of (GUI), (GUG), (=In), (=Out), and principles for a propositional modal logic S.

The quantifier rules of GS appear to be very cumbersome. However,

GS has a simple and natural reformulation in natural deduction format. The propositional modal logic *K* may be formulated by introducing boxed subproofs:

□|

Together with introduction and elimination rules for □:

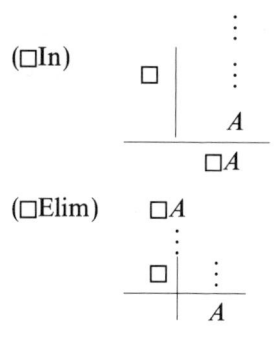

(□In)   □|
           | *A*
         ─────
          □*A*

(□Elim) □*A*
          □| ⋮
         ─────
           | *A*

When natural deduction rules are employed, GS may be formulated using the simpler FS rules (FUI) and (FUG), with the understanding that these apply in any subproof. It is a straightforward matter to show that this natural deduction formulation NDFS is equivalent to GS.[13]

Another feature of GS is evidence for its naturalness. One would hope to construct a quantified modal logic with *fixed* domains by adding *Et* as an axiom, thus insuring that the free logic rules collapse to their classical counterparts. In FS, the addition of *Et* entails (CBF), but (BF) is independent, and must be added as a separate axiom. However, when *Et* is added to GS, both the Barcan Formula (BF) and its converse (CBF) are provable. It is pleasing that the generalized rules are symmetrical with respect to the adoption of the Barcan Formula and its converse.

Completeness of GS. We will now turn to the completeness proof for GS. As we have already pointed out, w^* is not omega complete. However, in GS, w^* has a weaker property called generalized omega completeness (GOC) that insures that w^* can be extended to a set that has correspondingly weaker form of saturation, a form that insures a proof of the quantifier case of the Truth Lemma:

(GOC)   If $w \vdash G(Et \to Pt)$ for every term t of L, then $w \vdash G(\forall x Px)$

A *GOC set* is just a set with property (GOC), and a set is *generally saturated* (for language L) just in case it is a maximally consistent GOC set.

The main portions of the completeness proof are given in the following lemmas.

Lemma 1. *If w is GOC, then so is $w \cup f$, provided that f is finite.*

Proof. Suppose that w is GOC. To show that $w \cup f$ is also GOC, let us assume that $w \cup f \vdash G(Et \to Pt)$ for all terms t. It follows that $w \vdash \&f \to G(Et \to Pt)$ for all terms t, where $\&f$ is the conjunction of the (finitely many) members of f. It follows that $w \vdash \&f \to G(Et \to Pt)$. By principles of propositional logic, this sentence is equivalent to one with the shape (GF), so we know that $w \vdash \&f \to G(\forall xPx)$, and hence that $w \cup f \vdash G(\forall xPx)$. ●

Lemma 2. *Any consistent GOC set can be extended to a generally saturated set written in the same language.*

Proof. Let w be a GOC set. We construct a generally saturated extension of w using a variant of the method described in Section 3.A.(2). Suppose that the set $M_i \cup \{\sim G(\forall xPx)\}$ is consistent, so that $\sim G(\forall xPx)$ is the candidate for addition to M_i in the Lindenbaum construction. Then we add both $\sim G(\forall xPx)$ and $\sim G(Et \to Pt)$ to M_i to form M_{i+1}, where t is a term that leaves M_{i+1} consistent. There is such a term because w is GOC and so, by Lemma 1, M_i is GOC. This construction preserves consistency, and results in a GOC set, and so it yields a generally saturated set. ●

Lemma 3. *If w is a generally saturated set that contains $\sim \Box B$, then $w^* = \{A : \Box A \in w\} \cup \{\sim B\}$ is consistent and GOC.*

Proof. We can easily show that w^* is consistent. To show that w^* is GOC, assume that $w^* \vdash G(Et \to Pt)$ for any term t of L. It follows that $\{A : \Box A \in w\} \vdash \sim B \to G(Et \to Pt)$. By principles of propositional modal logic K, $w \vdash \Box(\sim B \to G(Et \to Pt))$, and so $w \vdash \sim B \dashv G(Et \to Pt)$ for every term t of L. Since w is GOC and $\sim B \dashv G(Et \to Pt)$ has the shape (GF), $w \vdash \sim B \dashv G(\forall xPx)$. Since w is maximal, $\sim B \dashv G(\forall xPx) \in w$. As a result, $\sim B \to G(\forall xPx) \in \{A : \Box A \in w\}$, and $w^* \vdash G(\forall xPx)$. ●

Lemma 4. *If w is generally saturated and contains $\sim \Box B$, then $w^* = \{A : \Box A \in w\} \cup \{-B\}$ can be extended to a generally saturated set written in the same language.*

Proof. By Lemmas 2 and 3. ●

Now that we have proved Lemmas 1–4, only a few details need to be mentioned to finish a completeness proof for GS. We begin with a

GS-consistent set, and we extend it to a generally saturated set M written in language L.[14] We define the canonical QS-model $\langle W, R, D, QS, a \rangle$ so that W is the set of all generally saturated sets for L. R is defined by (Def R). The extension $a_w(t)$ of a term t is $\{t': t = t' \in w\}$, and D is the set of all term intensions. The extension of an n-ary predicate letter P is defined so that $\langle d_1, \ldots, d_n \rangle \in a_w(P)$ iff each of the $a_w(t_i)$ is d_i, for $1 \leqslant i \leqslant n$, and $Pt_1 \cdots t_n \in w$.

We may now prove the stronger truth lemma (TL) in a straightforward way. The case for $\square$ requires that we show that if $\sim \square A \in w$ then there is a w' in W such that $w R w'$ and $\sim A \in w'$, but this is easily demonstrated using Lemma 4.

To prove the case for $\forall$, notice first that all generally saturated sets are free saturated, because free omega completeness (FOC) is a special case of (GOC). So we will have no difficulty proving that $a_w(\forall xPx)$ is T iff $\forall xPx \in w$ as long as we can show (EQS):

(EQS) $a(t) \in QS(w)$ iff $Et \in w$

In order to show (EQS) in the completeness proof for FS (Section 3.B(1)), we proved that if t and t' are distinct, then so are their intensions $a(t)$ and $a(t')$. We can show this is true for the canonical model of this section as follows. If t and t' are distinct, $\{\sim t = t'\}$ is GS-consistent and GOC. So by Lemma 2 there is a generally saturated set w in W that contains $\sim t = t'$. By the truth clause for identity $a_w(t) \neq a_w(t')$ and so the intensions $a(t)$ and $a(t')$ are not identical.

Since all sets in W on the canonical model are generally saturated in the same language, we no longer face the difficulties noted in Section 3.B(2) in showing that $\langle W, R \rangle \in R(S)$. Property $(\Diamond)$ now holds, and so the proof proceeds exactly the way it does in propositional modal logics. However, there are still modal logics for which strategy 2 does not apply.[15] The determination of exactly which choices of S lead to completeness of GS is an important question.[16]

Completeness for Q1R. One advantage of strategy 2 is that it can be used to obtain completeness proofs for a variety of less general logics. To illustrate, we will explain how to exploit the proof of the completeness of GS to obtain results for systems that use the objectual interpretation. Our first project will be to sketch the completeness for Q1R, objectual semantics with rigid terms. This will pave the way for a discussion of Q3, which is more complicated. The system we will show complete for Q1R is called GQ1R, and formed from GS by adding the following principles to insure that terms are rigid:

(RTE) $\quad \dfrac{t = t'}{\Box t = t'} \qquad \dfrac{\sim t = t'}{\Box \sim t = t'} \qquad \dfrac{t = t'}{Et \to Et'}$

A change in the definition of the canonical model is required to show completeness of GQ1R, along with an adjustment to the proof of (TL). We begin with a consistent set H, which we extend to a generally saturated set M. We insure rigidity of terms in the canonical model by restricting W to sets that contain exactly the identities of M. We must adjust the proof for the $\Box$ case of (TL) because we will need to know that w^* can be extended to a set that contains the same identities as M. This may be established using the fact that (RTE) is present in GQ1R. Since substitution now holds behind E, the proof of (EQS) is simplified. Since all term intensions are rigid on the canonical model, and since the domains only contain term intensions, we can create a Q1R-model from the canonical model by replacing constant term intensions with their values. The result is a Q1R-model that satisfies M, hence H.

Completeness for Q3. A variation on strategy 2 was invented by Thomason to prove completeness of Q3–S4. We will use strategy 2 here to prove completeness of more general Q3 logics. In our discussion of systems that use the objectual interpretation and nonrigid terms (Section 2.B), we pointed out that quantifier rules are quite complicated unless we introduce a *primitive* existence predicate E that expresses that a term intension is a constant function. So we will begin with Q3 systems that contain E. Later we will show how to eliminate E in systems as strong as S4.

The main problem that arises in the completeness proof for systems that allow nonrigid terms with the objectual interpretation concerns the substitution lemma. In GS the intensional substitution lemma (ISL) was used in establishing the $\forall$ case for the truth lemma (TL). Because our quantifiers range over objects in Q3, the $\forall$ case requires proof of the extensional version of the substitution lemma (ESL):

(ESL) $\quad a^{a_w(t)}/x_w(Ax)$ is $a_w(At)$

Unfortunately, (ESL) is not always true if t is nonrigid.[17]

We did not face this problem for systems with rigid terms, because (ESL) is true when $a(t)$ is a constant function. The problem did not arise with the substantial interpretation because there the lemma needed (ISL) concerns substitution of *intensions* and is readily proven:

(ISL) $\quad a^{a(t)}/x_w(Ax)$ is $a_w(At)$

Thomason tackles the problem posed by the failure of (ESL) by stipulating that variables are rigid designators. He then uses variables, not terms, to fix the domains of his canonical model. The extension $a_w(t)$ is set to $\{x: x = t \in w\}$, and the domain Q3(w) contains the extensions of all terms t such that $Et \in w$. By adding the rules (RV), to the system, he can insure that the canonical model has rigid variables:

(RV) $\quad \dfrac{x = y}{\Box x = y} \qquad \dfrac{\sim x = y}{\Box \sim x = y} \qquad \dfrac{x = y}{Ex \to Ey}$

However, the use of rigid variables leads to further complications. In order to establish the case for identity in (TL), we need to know that if $a_w(t)$ is $a_w(t')$ then $t = t' \in w$. The identity of $a_w(t)$ and $a_w(t')$ only establishes that $x = t \in w$ iff $x = t' \in w$, for all variables x. To show that $t = t' \in w$, we need to know that there is some variable y such that $y = t \in w$. This requires us to restrict the set W of possible worlds of our model to those that meet condition (V):

(V) $\quad$ For all $w \in W$, and all terms t of L, there is a y such that $y = t \in w$

In order to meet condition (V) when it comes times to extend w^* to a set in W, the following rule is needed:

(G=) $\quad \dfrac{G(\sim y = t)}{G(p \wedge \sim p)}$

The rule (G=) insures that we can consistently add a sentence of the form $y = t$ for each of the terms t during the construction of a saturated set, and to do so without extending the language.

The system Q3 that can be shown complete using this method is composed of GS, (RV), and (G=). The Q3 system Thomason (1970) discusses is similar to this, but it lacked the primitive existence predicate E, and was built on S4. In S4, the sentence $\exists x \Box x = t$ is true in the canonical model just in case the intension of t is rigid. Also, the replacement of Et with $\exists x \Box x = t$ in the rules of free logic results in valid quantifier rules. It follows that if S is S4 or stronger, we can formulate a complete system for Q3S without a primitive existence predicate by replacing Et with $\exists x \Box x = t$ in the rules of Q3S.

References

Bressan, A., *A General Interpreted Modal Calculus*, Yale University Press (1973).
Carnap, R., *Meaning and Necessity*, University of Chicago Press (1947).

Gabbay, D., *Investigations in Modal and Tense Logic with Applications to Problems in Philosophy and Linguistics*, Reidel, Dordrecht (1976).

Gabbay, D. and Guenthner, F. (eds.), *Handbook of Philosophical Logic*, vol. 2, Reidel, Dordrecht (1984).

Garson, J. W., "Completeness of some quantified modal logics," *Logique et Analyse*, vol. 21 (1978) pp. 153–64.

Garson, J. W., "Quantification in modal logic," in Gabbay and Guenthner (1984, pp. 249–307).

Garson, J. W., "Modularity in quantified modal logic" (unpublished manuscript).

Hintikka, J., "Existential and uniqueness presuppositions," in Lambert (1970, pp. 20–55).

Hughes, G. and Cresswell, H., *An Introduction to Modal Logic*, Methuen, London (1968).

Kripke, S., "Semantical considerations in modal logic," *Acta Philosophica Fennica*, vol. 16 (1963) pp. 83–94.

Lambert, K. (ed.), *The Logical Way of Doing Things*, Yale University Press (1969).

Lambert, K. (ed.), *Philosophical Problems in Logic*, Reidel, Dordrecht (1970).

Lewis, D., "Counterpart theory and quantified modal logic," *Journal of Philosophy*, vol. 65 (1968) pp. 113–26.

Lambert, K. and van Fraassen, B., *Derivation and Counterexample*, Dickenson Publishing, Encino, Calif. (1972).

Thomason, R., "Modal logic and metaphysics," in Lambert (1969, pp. 119–46).

Thomason, R., "Some completeness results for modal predicate calculi," in Lambert (1970, pp. 56–76).

Notes

1. Exactly how the strength of the underlying modality affects completeness results in such systems is not known, although we have completeness results for many of the standard systems.
2. This is reflected in the fact that the Barcan Formula is provable in systems that use the standard quantifier rules and are based on propositional modal logics as strong as B.
3. We cannot write $At \to \exists x Ax$ for arbitrary sentences At, because some of the instances of this sentence are valid, and others are not. In case we are using axioms with a rule of substitution of formulas for atoms, the problem reemerges in the failure of the rule of substitution. Either way, the use of local predicates leads to serious formal difficulties.
4. See Garson (1984, pp. 273 ff.).
5. In naming systems, we follow the nomenclature of Thomason (1970) where we can.
6. See Garson (1984, pp. 289 ff.) for the proof.
7. We may establish this as follows. Let f be a choice function that picks out an object $f(w) \in Q3(w)$ for every world w. Then $f(w) \in Q3(w)$ and $g(w)$ is $g(w)$. It follows that there is a function g (namely f itself!) such that $g(w) \in Q3(w)$ and for every world w, $f(w) \in Q3(w)$ and $f(w)$ is $g(w)$. This is the condition under which $\exists x \Box \exists y (y = x)$ is true.

8. $\exists x \Box \exists y(y = x)$ would only be true in world w if there were one substance f in QS(w') in every world w' accessible from w.
9. A set is S-consistent iff there is no proof of a contradiction from the sentences in that set.
10. The reason is that if $M_{i+1} = M_i \cup \{\sim \forall x A x, \sim (Et \to At)\}$ were inconsistent, then $M_i \cup \{\sim \forall x A x\} \vdash At$. Since t is foreign to both M_i and $\sim \forall x P x$, it follows by (FUG), the rule of free universal generalization, that $M_i \cup \{\sim \forall x A x\} \vdash \forall x A x$, which entails that $M_i \cup \{\sim \forall x A x\}$ is inconsistent, contrary to our assumption.
11. Notice that use of the hypothesis of the induction in this reasoning requires knowing that At is a sentence of L_w if $Et \in w$. This is true because when Et is in w, Et must be a sentence of L_w and t must be a term of L_w. Since we know $\forall x A x$ is also a sentence of L_w, it follows that so is At.
12. To see why, assume t is in L_w. It follows that $t = t$ is a sentence of L_w. So $t = t \in w$ and $t \in a_w(t)$. By rigidity of t, $a_w(t) = a_{w'}(t)$ for any saturated set w'. Hence $t \in a_{w'}(t)$, $t = t \in w'$, and so t is in $L_{w'}$.
13. To show that every argument provable in NDFS is provable in GS, we define the corresponding sentence to a line of a natural deduction proof to be the result of applying □In and Conditional Proof to the line repeatedly until the result lies outside all subproofs. Notice that when A is a line of a natural deduction proof, the corresponding sentence is equivalent to a formula of the form $G(A)$. We then establish that the sequence of corresponding lines of a natural deduction proof has the property that each line follows from the preceding lines in GS. To show that whatever is provable in GS is provable in NDFS, we show that (GUI) and (GUG) are derivable in NDFS by induction on the complexity of $G(\forall x A x)$.
14. To do so, we merely generalize the standard construction so that when $\sim G(\forall x P x)$ is added then so is $\sim G(Et \to Pt)$, where t is new to the construction.
15. We illustrate the problem with modal logics where R is convergent. In proving that the canonical model is convergent for propositional modal logics one assumes $w R w'$ and $w R w''$, establishes the consistency of $\{A: \Box A \in w'\} \cup \{A: \Box A \in w''\}$, and extends this set to a maximally consistent set w''' such that $w' R w'''$ and $w'' R w'''$. In the case of a quantified modal logic, we must know that $\{A: \Box A \in w'\} \cup \{A: \Box A \in w''\}$ is GOC as well as consistent before Lemma 4 can be used to extend it to a saturated set. However, there is no guarantee that $\{A: \Box A \in w'\} \cup \{A: \Box A \in w''\}$ will be GOC. It will not be GOC, for example, if $\{A: \Box A \in w'\}$ contains each of $Et_1 \to Pt_1$, $Et_3 \to Pt_3, \ldots$, and $\{A: \Box A \in w''\}$ contains $\sim \forall x P x$, $Et_2 \to Pt_2$, $Et_4 \to Pt_4, \ldots$, and $t_1, t_2, \ldots$ is a list of all terms of L. Under these circumstances $\{A: \Box A \in w'\} \cup \{A: \Box A \in w''\}$ contains $\{\sim \forall x P x, Et_1 \to Pt_1, Et_2 \to Pt_2, Et_3 \to Pt_2, \ldots\}$ and so is not even FOC.
16. Garson (unpublished manuscript) shows that QS is complete for systems S whose corresponding conditions on R are preserved under subsets. This includes universal conditions; that is, conditions that can be expressed with universal quantifiers alone. There are many interesting systems, however, whose conditions are not preserved under subsets, so the question invites further research.
17. (ESL) is false, for example, for $At = \Box Pt$ on the following model. The set of

worlds W contains (the real) world r, and (an unreal) world u, and they are both accessible from themselves and each other. The domain D contains two objects d (for David Lewis) and s (for Saul Kripke). The term t (read 'is author of "Counterpart Theory" ') has d as its extension in the real world, and s as its extension in the unreal world u. The extension of P (read 'is author of "Counterpart Theory" ') contains d in r, and s in u. Notice now that $a^{a_u(t)}/x_u(Ax)$ is $a^s/x_u(\Box Px)$, which is false, since s is not in the extension of P in both worlds. However, $a_u(\Box Pt)$ is true because the extension of t is in the extension of P in each world. We see that (ESL) fails for reasons closely related to the fact that substitution of identities fails for nonrigid terms.

PART III
Knowledge and Truth

Hintikka's essay "*Cogito, ergo sum*: Inference or performance?", a classic in the modern literature on Descartes, includes one of the earliest applications of bivalent positive free logic to an enduring philosophical debate, namely, the debate on the status of the Cartesian dictum: "I think, therefore, I am." It was published originally under the same title in *The Philosophical Review*, vol. 71 (1962) pp. 3–32, and is reprinted here with the kind permission of the editors. The essays by Grandy ("A definition of truth for theories with intensional description operators") and Burge ("Truth and singular terms") are contrasting views about the interplay between free logic and Tarskian truth theory. They appeared initially under the same titles, respectively, in *The Journal of Philosophical Logic*, vol. 1 (1972) pp. 137–35 (and reprinted here by permission of Kluwer Academic Publishers), and in *Noûs*, vol. VIII (1974) pp. 309–25 (and reprinted here with amendments by the permission of the editors of *Noûs*). Van Fraassen's essay, "Presuppositions, implication and self-reference," reprinted here from the article of the same title in *The Journal of Philosophy*, vol. 65 (1968) pp. 136–52, with the kind permission of the editors, utilizes the method of supervaluations to analyze the notion of existential presupposition, revealing, by the way, his own development of positive free logic in Part 1 of this book to be a paradigm presuppositional language, and exploits that analysis in a treatment of the famous paradox of the liar. Finally, Scales's essay, "A Russellian approach to truth," shows how to develop Russell's intuitions about truth in a bivalent negative free logic, a logic founded on a conception of logical form differing radically from Russell's own conception. Scales's essay appeared originally under the same title in *Noûs*, vol. 11 (1977) pp. 169–74, and is reprinted here with the permission of the editors.

7
Cogito, Ergo Sum: Inference or Performance?

Jaakko Hintikka

1. *Cogito, Ergo Sum* as a Problem

The fame (some would say the notoriety) of the adage *cogito, ergo sum* makes one expect that scholarly industry has long since exhausted whatever interest it may have historically or topically. A perusal of the relevant literature, however, fails to satisfy this expectation. After hundreds of discussions of Descartes's famed principle we still do not seem to have any way of expressing his alleged insight in terms that would be general and precise enough to enable us to judge its validity or its relevance to the consequences he claimed to draw from it. Thirty years ago Heinrich Scholz wrote that there not only remain many important questions concerning the Cartesian dictum unanswered but that there also remain important questions unasked.[1] Several illuminating papers later, the situation still seems essentially the same today.

2. Some Historical Aspects of the Problem

This uncertainty of the topical significance of Descartes's dictum cannot but reflect on the discussions of its historical status. The contemporaries were not slow to point out that Descartes's principle had been strikingly anticipated by St. Augustine. Although later studies have unearthed other anticipations,[2] notably in Campanella and in Schoolmen, scholars still seem to be especially impressed by Descartes's affinity with St. Augustine, in spite of his unmistakable attempts to minimize the significance of Augustine's anticipation. It cannot be denied, of course, that the similarities are striking. One may wonder, however, whether they are all there is to the matter. Perhaps there are also dissimilarities between Descartes and Augustine important enough to justify or at least to explain the one's reluctance to acknowledge the extent of the other's anticipation. But we cannot tell whether there is more to Descartes's *cogito, ergo sum* than there is to St. Augustine's similar argument before we can tell exactly what there is to the *cogito* argument.

If there are important differences between Descartes and his predecessors, the question will also arise whether some of the anticipations are closer than others. For instance, Descartes could have found the principle in St. Thomas Aquinas as well as in St. Augustine. Which of the two saints comes closer to the *cogito, ergo sum*?

3. What is the Relation of *Cogito* to *Sum*?

What kind of topical questions does *cogito, ergo sum* give rise to? One of the most important questions is undoubtedly that of the logical form of Descartes's inference. Is it a formally valid inference? If not, what is logically wrong about it?

But there is an even more fundamental question than these. Does Descartes's dictum really express an inference? That it does is suggested by the particle *ergo*. According to Descartes, however, by saying *cogito, ergo sum* he does not logically (syllogistically) deduce *sum* from *cogito* but rather perceives intuitively ("by a simple act of mental vision") the self-evidence of *sum*.[3] Similarly, Descartes occasionally says that one's own existence is intuitively obvious without bringing in *cogito* as a premiss.[4] Sometimes he intimates that his "first principle" is really the existence of his mind—and not the principle *cogito, ergo sum*, by means of which this existence is apparently deduced.[5] Once he formulates the *cogito* principle as *ego cogitans existo* without using the word *ergo* at all.[6]

But if it is true that the Cartesian dictum does not express an inference, equally perplexing questions are bound to arise. Not only is the particle *ergo* then misplaced; the word *cogito* is likewise out of place in a sentence that only serves to call attention to the self-evidence of *sum*.

But is the word *cogito* perhaps calculated to express the fact that thought is needed for grasping that *sum* is intuitively evident? Was it perhaps an indication of the fact that intuition was not for Descartes an irrational event but an act of the thinking mind, an "intellectual intuition," as it has been aptly expressed?[7] Even if this is part of the meaning of the word, the question will remain why Descartes wanted to stress the fact in connection with this particular insight. The same point would equally well apply to most of the other propositions of the Cartesian system; and yet Descartes does not say, for example, *cogito, ergo Deus est* in the way he says *cogito, ergo sum*.

Clearly the word *cogito* must have some further function in Descartes's sentence. Even if the sentence did not express a syllogistic inferences, it expressed something sufficiently like an inference to make Descartes call his sentence a reasoning (*ratiocinium*),[8] refer to expressing it as inferring (*inferre*),[9] and call *sum* a conclusion (*conclusio*).[10] As Martial Gueroult has trenchantly summed up the problem: "1° Descartes se refuse à considérer le *Cogito* comme un raisonnement.... 2° Pourquoi s'obstine-t-il alors au moins à trois reprises (*Inquisitio veritatis, Discours, Principes*) à présenter le *Cogito* sous la forme qu'il lui dénie?"[11]

Since the word *cogito* is not dispensable and since it is not just a premiss from which the conclusion *sum* is deduced, the relation of the two becomes a problem. One of the main objectives of this essay is to clear up their relation.

4. *Cogito, ergo sum* as a Logical Inference

But can we be sure that Descartes's dictum does not express a logical inference? In many respects it seems plausible to think that it does. Its logical form seems quite easy to define. In the sentence 'I think' an individual receives an attribute; for a modern logician it is therefore of the form '$B(a)$'. In the sentence 'I am', or 'I exist', this same individual is said to exist. How can one represent such a sentence formally? If Quine is right in claiming that "to be is to be a value of a bound variable," the formula '$(Ex)(x = a)$' serves the purpose. And even if he is not right in general, in this particular case his claim is obviously justified: 'a exists' and 'there exists at least one individual identical with a' are clearly synonymous. Descartes's dictum therefore seems to be concerned with an implication of the form

(1) $B(a) \supset (Ex)(x = a)$

Descartes perceives that he thinks; hence he obtains the premiss $B(a)$. If (1) is true, he can use *modus ponens* to conclude that he exists. Those who want to interpret the *Cogito* as a logical inference may now claim that (1) is in fact true, and even logically provable; for is not

$B(a) \supset (Ex)(x = a \;\&\; B(x))$

a provable formula of our lower functional calculi? And does not this formula entail (1) in virtue of completely unproblematic principles? It may seem that an affirmative answer must be given to these questions, and that Descartes's argument is thus easily construed as a valid logical inference.

Views of this general type have a long ancestry. Gassendi already claims that *ambuli, ergo sum*, "I walk, therefore I am," is as good an inference as *cogito, ergo sum*.[12] It is obvious that by the interpretation just suggested, Gassendi will be right. The alleged provability of (1) does not depend on the attribute '*B*' at all. The gist of Descartes's argument is on the present view expressible by saying that one cannot think without existing; and if (1) is an adequate representation of the logical form of this principle, one can indeed equally well say that one cannot walk without existing.

This already makes the interpretation (1) suspect. In his reply to Gassendi, Descartes denies that *ambulo, ergo sum* is comparable with *cogito, ergo sum*.[13] The reasons he gave are not very clear, however. A part of the burden of his remarks is perhaps that although the inferences *ambulo, ergo sum* and *cogito, ergo sum* are parallel—as being both of the form (1)—their premisses are essentially different. *Ambulo* is not an indubitable premiss in the way *cogito* may be claimed to be.

But even if we make this allowance, there remain plenty of difficulties. As we say, Descartes sometimes denies that in the *cogito* argument *sum* is deduced from *cogito*. But, on the view we are criticizing, the argument is a deduction. This view is therefore unsatisfactory.

It is also unsatisfactory because it does not help us to understand the role of the *cogito* argument in the Cartesian system. Insofar as I can see, it does not, for example, help us to appreciate the consequences Descartes wanted to draw from his first and foremost insight.

The gravest objection, however, still remains to be made. It may be shown that the provability of (1) in the usual systems of functional calculus (quantification theory) has nothing to do with the question whether thinking entails existence. An attempt to interpret Descartes's argument in terms of the provability of (1) is therefore bound to remain fruitless.

By this I mean the following: if we have a closer look at the systems of logic in which (1) can be proved, we soon discover that they are based on important *existential presuppositions*, as I have elsewhere called them.[14] They make more or less tacit use of the assumption that all the singular terms with which we have to deal really refer to (designate) some actually existing individual.[15] In our example this amounts to assuming that the term that replaces *a* in (1) must not be empty. But since the term in question is 'I', this is just another way of saying that I exist. It turns out, therefore, that we in fact decided that the sentence 'I exist' is true when we decided that the sentence 'I think' is of the form $B(a)$ (for the purposes of the usual systems of functional logic).[16] That we were then able to infer $(Ex)(x = a)$ from $B(a)$ is undoubtedly true, but completely beside the point.

It is possible to develop a system of logic that dispenses with the existential presuppositions.[17] If in such a system we could infer 'I exist' from 'I think'—that is, $(Ex)(x = a)$ from $B(a)$—it would be highly relevant to the question whether thinking implies existence in Descartes's sense. But this we cannot do. The truth of a sentence of the form (1) turns entirely on existential presuppositions. If they are given up, the provability of (1) goes by the board.

My point may perhaps be illustrated by means of an example constructed for us by Shakespeare. Hamlet did think a great many things; does it follow that he existed?

5. Descartes's Temptation

In spite of all this, there are passages in Descartes that seem to support the interpretation under criticism. I do not want to deny that it expresses *one* of the things Descartes had more or less confusedly in mind when he formulated his famous dictum. But it is important to realize that this interpretation is defective in important respects. It does not help to elucidate in any way some of Descartes's most explicit and more careful formulations. It is at best a partial interpretation.

One can see why some interpretations like the one we have been criticizing attracted Descartes. It gave him what must have seemed a very useful way of defending his own doctrines and of silencing criticism. He could always ask: How can it possibly be true of someone that he thinks unless he exists? And if you challenge the premiss that he is thinking (why cannot the all-powerful *malin génie* make it appear to him that he is thinking?), Descartes could have replied that in a sense the premiss is redundant. He could have resorted to some such argument as the following: If I am right in thinking that I exist, then of course I exist. If I err in thinking that I exist or if I as much as doubt whether I exist, then I must likewise exist, for no one can err or doubt without existing. In any case I must therefore exist: *ergo sum*.

However, this neat argument is a *petitio principii*, as you may perhaps see by comparing it with the following similar argument: Homer was either a Greek or a barbarian. If he was a Greek, he must have existed;

for how could one be a Greek without existing? But if he was a barbarian, he likewise must have existed. Hence he must have existed in any case.

The latter argument is obviously fallacious; the celebrated Homeric question cannot be solved on paper. By the same token, the former argument is also fallacious.[18]

Did Descartes realize that it is misguided to represent his insight in the way we have been discussing? It is very difficult to tell. Certainly he never realized it fully. He seems to have realized, however, that on this interpretation the validity of his argument depends essentially on existential presuppositions. For when he tried to present his fundamental doctrines in a deductive or "geometrical" form, he tried to formulate these presuppositions in so many words by saying that "we can conceive nothing except as existent (*nisi sub ratione existentis*)" (AT VII, 166; HR II, 57). This statement is all the more remarkable since it prima facie contradicts what Descartes says in the *Third Meditation* about "ideas ... considered only in themselves, and not as referred to some other thing," namely that "they cannot, strictly speaking, be false." It also contradicts the plain fact that we can think of (mentally consider) unicorns, or Prince Hamlet, without thereby committing ourselves to maintaining that they exist.

The fact also remains that Descartes resorted to the interpretation we have been criticizing mainly in his more popular writings. As Gueroult noticed, he does not resort to it in the *Meditationes*. His most explicit use of it occurs in *Recherche de la vérité*, in a dialogue whose didactic character has been particularly emphasized by Ernst Cassirer.[19] Descartes's most careful formulations of the *cogito* argument, notably those in the *Meditationes de prima philosophia*, seem to presuppose a different interpretation of the argument.

6. Existential Inconsistency

In order to understand this second interpretation of the *Cogito* we have to have a closer look at the logic of Descartes's famed argument. Descartes's formulations in the *Meditationes* and elsewhere suggest that his result may be expressed by saying that it was impossible for him to deny his existence. One way in which Descartes could have tried to (but did not) deny this would have been to say 'Descartes does not exist'. As a preliminary to our study of Descartes's first-person sentence *cogito, ergo sum* we shall inquire into the character of this third-person sentence. The reasons why Descartes could not have maintained the latter will turn out to be closely related to the reasons why he asserted the former, if I am right.

Cogito, Ergo Sum: Inference or Performance?

What, then, are these reasons? What general characteristic of the sentence 'De Gaulle does not exist' makes it awkward for De Gaulle to assert it?[20] I shall try to formulate this general characteristic by saying that it is *existentially inconsistent* for De Gaulle to assert (to utter) this sentence. The notion of existential inconsistency may be defined as follows: Let p be a sentence and a a singular term (e.g., a name, a pronoun, or a definite description). We shall say that p is *existentially inconsistent for the person referred to by a to utter* if and only if the longer sentence

(2) 'p; and a exists'

is inconsistent (in the ordinary sense of the word). In order to avoid our own objections we must of course require that the notion of ordinary inconsistency which is used in the definition involves no existential presuppositions. Provided that this is the case, we may write (2) more formally as

(2)' 'p & $(Ex)(x = a)$'

(As the informed reader has no doubt already noticed, we should really use quasi quotes instead of single quotes in (2) and (2)'.)

A trivial reformulation of the definition shows that the notion of existential inconsistency really formulates a general reason why certain statements are impossible to defend although the sentences by means of which they are made may be consistent and intelligible. Instead of saying that (2) is inconsistent, we could have said that p entails 'a does not exist' (without the use of any existential presuppositions but otherwise in the ordinary sense of entailment). Uttering such a sentence, p, will be very awkward for the bearer of a: it means making a statement which, if true, entails that its maker does not exist.

It is important to realize that the ills of such *statements* cannot be blamed on the *sentences* by means of which they are made.[21] In fact, the notion of existential inconsistency cannot be applied at all to sentences. As we defined the notion, it is a relation between a sentence and a singular term rather than a property of sentences. The notion of existential inconsistency, however, can often be applied to statements in a fairly natural sense. In order to specify a statement, we have to specify (inter alia) the sentence uttered (say, q) and its utterer. If the latter refers to himself by means of the singular term b when he makes his statement, we may say that the notion applies to the statement if and only if it applies to q in relation to b.

A simple example will make the situation clear. The *sentences* 'De Gaulle does not exist' and 'Descartes does not exist' are not inconsistent or otherwise objectionable any more than the moot sentence 'Homer does not exist'. None of them is false for logical reasons alone. What would be (existentially) inconsistent would be the attempt of a certain man (De Gaulle, Descartes, or Homer, respectively) to use one of these sentences to make a statement. Uttered by somebody else, the sentences in question need not have anything wrong or even strange about them.

It lies close at hand to express this important feature of the notion of existential inconsistency by means of a term that has recently enjoyed wide currency. The inconsistency (absurdity) of an existentially inconsistent statement can in a sense be said to be of *performatory* (performative) character. It depends on an act or "performance," namely on a certain person's act of uttering a sentence (or of otherwise making a statement); it does not depend solely on the means used for the purpose, that is, on the sentence which is being uttered. The sentence is perfectly correct as a sentence, but the attempt of a certain man to utter it assertively is curiously pointless. If one of these days I should read in the morning paper, "There is no De Gaulle any more," I could understand what is being said. But no one who knows Charles De Gaulle could help being puzzled by these words if they were uttered by De Gaulle himself; the only way of making sense of them would be to give them a nonliteral meaning.

We can here see how the existential inconsistency of De Gaulle's fictional utterance (as well as the inconsistency of other existentially inconsistent statements) manifests itself. Normally a speaker wants his hearer to believe what he says. The whole "language-game" of fact-stating discourse is based on the assumption that this is normally the case. But nobody can make his hearer believe that he does not exist by telling him so; such an attempt is likely to have the opposite result. The pointlessness of existentially inconsistent statements is therefore due to the fact that they automatically destroy one of the major purposes which the act of uttering a declarative sentence normally has. ("Automatically" means here something like "for merely logical reasons.") This destructive effect is of course conditional on the fact that the hearer knows who the maker of the statement is, that is, that he identifies the speaker as the same man the uttered sentence is about.

In a special case a self-defeating attempt of this kind can be made without saying or writing anything or doing anything comparable. In trying to make *others* believe something I must normally do something which can be heard or seen or felt. But in trying to make *myself* believe something there is no need to say anything aloud or to write anything on paper. The performance through which existential inconsistency arises can in this case be merely an attempt to think—more accurately, an attempt to make oneself believe—that one does not exist.[22]

This transition from "public" speech-acts to "private" thought-acts, however, does not affect the essential features of their logic. The reason why Descartes's attempt to *think* that he does not exist necessarily fails is for a logician exactly the same as the reason why his attempt to tell one of his contemporaries that Descartes did not exist would have been bound to fail as soon as the hearer realized who the speaker was.

7. Existentially Inconsistent Sentences

It can be seen that we are approaching Descartes's famous dictum. In order to reach it we have to take one more step. We have found that the notion of existential inconsistency is primarily applicable to statements (e.g., declarative utterances) rather than to sentences. In a sense, it may of course be defined for sentences, too, namely by making it relative to a term (name, pronoun, or definite description) occurring therein. This is in fact what we did when we first introduced the notion; we said inter alia that the *sentence* 'De Gaulle does not exist' is existentially inconsistent for De Gaulle (i.e., for the person referred to by 'De Gaulle') to utter. Sometimes it may even be possible to omit the specification 'for . . . to utter," namely when the intended speaker can be gathered from the context.

In a frequently occurring special case such an omission is not only natural but almost inevitable. It is the case in which the speaker refers to himself by means of the first-person singular pronoun "I." This pronoun inevitably refers to whoever happens to be speaking. The specification "inconsistent for . . . to utter" therefore reduces to the tautology "inconsistent for whoever happens to be speaking to utter," and may therefore be omitted almost always. In a special case, the notion of existential inconsistency may therefore be defined for sentences *simpliciter* and not only for sentences thought of as being uttered by some particular speaker. These are the sentences which contain a first-person singular pronoun. The existential inconsistency of such a sentence will mean that its utterer cannot add 'and I exist' without contradicting himself implicitly or explicitly.

There are purposes, however, for which it may be misleading to forget the specification. Forgetting it may be dangerous since it leads one to overlook the important similarities that obtain between existentially inconsistent *sentences* and existentially inconsistent *statements*. In a perfectly good sense, existentially inconsistent sentences are all right as sentences. They may be said to be consistent and sometimes even significant (e.g., when they occur as parts of more complicated sentences). According to their very definition, existentially inconsistent sentences are not so much inconsistent as such as absurd for anyone to utter. Their (existential) inconsistency is therefore of performatory character exactly in the same sense as that of the existentially inconsistent statements. The only difference between the two lies in the fact that the latter are inconsistent for some particular man to make while the former are inconsistent for anyone to utter. The inconsistency of existentially inconsistent sentences means that whoever tries to make somebody (anybody) believe them, by so doing, helps to defeat his own purpose.[23] Such an attempt may take the form of uttering the sentence assertively; or it may take the form of trying to persuade oneself of the truth of the sentence in question.

In the same way as existentially inconsistent sentences defeat themselves when they are uttered or thought of, their negations verify themselves when they are expressly uttered or otherwise professed. Such sentences may therefore be called existentially self-verifying. The simplest example of a sentence of this kind is 'I am', in Descartes's Latin *ego sum, ego existo*.

8. Descartes's Insight

Now we have reached a point where we can express precisely the import of Descartes's insight (or at least one of its most important aspects). It seems to me that the most interesting interpretation one can give to it is to say that Descartes realized, however dimly, the existential inconsistency of the sentence 'I do not exist' and therefore the existential self-verifiability of 'I exist'. *Cogito, ergo sum* is only one possible way of expressing this insight. Another way actually employed by Descartes is to say that the sentence *ego sum* is intuitively self-evident.

We can now understand the relation of the two parts of the *cogito, ergo sum* and appreciate the reasons why it cannot be a logical inference in the ordinary sense of the word. What is at stake in Descartes's dictum is the status (the indubitability) of the sentence 'I am'. (This is shown particularly clearly by the formulations of the *Second Meditation*.) Contrary appearances notwithstanding, Descartes does not demonstrate this indubitability by deducing *sum* from *cogito*. On the other hand, the sentence 'I am' ('I exist') is not by itself logically true, either. Descartes realizes that its indubitability results from an act of thinking, namely from an attempt to think the contrary. The function of the word *cogito* in Descartes's dictum is to refer to the thought-act through which the existential self-verifiability of 'I exist' manifests itself. Hence the indubitability of this sentence is not strictly speaking perceived *by means of* thinking (in the way the indubitability of a demonstrable truth may be said to be); rather, it is indubitable *because* and *insofar as* it is actively thought of. In Descartes's argument the relation of *cogito* to *sum* is not that of a premiss to a conclusion. Their relation is rather comparable with that of a *process* to its *product*. The indubitability of my own existence results from my thinking of it almost as the sound of music results from playing it or (to use Descartes's own metaphor[24]) light in the sense of illumination (*lux*) results from the presence of a source of light (*lumen*).

The relation which the particle *ergo* serves to express in Descartes's sentence is therefore rather peculiar.[25] Perhaps it would have been less misleading for Descartes to say, 'I am in that I think', or 'By thinking I perceive my existence', than to say, 'I think, therefore I am'. It may be worth noting that one of our formulations was closely anticipated by St. Thomas Aquinas when he wrote: "Nullus potent cogitare se non esse cum assensu: in hoc enim quod cogitat aliquid, percipit se esse" (*De veritate*, X, 12, *ad* 7). The peculiarity of this relation explains Descartes's vacillation in expressing it in that he sometimes speaks of the *cogito* as an inference and sometimes as a realization of the intuitive self-evidence of its latter half.

Similarly we may now appreciate the function of the word *cogito* in Descartes's sentence as well as his motives in employing it. It serves to express the performatory character of Descartes's insight; it refers to the "performance" (to the act of thinking) through which the sentence 'I exist' may be said to verify itself. For this reason, it has a most important function in Descartes's sentence. It cannot be replaced by any arbitrary verb. The performance (act) through which the existential self-verifiability is manifested cannot be any arbitrary human activity, contrary to what Gassendi claimed. It cannot be an act of walking or an act of seeing. It cannot even be an instance of arbitrary mental activity, say of willing or of feeling. It must be just what we said it is: an attempt to think in the sense of making myself believe (an attempt to think *cum assensu*, as Aquinas put it) that I do not exist. Hence Descartes's choice of the word *cogito*. This particular word is not absolutely indispensable, however, for the act of thinking to which it refers could also be called an act of doubting; and Descartes does admit that his insight is also expressible by *dubito, ergo sum* (in *Recherche de la vérité*, AT X, 523; HR I, 324; cf. also *Principia philosophiae*, I, 7).

But did I not say that the performance through which an existentially self-verifying sentence verifies itself may also be an act uttering it? Is this not incompatible with Descartes's use of the word *cogito*? There is no incompatibility, for Descartes says exactly the same. In his second meditation on first philosophy he says in so many words that the sentence 'I exist' is necessarily true "whenever I utter it or conceive it in my mind"—"quoties a me profertur, vel mente concipitur" (AT VII, 25; HR I, 150).[26]

The performatory character of Descartes's insight presupposes a characteristic feature of his famous method of doubt that has frequently been commented on in other contexts. Descartes's doubt does not consist in the giving up of all opinions, as a skeptic's doubt might. Nor is it an attempt to remove certain specific sources of mistakes from our thinking, like Francis Bacon's. It amounts to an active attempt to think the contrary of what we usually believe. For this reason Descartes could claim that in an important point this rather doctrinaire doubt of his defeats itself. A skeptic's passive doubt could never do so.

9. The *Cogito* and Introspection

The attempt to see the *cogito* as a logical inference is not the only one-sided interpretation of Descartes's insight. Sometimes it has been understood, on the contrary, as a more or less purely factual statement, as a mere *Tatsachenwahrheit*.[27] This interpretation is often combined with a definite view as to how this particular truth is ascertained, namely by introspection. The function of the *cogito*, on this view, is to call our attention to something every one of us can ascertain when he "gazes within himself."

It is very misleading, however, to appeal to introspection in explaining the meaning of the *cogito*, although there is likely to be a connection between the notion of introspection and the peculiarities of the Cartesian argument. We have seen that an existentially inconsistent sentence may also defeat itself through an "external" speech-act. The reason why Descartes could not doubt his own existence is in principle exactly the same as the reason why he could not hope to mislead anybody by saying "I do not exist." The one does not presuppose introspection any more than the other. What the philosophers who have spoken of introspection are likely to have had in mind here is often performatoriness rather than introspectiveness.

The independence of Descartes's insight of introspection is illustrated by the fact that there is a peculiarity about certain sentences in the *second* person which is closely related to the peculiarities of Descartes's *ego sum*, *ego existo.* In the same way as it is self-defeating to say "I do not exist," it is usually absurd to say "You do not exist." If the latter sentence is true, it is *ipso facto* empty in that there is no one to whom it could conceivably be addressed.

What makes us connect the *cogito* with introspection is the "spiritualization" which takes place when an "external" speech-act is replaced by a thought-act and on which we commented above. In the *cogito* it is presupposed that a man not only can converse with his fellow men but is also able to "discourse with himself without spoken sound" in a way closely reminiscent of Plato's famous definition of thinking "as a discourse that the mind carries on with itself" (and also reminiscent of Peirce's pertinent remarks on the dialogical character of thought[28]).

Another reason why it is natural to connect the *cogito* with one's self-knowledge is implicit in what was said above. In order to ascertain that a statement like 'De Gaulle does not exist' (supposing that it is made by De Gaulle himself) is existentially inconsistent, I have to know the speaker; I have to identify him as the selfsame man whom his statement is about. In the same way, appreciating the existential inconsistency of an utterance of the form 'I do not exist' presupposes realizing that the man whom it is about is necessarily the speaker himself. Descartes's *cogito* insight therefore depends on "knowing oneself" in the same literal sense in which the insight into the self-defeating character of the statement 'De Gaulle does not exist' depends on knowing De Gaulle. Expressed in less paradoxical terms, appreciating the *cogito* argument presupposes an ability to appreciate the logic of the first-person pronoun 'I'. And although mastering the latter is not the same thing as the capacity for introspection, the two are likely to be connected with each other conceptually (logically). The *cogito* insight is essentially connected with one's own case in the same way introspection is, we might say.

10. The Singularity of the *Cogito*

Descartes realized that his *cogito* argument deals with a particular case, namely with his own. This is in fact typical of his whole procedure; it is typical of a man who asked "What can *I* know?" rather than "What can men know?" Descartes denied that his argument is an enthymeme whose suppressed major premiss is 'Everybody who thinks, exists.' He seems to have thought, nevertheless, that this general sentence is a genuine generalization of the insight expressed by his singular sentence.[29]

The general sentence cannot be such a generalization of the *cogito*, however; it cannot serve as a general truth from which the sentence *cogito, ergo sum* could be inferred, as Descartes seems to have thought. This is perhaps seen most readily by making explicit the existential presuppositions that are implicit in the general sentence. If they are removed, the sentence takes the form 'Every actually existing individual that thinks, exists' and becomes a tautology. This tautology is useless for the purpose Descartes had in mind; it can entail 'I think, therefore I exist' only in conjunction with the further premiss 'I exist'. This further premiss, however, is exactly the conclusion that Descartes ultimately wanted to draw by means of the *cogito* argument. Hence the alleged deduction becomes a *petitio principii*.

Alternatively, we might try to interpret the word 'everybody' which occurs in the general sentence as somehow ranging over all *thinkable* individuals rather than all *actually existing* individuals. I am sure that such a procedure is illicit unless further explanations are given. But even if it were legitimate, it would not help us to formulate a true generalization of the Cartesian sentence. For then our generalization would take the form 'Every thinkable individual that thinks, exists' and become false, as witnessed by Shakespeare's meditative Prince of Denmark.

In a sense, therefore, Descartes's insight is not generalizable. This is of course due to its performatory character. Each of us can formulate "for himself" a sentence in the first-person singular that is true and indubitable, namely the Cartesian sentence *ego sum, ego existo*. But since its indubitability is due to a thought-act that each man has to perform himself, there cannot be any general sentence that would be indubitable in the same way without being trivial. The *cogito* insight of each of us is tied to his own case even more closely than Descartes realized.

11. The Role of the *Cogito* in Descartes's System

Our interpretation is supported by the fact that it enables us to appreciate the role of Descartes's first and foremost insight in his system, that is, to understand the conclusions he thought he could draw from the *cogito*. For one thing, we can now see the reason why Descartes's insight emerges from his own descriptions as a curiously *momentary* affair. It is a consequence of the performatoriness of his insight. Since the certainty of my existence results from my thinking of it in a sense not unlike that in which light results from the presence of a source of light, it is natural to assume (rightly or wrongly) that I can be really sure of my existence only as long as I actively contemplate it. A property that a proposition has *because* and *insofar as* it is actually thought of easily becomes a property that belongs to it only *as long as* it is thought of. In any case, this is what Descartes says of the certainty of his own existence. I can be sure of my existence, he says, "while" or "at the same time as" I think of it or "whenever" or "as often as" I do so.[30] "Whereas I had only to cease to think for an instant," he says, "and I should then (even although all the other things I had imagined still remained true) have no grounds for believing that I can have existed in that instant" (*Discours*, Part IV; AT VI, 32–33; HR I, 101).

This shows, incidentally, that the sole function of the word *cogito* in Descartes's dictum cannot be to call attention to the fact that his insight is obtained *by means of* thinking. For of an ordinary insight of this kind (e.g., of a demonstrative truth) we may of course continue to be sure once we have gained it.

In the same way we can perhaps see why Descartes's insight *cogito, ergo sum* suggested to him a definite view of the nature of this existing *ego*, namely that its nature consists entirely of thinking. We have seen that Descartes's insight is not comparable with one's becoming aware of the sound of music by pausing to listen to it but rather with making sure that music is to be heard by playing it oneself. Ceasing to play would not only stop one's hearing the music, in the way ceasing to listen could; it would put an end to the music itself. In the same way, it must have seemed to Descartes, his ceasing to think would not only mean ceasing to be aware of his own existence; it would put an end to the particular way in which his existence was found to manifest itself. To change the metaphor, ceasing to think would not be like closing one's eyes but like putting out the lamp. For this reason, thinking was for Descartes something that could not be disentangled from his existence; it was the very essence of his nature. We may thus surmise that the original reason why Descartes made the (illicit but natural) transition from *cogito, ergo sum* to *sum res cogitans* was exactly the same as the reason for the curious momentariness of the former which we noted above, namely the performativeness of the *cogito* insight. In any case, the two ideas were introduced by Descartes in one and the same breath. The passage we just quoted from the *Discours* continues as follows: "From this I knew that I was a substance whose whole essence or nature consists entirely in thinking." In the *Meditationes* Descartes is more reserved. He has already become aware of the difficulty of converting his intuitive idea of the dependence of his existence on his thinking into a genuine proof. The way in which the idea of the dependence is introduced is, nevertheless, exactly the same: '*Ego sum, ego existo*. This is certain. How long? As long as I think. For it might indeed be that if

I entirely ceased to think, I should thereupon altogether cease to exist. I am not at present admitting anything which is not necessarily true; and, accurately speaking, I am therefore only a thinking thing" (AT VII, 27; HR I, 151–2).

The transition from *cogito, ergo sum* directly to *sum res cogitans* remains inexplicable as long as we interpret the *cogito* in terms of the logical truth of (1). For then the blunt objections of Hobbes carry weight: Even if it were true that we can validly infer *ambulo, ergo sum* or *video, ergo sum*, there would not be the slightest temptation to take this to suggest that one's nature consists entirely of walking or of seeing in the way Descartes thought he could move from *cogito, ergo sum* to *sum res cogitans* (Cf. AT VII, 172; HR II, 61).

12. Descartes and His Predecessors

It seems to me that Descartes is distinguished from most of his predecessors by his awareness of the performatory character of his first and foremost insight.[31] In spite of all the similarities that there obtain between Descartes and St. Augustine, there are also clear-cut differences. Insofar as I know, there is no indication that Augustine was ever alive to the possibility of interpreting his version of the *cogito* as a performance rather than as an inference or as a factual observation.[32] As far as Augustine is concerned, it would be quite difficult to disprove a "logical" interpretation such as Gassendi and others have given of the Cartesian *cogito* argument. What he dwells on is merely the "impossibility of thinking without existing." I do not see any way in which Augustine could have denied that *ambulo, ergo sum* or *video, ergo sum* are as good inferences as *cogito, ergo sum* and that the sole difference between them lies in the different degree of certainty of their premisses.

In this respect, there is an essentially new element present, however implicitly, in Descartes's formulations. This difference also shows in the conclusions which Descartes and Augustine drew from their respective insights. For instance, Augustine used his principle as a part of an argument which was designed to show that the human soul is tripartite, consisting of being, knowing, and willing. We have already seen that Descartes's insight was for him intimately connected with the notion of thinking (rather than, say, of willing or feeling): the performance through which an existentially inconsistent sentence defeats itself can be an act of thinking of it, but it cannot possibly be an act of willing or of feeling. Hence Descartes could use the performatorily interpreted *cogito* insight to argue that the human soul is a *res cogitans*, but not to argue that it is essentially a willing or feeling being. In view of such differences, is it at all surprising that Descartes should have emphasized his independence of Augustine?

If there is a predecessor who comes close to Descartes, he is likelier to be St. Thomas than St. Augustine. We have already quoted a passage in Aquinas which shows much more appreciation of the performatory aspect of the *cogito* than anything in Augustine. The agreement is not fortuitous; Aquinas's ability to appreciate the performatoriness of the *cogito* was part and parcel of his more general view that "the intellect knows itself not by its essence but by its act."[33] I should go as far as to wonder whether there is more than a coincidence to the fact that Descartes was particularly close to Aquinas (as far as the *cogito* insight is concerned) in that work of his, in the *Meditationes*, in which the Thomistic influence on him is in many other respects most conspicuous.

13. Summing up

Some of the main points of our analysis of the *cogito* may be summed up as follows: Whatever he may have thought himself, Descartes's insight is *clear* but not *distinct*,[34] to use his own terminology. That is to say, there are several different arguments compressed into the apparently simple formulation *cogito, ergo sum* that he does not clearly distinguish from each other.

(i) Sometimes Descartes dealt with the *cogito* as if it were an expression of the logical truth of sentences of the form (1) or at least of the indubitable truth of a particular sentence of this form. On this interpretation the argument *cogito, ergo sum* is on the same footing with such arguments as *volo, ergo sum*. Arguments like *video, ergo sum* or *ambulo, ergo sum* can be said to be less convincing than the *cogito* merely because their premises are not as indubitable as that of Descartes's argument. The word *cogito* may thus be replaced by any other word that refers to one of my acts of consciousness.

(ii) Descartes realized, however, that there is more to the *cogito* than interpretation (i). He realized, albeit dimly, that it can also serve to express the existential self-verifiability of the sentence 'I exist' (or the existential inconsistency of 'I do not exist'). On this interpretation the peculiarity of the sentence *ego sum* is of performatory character. The verb *cogitare* now has to be interpreted rather narrowly. The word *cogito* may still be replaced by such "verbs of intellection" as *dubito* (or *profero*) but not any longer by verbs referring to arbitrary mental acts, such as *volo* or *sentio*. This interpretation, and only this one, makes it possible to understand Descartes's rash transition from *cogito, ergo sum* to *sum res cogitans*.

By comparing the two interpretations we can further elucidate certain peculiarities of Descartes's thought. We shall mainly be concerned with the following two points:

(A) Descartes does not distinguish the two interpretations very clearly. We cannot always expect a clear answer to the question whether a particular instance of the *cogito* argument is for him an inference or a performance. The two types of interpretation merge into each other in his writings in a confusing manner.

(B) Nevertheless, the relation of these two possible interpretations of the Cartesian *cogito* throws light on the meaning of the critical verb *cogitare* in the different parts of Descartes's philosophy.

14. The Ambiguity of the Cartesian *Cogito*

Interpretation (ii) easily gives rise to an expectation that is going to be partly disappointed. It easily leads us to expect a definite answer to the question: What was Descartes thinking *of* in that thought-act that to him revealed the indubitability of his own existence? Interpretation (ii) suggests that Descartes should have been thinking of *his own existence*. This agrees very well with some of Descartes's most explicit pronouncements. One of them was already quoted above (in the penultimate paragraph of Section 8). In the same connection Descartes writes: "Let him [viz. Descartes's *malin génie*] deceive me as much as he will, he can never cause me to be nothing so long as I shall be thinking that I am something." The same point is repeated in the *Third Meditation* (AT VII, 36; HR I, 158–9).

Elsewhere, however, Descartes often uses formulations that clearly presuppose that his crucial thought-act pertains to something different from his mere existence. These formulations can be understood, it seems to me, as hybrids between the two arguments (i) and (ii). This hybridization was undoubtedly encouraged by the following (correct) observation: If the sentence 'I do not exist' is existentially self-defeating, then so are a fortiori such sentences as 'I think, but I do not exist' or 'I doubt, but I do not exist'. In other words, there are no objections in principle to saying that what is at stake in the *cogito* is the status of these latter sentences rather than that of the sentence 'I do not exist'.

On this intermediate interpretation the word *cogito* has a curious double role in Descartes's dictum. On one hand, it is a part of the proposition whose status (indubitability) is at stake. On the other hand, it refers to the performance through which the indubitability of this proposition is revealed. If we are on the right track, we may expect that this duality of functions will sometimes be betrayed by Descartes's formulations; that is, that he will sometimes use two "verbs of intellection" (such as *think*, *doubt*, *conceive*, and the like) where on interpretation (i) there should be only one. This expectation turns out to be justified: "From this very circumstance that I *thought to doubt* [*je pensais à douter*] the truth of those other things, it very evidently and very certainly followed that I was ..." (*Discours*, Part IV; my italics); "but we cannot in the same way *conceive* that we who *doubt* these things are not..." (*Principia philosophiae* I, 7; my italics).

This duplication of verbs of intellection[35] shows that we still have to do with a performatory insight. Where Augustine would have said that nobody can doubt anything without existing, Descartes in effect says that one cannot think that one doubts anything without thereby demonstrating to oneself that one exists. But he does not clearly distinguish the two arguments from each other. He thinks that interpretation (ii), thus expanded, is tantamount to interpretation (i). For instance, the passage that we just quoted from the *Principia* continues as follows: "For there is a contradiction in conceiving that what thinks does not, at the same time as it thinks, exist." The change may seem small, but it makes all the difference. In the first passage Descartes is saying that it is impossible for him to think that *he himself* should not exist while he doubts something. In the second passage he says that it is impossible for him to think that *anybody else* should not exist while he (the other man) doubts something. The former passage expresses a performatory insight, whereas the latter cannot do so. We have moved from the ambit of interpretation (ii) to that of interpretation (i).[36]

15. The Ambiguity of the Cartesian *Cogitatio*

To tell what Descartes meant by the verb *cogitare* is largely tantamount to telling what is meant by his dictum: *sum res cogitans*. We saw that this dictum originally was for Descartes a consequence (a fallacious, albeit natural one) of the principle *cogito, ergo sum*, which for this purpose had to be given interpretation (ii). From this it follows that the word *cogitans* has to be interpreted as referring to thinking in the ordinary sense of the word. It is not surprising, however, that Descartes should have included more in his alleged conclusion *sum res cogitans* than it would have contained on the basis of the way in which he arrived at it even if this way had amounted to a demonstration.

Descartes had to reconcile his "conclusion" that the essence of a human being consists entirely of thinking (in the ordinary sense of the word) and the obvious fact that there are genuine acts of consciousness other than those of thinking, for example, those of willing, sensing, feeling, and the like. This he sought to accomplish by extending the meaning of the verb *cogitare*. He tried to interpret all the other acts of consciousness as so many modes of thinking.[37] In this attempt he was helped by the following two facts:

(a) The meaning of the verb *cogitare* was traditionally very wide. According to Alexander Koyré, "it embraced not only 'thought' as it is now understood, but all mental acts and data: will, feeling, judgment, perception, and so on."[38] Because of this traditionally wide range of senses of the word Descartes was able to smuggle more content into his "result" *sum res cogitans* than the way in which he reached it would, in any case, have justified.

It is significant that nonintellectual acts of consciousness enter into the argument of the *Meditationes* at the moment when Descartes pauses to ask what a *res cogitans* really is, that is, what is meant by the *cogitatio* of a *res cogitans*:

> What then am I? A thinking thing [*res cogitans*]. What is a thinking thing? It is a thing that doubts, understands, asserts, denies, wills, abstains from willing, that also has sense and imagination. These are a good many properties—if only they all belong to me. But how could they fail to? [AT VII, 28; HR I, 153]..

Descartes is not here simply stating what is meant by a *res cogitans*. He is not merely formulating the conclusion of an argument; he is proceeding to interpret it.[39] This is shown by the last two quoted sentences. For if willing and sensation were included in Descartes's thinking *ego* already in virtue of the argument which led him to conclude *sum res cogitans*, there would not be any point in asking whether they really belong to his nature.

(b) However, the wide range of senses of the verb *cogitare* in Descartes is not all due to external influence. There are factors in his own thinking which tend in the same direction. Among other things, the confusion between the two interpretations is operative here. Descartes can hope (as we saw) to be able to jump from *cogito, ergo sum* to *sum res cogitans* only if interpretation (ii) is presupposed. This interpretation in turn presupposes a narrowly "intellectual" meaning of the verb *cogitate* in that it cannot be replaced by any arbitrary verb that refers to some act of one's immediate consciousness. In contrast, on interpretation (i) the verb *cogitare* could be understood in this wide sense. The confusion between the two interpretations made it possible for Descartes to deal with the "conclusion" *sum res cogitans* as if it were based on a *cogito* argument in which *cogitatio* covers all one's acts of consciousness—as he strictly speaking is not justified in doing.

This explains Descartes's apparent inconsistency in using the verb *cogitare*. It is interesting to note that some of the critics (e.g., Anscombe and Geach; see Note 38, *op. cit.*, p. xlvii) who have most strongly stressed the wide extent of this verb in Descartes have nevertheless been forced to say that in the *cogito* argument the verb is used in a rather narrow sense to refer to what we nowadays call thinking. This may seem paradoxical in view of the fact that the broad interpretation is applied in the first place to the sentence *sum res cogitans* to which Descartes moved directly from the *cogito* argument. In our view, this prima-facie paradox disappears if we realize the ambiguity of the *cogito* argument.

The close connection between this argument and the notion of *cogitatio* in Descartes is amply demonstrated by his formulations. In our last quotation, Descartes was left asking whether doubt, understanding, will, sense, imagination, and the like belong to his nature. He reformulates this question successively as follows: "How can any of these things be less true than my existence? Is any of these something distinct from my thinking [*cogitatione*]? Can any of them be called a separate thing from myself?" Only such things could belong to Descartes's nature as were as certain as his existence. Why? The reason is seen from the context of the quotation. Descartes had already pronounced his *cogito*; he had already ascertained the indubitability of his existence. He held that nothing he did not have to know in order to ascertain this could, in the objective order of things, constitute a necessary condition of his existence.[40] Such things could not belong to his essence, for "nothing without which a thing can still exist is comprised in its essence."[41] Hence nothing could belong to his essence or nature that he could not be sure of already at the present stage of his argument, that is, nothing that he could not ascertain in the same way and at the same time as he ascertained his own existence. For this reason, nothing that belonged to his nature could be "less true than his existence."

What this requirement amounts to is that everything that Descartes was willing to accept as a part of his nature (even in the sense of being a mere mode of his basic nature of thinking) had to be shown to belong to him by means of the *cogito* argument in the same way in which he "demonstrated" that thinking belonged to him by "deducing" *sum res cogitans* from *cogito, ergo sum*. A mental activity was for Descartes a part of his nature if and only if the corresponding verb could function as the premiss of a variant of the *cogito* argument. For instance, the sense in which apparent sensation can be said to belong to his nature (as a mode of thinking) is for Descartes exactly the same as the sense in which he could infer *sentio, ergo sum*. The former is explained by Descartes as follows:

> Finally, it is I who have sensations, or who perceive corporeal objects as it were by the senses. Thus, I am now seeing light, hearing a noise, feeling heat. These things are false [it may be said], for I am asleep; but at least I seem to see, to hear, to be warmed. This cannot be false; and this is what is properly called my sensation; further, sensation, precisely so regarded, is nothing but thinking [*cogitare*] [AT VII, 29; HR I, 153].

The latter is explained in a strikingly similar way:

> Suppose I say *I see* or *I am walking, therefore I exist*. If I take this to refer to vision or walking as corporeal action, the conclusion is not absolutely certain; for, as often happens during sleep, I may think I am seeing though I do not open my eyes, or think that I am walking although I do not change my place; and it may even be that I have no body. But if I take it to refer to the actual sensation or awareness [*sensu sive conscientia*] of seeing or walking, then it is quite certain; for in that case it has regard to the mind, and it is the mind alone that has sense or thought [*sentit sive cogitat*] of itself seeing or walking [*Principia* I, 9; cf. Descartes's similar reply to Gassendi's objections to the *cogito*].

In short, the reason why sensation belonged to Descartes's nature was for him exactly the same as the reason why he could argue *sentio, ergo sum*. For him, doubting, willing, and seeing were modes of his basic nature of thinking exactly in the same sense in which the arguments *dubito, ergo sum*, *volo, ergo sum*, and *video, ergo sum* were variants or "modes" of the argument *cogito, ergo sum*.

Why, then, is one of these arguments a privileged one? If Descartes could argue *volo, ergo sum* and *sentio, ergo sum* as well as *cogito, ergo sum*, why did he refuse to infer that his nature consists of *Wille und Vorstellung*, claiming as he did that it consists entirely of thinking? The answer is again implicit in the ambiguity of the *cogito* argument. Such parallel arguments as *volo, ergo sum* presuppose interpretation (i). Now there was more to the Cartesian *cogito* than this interpretation; Descartes was also aware of the "performatory" interpretation (ii). It is the latter interpretation that gives the verb *cogitare* a privileged position vis-à-vis such verbs as *velle* or *videre*. Descartes could replace the word *cogito* by other words in the *cogito, ergo sum*; but he could not replace the performance which for him revealed the indubitability of any such sentence. This performance could be described only by a "verb of intellection" like *cogitare*. For this reason, the verb *cogitare* was for Descartes a privileged one; for this reason nothing could for him belong to his nature that was "something distinct from his thinking."

This special role of the verb *cogitare* seems to me difficult to explain otherwise. If I am right, the conspicuous privileges of this verb in Descartes therefore constitute one more piece of evidence to show that he was aware of interpretation (ii).

There is a further point worth making here. We have already pointed out that the verb *cogitare* is not the most accurate one for the purpose of describing the performance which for Descartes revealed the certainty of his existence (see Note 26). This inaccuracy led Descartes to assimilate the peculiarities of the existentially self-defeating sentence 'I do not exist' to the peculiarities of such sentences as 'I doubt everything' or 'I am not thinking anything'. There is an important difference here, however. The latter sentences are not instances of existential inconsistency. They are instances of certain related notions; they are literally impossible to believe or to think in a sense in which 'I do not exist' is not. I have studied the peculiarities of some such sentences elsewhere.[42] In many respects, their properties are analogous to those of existentially self-defeating sentences.[43]

Notes

1. Heinrich Scholz, "Über das Cogito, ergo sum," *Kant-Studien*, vol. XXXVI (1931) pp. 126–47.
2. See, for example, L. Blanchet, *Les antécédents du 'Je pense, donc je suis'* (Paris, 1920); Étienne Gilson, *Études sur le rôle de la pensée médiévale dans la formation du système cartésien* (*Études de philosophie médiévale*, XIII) (Paris, 1930), 2d pt., ch. ii, and the first appendix; Heinrich Scholz, "Augustinus und Descartes," *Blätter für deutsche Philosophie*, vol. V (1932) pp. 406–23.
3. *Œuvres de Descartes*, published by C. Adam and P. Tannery (Paris, 1897–1913), VII, 140; *The Philosophical Works of Descartes*, trans. by E. S. Haldane and G. R. T. Ross Cambridge (1931), II, p. 38. In the sequel, these editions will be referred to as AT and HR, respectively, with Roman numerals referring to volumes. Normally I shall not follow Haldane and Ross's translation, however; I shall make use of the existing translations (notably of those by N. Kemp Smith and by Anscombe and Geach) rather eclectically.
4. AT X, 368; HR I, 7.
5. AT IV, 444; AT VII, 12; HR I, 140.
6. AT VII, 481; HR II, 282.
7. L. J. Beck, *The Method of Descartes*, Oxford (1952) ch. iv.
8. AT X, 523; HR I, 324.
9. AT VII, 352; HR II, 207; cf. AT III, 248.
10. *Principia philosophiae* I, 9; AT VIII, 7; HR I, 222; cf. AT II, 37, and at AT V, 147.

11. Martial Gueroult, "Le *Cogito* et la notion 'pour penser il faut être'," *Travaux du IXe Congrès International de philosophie* (*Congrès Descartes*) (Paris, 1937; reprinted as the first appendix to Gueroult's *Descartes selon l'order des raisons*, Paris, 1953) vol. II, pp. 307–12, see p. 308.
12. In his objections to the *Second Meditation* (AT VII, 258–9; HR II, 137).
13. AT VII, 352; HR II, 207.
14. In "Existential presuppositions and existential commitments," *Journal of Philosophy*, vol. LVI (1959) pp. 125–37.
15. All the singular terms (e.g., names or pronouns) that in an application may be substituted for a free individual variable are assumed to do so; and as a consequence all the free individual variables have to behave like singular terms that really possess a reference (or "bearer," vulgarly "referent").
16. Cf. Leibniz' incisive remark: "And to say *I think, therefore I am*, is not properly to prove existence by thought, since to think and to be thinking is the same thing; and to say, I am thinking, is already to say, *I am*" (*Nouveaux Essais*, trans. by A. G. Langley (La Salle, Ill., 1949), IV, 7, sec. 7).
17. Such a system was outlined in the paper referred to in Note 14. Essentially the same system was developed independently by Hugues Leblanc and Theodore Hailperin in "Nondesignating singular terms," *Philosophical Review*, vol. LXVII (1959) pp. 239–43.
18. But maybe you are not convinced; maybe you feel that the question of Descartes's own existence is essentially different from the question of Homer's existence. If so, you are right. I have not wanted to deny that there is a difference, and an important one. All I am saying is that the reconstruction we are considering does not bring out this difference.
19. *Descartes: Lehre, Persönlichkeit, Wirkung*, Stockholm (1939) p. 126.
20. My example is inspired by his predilection for referring to himself in the third person.
21. It may be worth while to recall here the distinction between a sentence, an utterance, and a statement. A sentence is of course a grammatical entity that involves no reference to any particular utterer or any particular time of utterance. An utterance is an event (a speech-act) that may be specified by specifying the uttered sentence, the speaker, and the occasion on which he makes his utterance.

    Utterances of declarative sentences (with prima-facie fact-stating intent) are typical examples of *statements*. (The term does not seem especially happy, but I shall retain it because it appears to be rather widespread.) A statement is an event (an act) occurring in some particular context. Usually it is a speech-act of a certain kind, but we shall not insist on that. For our purposes a statement may equally well be made, for example, by writing a sentence. *Any* act will do that is prima facie designed to serve the same purposes as the act of uttering a declarative sentence with the intention of conveying bona fide information.
22. This means, in effect, that Descartes arrives at his first and foremost insight by playing for a moment a double role: he appears as his own audience. It is interesting and significant that Balz, who for his own purposes represents Descartes's quest as a dialogue between "Cartesius, who voices Reason itself," and "René Descartes the Everyman," finds that they both "conspire in effecting this renowned utterance," the *cogito ergo sum*, wherefore "in some sense,

23. For this reason it might be more appropriate to call them (existentially) *self-defeating* than (existentially) *inconsistent*.
24. See his letter to Morin, dated July 13, 1638 (AT II, 209).
25. Martial Gueroult has again neatly located the source of trouble by calling our attention to the peculiarities of this relation. He has realized that Descartes's dictum does not (merely) express a logical relation between thinking and existing but that it is concerned with an additional "fact" or "act" ("le fait ou l'acte," "le fait brut de l'existence donnée"), which is just what is needed to show the certainty of my existence. However, his explanations leave the status of this fact or act (which cannot be an ordinary fact given to us by our senses or by introspection) rather vague. Nor does Gueroult realize that the logical aspect of Descartes's insight is in principle completely dispensable. See Gueroult's *Descartes*, II, p. 310.
26. What we have said shows that Descartes's verbs *cogitare* and *dubitare* are not, in the last analysis, the most accurate ones for describing the act through which the sentence 'I do not exist' defeats itself. It is not strictly true to say that an inconsistency arises from Descartes's attempt to *think* that he does not exist or to *doubt* that he does. Somebody else may think so; why not Descartes himself? He can certainly think so in the sense of contemplating a "mere possibility." What he cannot do is to *persuade* anybody (including himself) that he does not exist; wherefore he cannot try to *profess* (to others or to himself) that he does not exist without defeating his own attempt. In fact, Descartes himself resorts to explanations of this kind when he gives his most explicit explanation of the moves that made him recognize the self-evidence of his own existence. In the passage just quoted he uses the Latin verb *proferre* and a little earlier the verb *persuadere* for the purpose. If you are very literal-minded, you may thus say that Descartes ought to have concluded *ego sum professor* rather than *sum res cogitans*.
27. For the history of this view as well as for an interesting argument for its importance, see P. Schrecker, "La méthode cartésienne et la logique," *Revue philosophique*, vol. CXXIII (1937) pp. 336–67, especially pp. 353–4.
28. *Collected Papers*, Cambridge, Mass. (1931–1958), VI, sec. 338; V, sec. 421.
29. See AT IX, 205–6; HR II, 127; cf. AT VII, 140–1; HR II, 38.
30. See, for example, *Principia philosophiae* I, 7; I, 8; I, 49.
31. The difference is marked even though Descartes himself was not fully aware in all respects of the nature of his insight.
32. That he was not aware of it is also suggested by his conspicuous use of the argument to combat the skeptics. We have seen that for this purpose (which in Descartes is a subordinate one) the performatorily interpreted *cogito* argument is of little use.
33. *Summa theologica*, I, Q.87, art. 1.
34. For the relation of the two notions in Descartes, see N. Kemp Smith, *New Studies in the Philosophy of Descartes*, London (1952) pp. 52 ff.
35. That a verb of intellection should in Descartes serve to describe the object of another thought-act is all the more remarkable as it is virtually inconsistent with his explicit doctrines. For Descartes held that "one thought [conscious

act, *cogitationem*] cannot be the object of another" (reply to Hobbes's second objection; cf. AT VII, 422; HR II, 241).
36. This is not strictly true, for the second passage is concerned with the alleged inconsistency of sentences of the form "*b* thinks that *a* does not exist while *a* doubts something," whereas interpretation (i) was concerned with the alleged inconsistency of sentences of the form "*a* does not exist while he doubts something." The difference is immaterial for our purposes, however, and was obviously neglected by Descartes.
37. Cf. N. Kemp Smith, *op. cit.*, pp. 324–31.
38. See his introduction to *Descartes, Philosophical Writings*, ed. and trans. by E. Anscombe and P. Geach, Edinburgh (1954) p. xxxvii.
39. A little earlier Descartes had written: "I am, then, a real thing.... What thing? I have said it, a thinking thing. And *what more* am I?" (my italics; AT VII, 27; HR I, 152).
40. This part of his doctrine was criticized by Arnauld and others. In the preface to the *Meditationes* and in his replies to objections Descartes sought to defend himself. The question whether he succeeded is not relevant here.
41. AT VII, 219; HR II, 97.
42. Hintikka, Jaakko, *Knowledge and Belief*, Ithaca, N.Y. (1962).
43. I am indebted to Professors Norman Malcolm and G. H. von Wright for several useful suggestions in connection with the present essay.

8
A Definition of Truth for Theories with Intensional Definite Description Operators

Richard E. Grandy

There is no dearth of theories about nondenoting singular terms and the creation of yet another requires some justification. The extant theories divide rather neatly into two types—those of Frege (1966) and Russell (1905) analyze nondenoting singular terms by paraphrasing discourse containing them into discourse that is already well understood semantically and contains no nonextensional terms. More recent proposals by Leonard (1956), Hintikka (1959a, b), Lambert (1962, 1964), Rescher (1959), Scott (1967), and others treat singular terms with more respect and suggest modified logical theories to accommodate them.[1]

The theories of Russell and Frege were motivated primarily by systematic semantic considerations rather than by intuitions about ordinary usage; the more recent contributions are intended to reflect better what is said in ordinary discourse, but in some cases they offer merely syntactic treatments without semantic backing. Part of the purpose of this paper is to present a theory with a coherent semantics for a nonextensional definite description operator. By a nonextensional description operator, I mean one such that the formula $(x)[Ax \leftrightarrow Bx] \to \imath xAx = \imath xBx$ is not valid. This objective can be trivially met by several means; for example, one could declare all identities between nondenoting descriptions false, as is done in the theories of Russell, Schock, and Hintikka. However, I want to also attempt to satisfy another objective with the theory to be presented, which is that for any singular term t, $t = t$ should be valid. Another respect in which one wants the theory to be stronger than a system in which any two nondenoting descriptions are nonidentical is that one would like to be able to give a definition of truth for the language in a metalanguage with the same logical apparatus.[2]

Philosophers have recently discovered the usefulness of Tarski's work on truth in relation to problems other than those directly related to defining truth itself. The demand that any analysis of the logical form of some type of discourse (e.g., belief contexts or counterfactuals), provide (or at least be compatible with) a definition of truth for such discourse is an apparently useful one, though perhaps somewhat vaguer than is sometimes recognized.[3] It is clear what Tarski did for quantification theory, but is is less clear how to generalize the statement of what he accomplished so that we can measure new theories against the standard. For example, the usual goal of truth theory is to be able to derive in a suitable metalanguage all statements of the form 'S is true iff p' where S is an arbitrary sentence of the object language and p is the translation of S into the metalanguage. Beyond the fashionable question of what a translation is, however, we must also consider the fact that without further restrictions the demand is trivially satisfiable, for we could simply take the desired biconditionals as axioms. Thus, Davidson (1965) is led to formulate the requirement that the theory of truth employ only finitely many semantical primitives, while Wallace (1970) has demanded that the truth theory be finitely axiomatizable.[4]

In order to treat languages with primitive nondenoting singular terms there is the further complication that some aspect of Tarski's construction must be modified. One possibility would be to derive, instead of the Tarski biconditionals, conditional biconditionals of the form 'If all singular terms in S denote, then S is true iff p'. Another alternative would be to modify some of the machinery that is used to derive the biconditionals (i.e., either the notion of a sequence or of satisfaction or possibly the notion of logical consequence).

Thus, the project we have set ourselves is to explore the possibilities of developing a system that is sufficiently weak that the principle of extensionality is not valid, but that is sufficiently strong that a definition of truth is forthcoming. As we shall see, the possibilities are rather narrowly circumscribed in that a number of plausible-seeming modifications of the system lead to extensionality or inconsistency. The resulting system seems to me to shed light both on certain kinds of nonextensional contexts and on the role that the theory of truth can play in philosophy of language and philosophical logic.

Section 1 sets out the motivation for the system in more detail and compares it briefly with other treatments. Section 2.A presents the syntax and semantics of the theory, and soundness and completeness are proved in 2.C and 2.D respectively. Section 3.A presents a theory adequate to define truth in a metalanguage with the same logic and 3.B shows that the theory does adequately define truth. Section 4.A discusses alternatives to the theory presented and 4.C gives a more detailed comparison with other systems. To facilitate the comparison a general theory of intensional one-variable binding term-forming operators is given in 4.B. In Section 5 some of the philosophical aspects of the theory concerning ontology, intensionality and the general program of formalizing ordinary language are discussed. Readers whose primary concern is philosophical may omit sections 2.C, 2.D, and 3.B if they are prepared to believe the theorems without proof.

1. Logic

The logic to be used is most similar to that of Scott (1967)[5]; it resembles free logics in general in that the inference from $(x)Ax$ to At or from At to $(\exists x)Ax$ is permitted only with the additional premiss that $(\exists x)x = t$.[6] Thus, singular terms are permitted to be at least potentially denotationless in the sense that they may have no denotation in the range of values of the variables, or, more formally, one cannot infer $(\exists x)x = t$ from $t = t$. We will be concerned with singular terms both of the functional type (i.e., those that result from the application of a function to a (n-tuple of) singular term(s),[7] such as 'eldest brother of') and definite descriptions. The functional terms could be eliminated in favor of predicates and identity in well known ways, but the elimination is less interesting than retaining the terms and giving the truth definition, since the elimination rather arbitrarily fills what may be truthvalue gaps with falsity.

All of the free logic theories mentioned agree that when a definite description denotes, the sentence containing that description is equivalent to its Russellization. The effect of this can be achieved with the single axiom schema $(x)(x = \iota y A y \leftrightarrow (z)[x = z \leftrightarrow Az])$. Problems arise when one considers what statements containing nondenoting descriptions should be taken as true on logical grounds. The theory to be presented includes the usual identity theory for all singular terms, that is, the axioms $s = s$ and the schema $s = t \to \cdot As \to At$ are valid for all terms s and t without restriction. Thus, we side on this point with Lambert, Scott, and Rescher against Hintikka and Schock for whom $s = s$ is true only if s denotes. Elaboration of the obvious appeals to ordinary usage to support the truth of $s = s$ can be found in Lambert (1962) and Meyer and Lambert (1968). My own view is that to the extent that we have any clear intuitions about sentences containing nondenoting expressions being sometimes true, these are the clearest examples. If no such sentences are taken to be true, then there is no motivation for free logic at all.

It might be wise to say a few words about motivation for free logic in general at this point. Motivation is usually cited as the obvious truth of some such sentences as 'Pegasus is a horse' or 'Hamlet loved Ophelia', but I find these examples not entirely convincing. One could argue that appeals to the truth of sentences of mythology or fiction are sufficient grounds for extending classical logic to handle such sentences. There is an evident, though not always clearly delineated, respect in which such sentences are being considered within a mythological or fictional context, and thus there is considerable plausibility to treating them as sentences about a possible world (partially) specified by the myth or work of art in question.[8] I take the serious motivation for extending logic (and truth theory) to include nondenoting singular terms to be given by sentences such as the original Russell example, 'The present King of France is bald'. In everyday talk we use singular terms that we believe, but do not know, denote; it seems desirable to have a logic that acknowledges this fact and it seems plausible that a theory of truth can be formulated that expresses the fact.

The question of what statements containing nondenoting definite descriptions should be taken as true becomes more difficult when we look for principles beyond self-identity. The strongest principle one might want to adopt is that an identity between any two nondenoting expressions is true. This is the alternative adopted by Scott and it can be most concisely characterized by the axiom $(x)[x \neq t \land x \neq s] \rightarrow s = t$, due to Lambert (1964) although the semantics for the system is more intuitive in Scott's presentation. The idea of the semantics is to have an object not in the domain of the quantified variables of the theory that is the denotation of any nondenoting expression.

This theory of descriptions is formally coherent, but it leads to implausible ordinary-language examples, 'The present King of France is the present Queen of France' comes out true, for example. Further, if one develops a theory of truth along the obvious lines one finds that one has as a general principle that 'If s and t do not denote, then As is true iff At is true'. In view of such consequences, it seems desirable to develop a subtler theory about the identity of nondenoting descriptions.[9] Similar remarks apply to any system with the principle of extensionality, $(x)[A \leftrightarrow B] \rightarrow \iota x A = \iota x B$, which is derivable in the Scott system, but which could serve as a weaker axiom. This theory would not necessarily identify the present King of France with the author of *Principia Mathematica* but it would still identify the present King of France with the present Queen of France.

What we have in fact chosen to adopt is the rule that from $C \to [A \leftrightarrow B]$ one may infer that $C \to [\iota x A = \iota x B]$ if x is not free in C. Some form of such a rule seems desirable on the ground that such identities as $\iota x A = \iota x [A \vee A]$ are only slightly, if any, less obvious than $\iota x A = \iota x A$. Also, it proves necessary in defining truth to be able to make substitutions that identify objects definitely described in different ways.

The semantics that naturally suggests itself for such a theory is an expansion of Scott (1967). The first step is to permit an arbitrary nonempty set of objects (disjoint from the domain of individuals, of course) to function as a pseudodomain for the nondenoting descriptions to pseudodenote. The second step is to permit the assignment of pseudodenotation to a description to depend on what objects in the pseudodomain satisfy the formula to which the description operator is applied. If the assignment of the pseudodenotation depends only on which objects in the domain satisfy the formula, then one obtains an extensional system. Thus, the theory treats the description operator in effect as having a broader range of values for the variables that it binds than the quantifiers. The evaluation of the result of applying a quantifier to a formula may depend on all of the objects in the domain, but the evaluation of a description may depend on the objects in the pseudodomain as well.

In connection with this point it should be noted that free variables may be denotationless; that is, an assignment need not assign objects in the domain to all of the free variables, though it must assign an object from either the domain or pseudodomain. This is not a matter of caprice, for this formulation of the semantics ensures that a formula Ax will differ in interpretation from its closure $(x)A$. This is necessary in formulating the rule for identifying description terms; if one requires free variables to be assigned denotations in the domain, the rule quickly yields the principle of extensionality.[10]

The discussion of pseudodomains and pseudodenotation will likely have conjured up visions of possible objects or worse in the reader's mind. It should be mentioned now that not all the objects in the pseudodomain are possible objects, for one of them will be the denotation of $\iota x(x \neq x)$. Further remarks about ontology and intensionality are better left until after the model theory and truth theory have been given in detail, for it will then be easier to argue that the second domain is respectable in the model-theoretic context and does not enter the truth definition.

It might be thought by some that the use of free logic in the metalanguage removes most philosophical interest from the truth theory. On this view the point of the truth theory would be to explicate the truth conditions of sentences containing nondenoting singular terms in a theory that was better understood, that is, had no nondenoting singular terms. I think rather that it is advisable to distinguish the problem of giving a reduction from that of defining truth. The definition given here is not in any interesting sense a reduction, but it does have the philosophical interest of making plausible the rule of identifying descriptions. This is particularly true since the use of the rule is localized in that it is required only in the derivation of the biconditionals for sentences containing definite descriptions. The use of principles about nondenoting terms in the metalanguage seems quite parallel to the use made of principles about propositional connectives in deriving sentences containing those connectives and quantifier principles for deriving biconditionals about sentences containing quantifiers.

2.A. Syntax

We are studying a language L which is a metalanguage of a second language OL. Details of the relation will be made explicit in Section 3, but for the purposes of this section they are irrelevant. The primitive symbols of L are the logical operators, $\sim$, $\rightarrow$, $\forall$, a finite number of predicate letters P_i^n, identity $(=)$, an infinite list of individual variables $x, y, z, x_0, y_0, z_0, \ldots$, a finite list of function constants f_i^n and the definite description operator ι. The formation rules for wffs and terms are the obvious ones and we assume as given some standard abbreviations of the other logical operators.

We will use $A, B, C, \ldots$ as variables ranging over wffs of L and $s, t, s', t', \ldots$ similarly for terms of L. The notation $A(x, y, z)$ will be used to indicate that x, y, and z are among the free variables occurring in $A(x, y, z)$; similar notation will be used with $s, t, \ldots$ to indicate occurrence of terms in a wff. Typographical replacement will be used to indicate substitution and we assume that the substitution operation is defined in such a way that formulas are alphabetically varied so as to avoid undue capture of free variables when terms are substituted. The axioms and rules of L are the following

(A1) If A is a tautology, $\vdash A$
(A2) $(x)[A \rightarrow B] \rightarrow [(x)A \rightarrow (x)B]$
(A3) $s = s$
(A4) $s = t \rightarrow [As \rightarrow At]$
(A5) $[(v)Av \wedge (\exists v)v = s] \rightarrow As$
(A6) $(y)(\exists x)[x = y]$
(A7) $(x)[x = \iota y A y \leftrightarrow (z)[Az \leftrightarrow z = x]]$

(A8) $\iota x A = \iota y B$, if y is not free in A and B is the result of substituting y for x in A.
(A9) $(\exists x)[x = x]$
(R1)   If $\vdash A$ and $\vdash A \to B$, then $\vdash B$
(R2)   If $\vdash A \to B$ and x is not free in A then $\vdash A \to (x)B$
(R3)   If $\vdash C \to [A \leftrightarrow B]$ and x is not free in C, then $\vdash C \to [\iota x A = \iota x B]$

Only a few brief comments on the syntax of L are required. The provable closed wffs without singular terms are exactly the closed theorems of ordinary quantification theory with identity. The rules, if $\vdash A$ then $\vdash (x)A$ and if $\vdash A \leftrightarrow B$ then $\vdash \iota x A = \iota x B$ are derived rules. A deduction theorem is provable with the restriction that the x in applications of (R2) and (R3) also not occur free in any assumption.

2.B. Model Theory

An interpretation of L is a quadruple $\langle \phi, D, D^*, \pi \rangle$, where D and D^* are disjoint nonempty sets; π is a function defined on all subsets of $D \cup D^*$ whose values are elements of $D \cup D^*$, and $\pi(x) \in D$ iff $x \cap D = \{\pi(x)\}$, ϕ is a function which is defined on all terms, wffs, predicate letters, and function symbols of L and is such that

(a)   For any wff A, $\phi(A) = T$ or $\phi(A) = F$.
(b)   For each variable v, $\phi(v) \in D \cup D^*$.
(c_0)   For each P_i^0, $\phi(P_i^0) = T$ or F.
(c_n)   For each P_i^n, $n > 0$, $\phi(P_i^n) \subseteq (D \cup D^*)^n$.
(d)   For each atomic wff $P^n(s_1{}^i, \ldots, s_n)$, $\phi(P^n(s_1, \ldots, s_n)) = T$ iff $\langle \phi(s_1), \ldots, \phi(s_n) \rangle \in \phi(P^n)$.
(e) $\phi(\sim A) = T$ iff $\phi(A) = F$.
(f) $\phi(A \to B) = F$ iff $\phi(A) = T \neq \phi(B)$.
(g) $\phi((v)A) = T$ iff for every interpretation $\langle \psi, D, D^*, \pi \rangle$ such that ϕ and ψ agree on all predicate and function letters and all variables except possibly v, $\psi(A) = T$.
(h_0)   For each f_i^0, $\phi(f_i^0) \in D \cup D^*$.
(h_n)[11]   For each f_i^n, $\phi(f_i^n)$ is a function with domain $(D \cup D^*)^n$ and range included in $D \cup D^*$.
(i) $\phi(f^n(s_1, \ldots, s_n)) = \phi(f^n)(\phi(s_1), \ldots, \phi(s_n))$.
(j) $\phi(s = t) = T$ iff $\phi(s) = \phi(t)$.
(k) $\phi(\iota x A) = \pi(\{d: \text{for all } \langle \psi, D, D^*, \pi \rangle, \text{ if } \psi \text{ agrees with } \phi \text{ on all predicate and function letters, and on all variables except } x, \text{ and } \psi(x) = d, \text{ then } \psi(A) = T\})$.

A wff is said to be valid if for every $\langle \phi, D, D^*, \pi \rangle$, $\phi(A) = T$.

2.C. Soundness

The soundness of most of the axioms and rules can be verified directly from the definition of an interpretation. The soundness of (A7) can be shown by noting that if only one object d in D is such that the result of assigning it to z provides a variant of ϕ such that $\phi_d^e(Az) = T$,[12] then the set x defined in clause k will have only d in common with D, and thus $\pi(x) = d$.

The soundness of (R3) is most easily shown by a reductio argument. Suppose $C \to [A \leftrightarrow B]$ valid, x not free in C and $C \to [\iota x A = \iota x B]$ not valid. There is a $\langle \phi, D, D^*, \pi \rangle$ such that $\phi(c) = T$ and $\phi(\iota x A = \iota x B) = F$. Thus, $\pi(\{d: \phi_d^x(A) = T\}) \neq \pi(\{d: \phi_d^x(B) = T\})$ and so for some d_0, $\phi_{d_0}^x(A) = T = \phi_{d_0}^x B$, or conversely. But since x is not free in C and $\phi(c) = T$, $\phi_{d_0}^x(c) = T$ and thus $\phi_{d_0}^x(C \to [A \leftrightarrow B]) = F$.

2.D. Completeness

We shall prove completeness by a slight modification of Henkin's method. We will define a sequence Γ_n of sets of formulas associated with each formula A, and will use Γ_ω, the union of the Γ_n to define an interpretation which satisfies A unless $\sim A$ is provable. We assume given an enumeration $B_1, B_2, \ldots$ of all wffs of L.

$\Gamma_0 = \{A\}$

$\Gamma_{n+1} = \Gamma_n \cup \{B_{n+1}\} \cup \{S_{n+1}\}$ or $= \Gamma_n \cup \{\sim B_{n+1}\} \cup \{S_{n+1}\}$

according as $\Gamma_n \cup \{B_{n+1}\}$ is or is not consistent. B_{n+1} or $\sim B_{n+1}$ is the *primary* formula of stage $n + 1$ and S_{n+1} the *secondary* formulas. If the primary formula is of the form $\sim(v)Cv$ then the secondary formulas are $\sim C x_k$ and $(Ev)v = x_k$; if the primary formula is $\iota x C x \neq \iota x D x$, then the secondary formula is $\sim[C x_k \leftrightarrow D x_k]$, where x_k is the first variable not free in Γ_n or the primary formula.

Lemma 1. $\Gamma_\omega = \bigcup_n \Gamma_n$ *is consistent if A is.*

If Γ_ω is inconsistent then some finite subset is, but every finite subset of Γ_ω is included in some Γ_n; thus, it suffices to show by induction that Γ_n is consistent if A is. To prove the induction step we note that B_{n+1} is added only if consistent, and that if $\Gamma_n \cup \{B_{n+1}\}$ and $\Gamma_n \cup \{\sim B_{n+1}\}$ are both inconsistent, then Γ_n is inconsistent. Thus, we need only prove that secondary formulas do not cause inconsistency. We will deal only with

Theories with Intensional Definite Description Operators

the description case, the treatment of the quantifier is analogous and more familiar.

Suppose $\sim(Cx_k \leftrightarrow Dx_k)$, $\iota xCx \neq \iota xDx$, $\Gamma_n \vdash p \cdot \sim p$. Then by the deduction theorem and propositional calculus $\iota xCx \neq \iota xDx$, $\Gamma_n \vdash Cx_k \leftrightarrow Dx_k$, but since x_k is not free in any assumption $\iota xCx \neq \iota xDx$, $\Gamma_n \vdash \iota xCx = \iota xDx$ follow by rule (R3). Thus, the set of wffs without the secondary formula was already inconsistent.

Lemma 2. *If A is consistent,*

(a) *For any wff B, $B \in \Gamma_w$ or $\sim B \in \Gamma_w$.*
(b) *$[B \to C] \in \Gamma_w$ iff $C \in \Gamma_w$ or $B \notin \Gamma_w$.*
(c) *$\forall vBv \in \Gamma_w$ iff $Bs \in \Gamma_w$ for all terms such that $(\exists x)[x = s] \in \Gamma_w$.*
(d) *$s = s \in \Gamma_w$ for all s.*
(e) *If $s = t \in \Gamma_w$ and $t = t_0 \in \Gamma_w$ then $s = t_0 \in \Gamma_w$.*
(f) *$\iota v A = \iota v B \in \Gamma_w$ iff $As \leftrightarrow Bs \in \Gamma_w$ for all terms s.*
(g) *If $(\exists x)[x = \iota y By] \in \Gamma_w$ and $(\exists x)x = s \in \Gamma_w$ and $Bs \in \Gamma_w$, then $s = \iota x By \in \Gamma_w$.*

(a)–(e) follow in the usual way; (f) follows from the definition of Γ_w and (a); (g) follows from (A4) and (A7).

We now define an interpretation $\langle \phi, D, D^*, \pi \rangle$ as follows: We form equivalence classes of terms $s/=$ by defining $s/= = \{t : s = t \in \Gamma_w\}$; this is an equivalence class by Lemma 2(d, e). We set $D = \{s/= : (\exists x)[x = s] \in \Gamma_w\}$ and $D^* = \{s/= : (\exists x)[x = s] \notin \Gamma_w\}$. D and D^* are clearly disjoint; D is nonempty because of (A6) and (A9); D^* is nonempty because $(\exists x)[x = \iota y [Fy \wedge \sim Fy]]$ belongs to no consistent set of formulas.

π is defined by cases; if there is a formula B such that $X = \{s/= : Bs \in \Gamma_w\}$ then $\pi(X) = \iota y By/=$. If there is no such B, $\pi(X) = x_0/=$. For terms, $\phi(s) = s/=$; for atomic predicates, $\phi(P^n) = \{\langle s_1/=, \ldots, s_n/= \rangle : P^n_{s_1}, \ldots, s_n \in \Gamma_w\}$; for function symbols, $\phi(f^n) = \{\langle s_1/=, \ldots, s_n/=, t/= \rangle : f(s_1, \ldots, s_n) = t \in \Gamma_w\}$; the remaining clauses for ϕ can be filled in directly from the definition of an interpretation in Section 2.B.

Using Lemma 2, one can now verify that $\langle \phi, D, D^*, \pi \rangle$ is an interpretation. Lemma 2(f) guarantees that the π defined above is a function, that is, that the choice of the fomula used in defining $\pi(X)$ does not matter; Lemma 2(g) is necessary to verify that if $X \cap D = \{s/=\}$, $\pi(X) = s/=$.

Lemma 3. *If A is consistent and $\langle \phi, D, D^*, \pi \rangle$ is the interpretation just defined, $\phi(B) = T \leftrightarrow B \in \Gamma_w$.*

Proof. The lemma follows immediately from the definition of ϕ for atomic cases, and can be shown in the general case by induction on complexity of B using Lemma 2.

Thus, by contraposing Lemma 3, appealing to the fact that $A \in \Gamma_w$ and using double negation elimination:

Theorem. *If A is valid,* $\vdash A$.

3.A. Definition of Truth

Before presenting the theory of truth in L for OL, we must be more explicit about what is meant by saying that L is a metalanguage of OL. The conditions are not intended to be a general characterization of the relation; indeed, I do not know whether such a general characterization is possible, but will merely be a statement of assumptions used in the proof.

We assume that for each predicate and function letter in OL there is a corresponding predicate or function letter in L, and that there is a one-place predicate D in L whose intended extension is the domain of OL. We assume also that L has adequate resources to describe the syntax of OL. No explicit theory of syntax will be given; instead we assume a correspondence between the vocabulary of OL and a subvocabulary of L. Thus, we will use $\underline{A}, \underline{x}, \underline{s}$, and so on as names of the L names of the OL strings corresponding to A, x, s, and so on where such exist, with one complication. Although the string A corresponds syntactically to the string named by $\underline{A}$, the sentence expressed by the latter is expressed by the result of relativizing the quantifiers and definite descriptions to the domain of OL. Rather than add more notation for the relativization of A, we will apply the underlining notation only to relativized wffs and take $\underline{A}$ to be an L name of the OL wff corresponding syntactically to the derelativization of A.[13]

In addition to the syntactic machinery, L must also contain a one-place predicate $\text{Seq}(x)$ ('x is a sequence'), a two-place satisfaction predicate set, and a three-place relation $\text{App}(x, y, z)$ for 'the result of applying sequence x to y is z'. In practice, it is convenient to foresake the purity of the official language and use a many-sorted theory with Greek letters for variables for sequences, and to write $\alpha(x)$ for the value of α at X. We will also use underlined Roman capitals for variables ranging over wffs of OL and $\underline{v}, \underline{v}_1, \ldots$ as variables ranging over variables of OL, that is, $\underline{x}, \underline{y}, \ldots$. One final abbreviation will save much repetition: $\alpha \approx \beta_{x_1 \cdots x_n}^{v_1 \cdots v_n}$ is an abbreviation of $(\underline{v}_{n+1})[\underline{v}_{n+1} \neq \underline{v}_1 \wedge \cdots \wedge \underline{v}_{n+1} \neq \underline{v}_n \rightarrow \alpha(\underline{v}_{n+1}) = \beta(\underline{v}_{n+1})] \wedge \beta(\underline{v}_1) = x_1 \wedge \cdots \wedge \beta(\underline{v}_n) = x_n$; that is, β is like α except that it assigns $x_1, \ldots, x_n$ to $\underline{v}_1, \ldots, \underline{v}_n$.

The theory of truth (TOL) has two standard axioms of sequence existence,

(T1) $(\exists \alpha)(\underline{v})(\exists x)[\alpha(\underline{v}) = x]$

(T2) $(\alpha)(\underline{v})(x)(\exists \beta) . Dx \to \alpha \approx \beta_x^v$

We must also have axioms that define $\alpha(\underline{s})$ for other $\underline{s}$ other than variables of OL. Thus for each f^n we have an axiom

(T3) $(\alpha)(\underline{s}_1) \cdots (\underline{s}_n) . \alpha(\underline{f^n(s_1, \ldots, s_n)}) = f^n(\alpha(\underline{s}_1), \ldots, \alpha(\underline{s}_n))$

There will also be a parallel axiom for ι,

(T4) $(\alpha)(\underline{v})(A)\alpha(\underline{\iota vA}) = \iota x[D \land (\exists \beta)[\alpha \approx \beta_x^v \land \beta \text{ sat } \underline{A}]]$

The axioms for satisfaction are the standard ones:

(T5) $\alpha \text{ sat } \underline{Ps_1 \cdots s_n} \leftrightarrow P\alpha(s_1) \cdots \alpha(s_n)$, for each P

(T6) $\alpha \text{ sat } \underline{s = t} \leftrightarrow \alpha(s) = \alpha(t)$

(T7) $\alpha \text{ sat } \underline{\sim A} \leftrightarrow \sim \alpha \text{ sat } \underline{A}$

(T8) $\alpha \text{ sat } \underline{A \to B} \leftrightarrow \alpha \text{ sat } \underline{A} \to \alpha \text{ sat } \underline{B}$

(T9) $\alpha \text{ sat } \underline{\forall vA} \leftrightarrow (\beta)(x) . Dx \to \cdot \alpha \approx \beta_x^v \to \beta \text{ sat } \underline{A}$

3.B. The Adequacy of *TOL*

The goal of this section is to show that $T(\underline{A}) \leftrightarrow A$ is provable for all closed $\underline{A}$ of OL, where $T(\underline{A}) =_{\text{df}} (\alpha)\alpha \text{ sat } \underline{A}$. This result will follow easily if we can prove

Lemma. $\vdash \alpha \text{ sat } \underline{Bv_1 \cdots v_n} \leftrightarrow B\alpha(\underline{v}_1) \cdots \alpha(\underline{v}_n)$, *where $\underline{B}$ is an arbitrary wff of OL and $\underline{v}_1 \cdots \underline{v}_n$ are all of the free variables of $\underline{B}$.*

Proof. It is in fact easier to prove the following more complicated statement: For all wffs $Bv_1 \cdots v_n$ of OL, $\vdash \alpha \text{ sat } \underline{Bv_1 \cdots v_n} \leftrightarrow B\alpha(\underline{v}_1) \cdots \alpha(\underline{v}_n)$ and for all terms $\underline{s(v_1, \ldots, v_n)}$ of OL, $\vdash \alpha(s(\underline{v_1 \cdots v_n})) = s(\alpha(\underline{v}_1) \cdots \alpha(\underline{v}_n))$, where $\underline{v}_1 \cdots \underline{v}_n$ are all the free variables of $\underline{B}$ or $\underline{s}$ respectively.

The proof proceeds by induction on the complexity of $\underline{B}$ and $\underline{s}$, where complexity is defined by counting all occurrences of logical operators, ι, identity, predicate letters and function letters.

$n = 0, 1$.   No wffs are of complexity 0; those of complexity 1 must be atomic predicates or identity applied to variables. For such wffs the statement reduces to an instance of (T5) or (T6) terms of complexity 0 must be variables, for which the statement is $\alpha(\underline{v}) = \alpha(\underline{v})$.

$n \neq 0$. The proof for wffs divides into four cases; those in which $\underline{B}$ is logically complex with principal operator $\sim$, $\rightarrow$, or ι respectively, and the case where B is the application of a predicate letter or identity to complex terms.

$\sim$. If $Bv_1 \cdots v_n = \sim \underline{Cv_1 \cdots v_n}$, by the induction hypothesis we know that $\vdash \alpha$ sat $\underline{Cv_1 \cdots v_n} \leftrightarrow C\alpha(\underline{v_1}) \cdots \alpha(\underline{v_n})$, from which the desired result follows by (T7).

$\rightarrow$. If $\underline{B} = C \rightarrow D$, the result follows from the induction hypothesis by (T8).

If $Bv_1 \cdots v_n$ is $\forall v C v v_1 \cdots v_n$, by hypothesis we know that $\vdash (\gamma)\gamma$ sat $Cvv_1 \cdots v_n \leftrightarrow C\gamma(v)\gamma(v_i) \cdots \gamma(v_n)$ and by (T9) $\vdash \alpha$ sat $Bv_1 \cdots v_n \leftrightarrow (\gamma)(x) . Dx \rightarrow \alpha \approx \gamma_x^v \rightarrow \gamma$ sat $Cvv_1 \cdots v_n$, so by substitution of equivalents $\vdash \alpha$ sat $Bv_1 \cdots v_n \leftrightarrow (\gamma)(x) . Dx \rightarrow \alpha \approx \gamma_x^v \rightarrow C\gamma(v) \cdots \gamma(v_n)$. Unpacking $\alpha \approx \gamma_x^v$ and using (T2), $\vdash \alpha$ sat $Bv_1 \cdots v_n \leftrightarrow (x) . D_x \rightarrow Cx\alpha(v_1) \cdots \alpha(v_n)$, which is the desired result.

Atomic wffs with complex terms: If $\underline{B}$ is $Ps_1 \cdots s_n$, then by (T5) $\vdash \alpha$ sat $Ps_1 \cdots s_n \leftrightarrow P\alpha(s_1) \cdots \alpha(s_n)$, and by the induction hypothesis $\vdash \alpha(s_i(v_i \cdots v_{in}) = s_1(\alpha(v_{i_1}), \ldots, \alpha(v_{i_n}))$, for all s_i in $\underline{B}$, so by substitution $\vdash \alpha$ sat $P_v \cdots v_m \leftrightarrow P\alpha(v_i) \cdots \alpha(v_m)$. Similarly for identity by (T6).

Terms. If $s(v_1 \cdots v_n) = f(t_1 \cdots t_m)$, then using (T3) $\vdash \alpha(s(v_1 \cdots v_n)) = f(\alpha(t_1) \cdots \alpha(t_m))$ and by the induction hypothesis $\vdash \alpha(t_i(v_i, \ldots, v_{i_k})) = t_i(\alpha(v_{i_1}) \cdots \alpha(v_{i_k}))$ for each t_i in $\underline{s}$, so the desired conclusion follows by substitution.

If $s(v_1 \cdots v_n)$ is $\forall v A v v_1 \cdots v_n$, then by (T4) we know that $\vdash \alpha(s(v_1 \cdots v_n)) = \iota x[Dx \wedge (\exists \beta) \alpha \approx_x^v \beta \wedge A\beta(v_1) \cdots \beta(v_n)]$. But by (T2), the definition of $\alpha \approx_x^v \beta$ and substitution, $[Dx \wedge (\exists \beta) \alpha \approx_x^v \beta \wedge A\beta(v)\beta(v_1) \cdots \beta(v_n)] \leftrightarrow [Dx \wedge Ax\alpha(v_1) \cdots \alpha(v_n)]$, so by (R3) and transitivity of identity

$$\vdash \alpha(s(v_1 \cdots v_n) = \iota x[Dx \wedge Ax\alpha(v_1) \cdots \alpha(v_n)]$$

which is the desired result.

Theorem. *For all closed $\underline{A}$, $\vdash T(\underline{A}) \leftrightarrow A$.*

Proof. For closed $\underline{A}$, the lemma gives $\vdash \alpha$ sat $\underline{A} \leftrightarrow A$, whence $\vdash [(\alpha)\alpha$ sat $\underline{A}] \leftrightarrow A$.

4.A. Alternative Formulations

One principle concerning descriptions that suggests itself is $A\iota xA$. It is well known that this general principle together with the reflexivity of identity leads to contradiction because by reflexivity $\iota x(x \neq x) = \iota x(x \neq x)$ but by the principle $\iota x(x \neq x) \neq \iota x(x \neq x)$. A second thought along these lines would be to add $A\iota xA$ for consistent formulas A.

This, too, is inconsistent, however. Consider the two descriptions $\iota x[x = \iota yFy \wedge x = \iota yGy]$ and $\iota z[x = \iota yFy \wedge x \neq \iota yGy]$. If F and G are atomic predicates both descriptions are consistent, but now by the principle $A\iota xA$ for consistent A we obtain $s = \iota yFx \wedge s = \iota yGy \wedge t = \iota yFx \wedge t \neq \iota xGy$ (abbreviating the long descriptions by 's' and 't'). But this last formula entails $s = t$ and $s \neq t$.

A second modification of the system would be to require all free variables to denote. This would render the principle $(x)Ax \to Ay$ is valid, and would lead to the extensionalization of the system in two steps:

$\vdash (x)[Fx \leftrightarrow Gx] \to [Fy \leftrightarrow Gy]$

$\vdash (x)[Fx \leftrightarrow Gx] \to \iota yFy = \iota yGy$

The principle of extensionality can be shown to be unprovable in the system by considering the following interpretation for a language containing two atomic predicates, F and G:

$D = \{0\}, \quad D^* = \{1, 2\}, \quad \phi(F) = \{1\}, \quad \phi(G) = \{2\}$

if X is a unit set $\{d\} \pi(X) = d$, if $0 \in X$, $\pi(X) = 0$, and $\pi(X) = 1$ otherwise. This $\langle \phi, D, D^*, \pi \rangle$ satisfies all of the axioms and $\phi(x)[Fx \leftrightarrow Gx]) = T$ but $\phi(\iota xFx) = 1 \neq 2 = \phi(\iota xGx)$.

4.B. General Intensional Operators

Scott (1967) discusses the general theory of operators that are extensional in the sense that if $X \cap D = Y \cap D$ then $\pi(X) = \pi(Y)$. It is shown there that the system obtained by adding an operator 0 and supplementing (A1)–(A6) and (R1), (R2) by $(x)[A \leftrightarrow B] \to 0xA = 0xB$ is complete for such a general operator.

A similar question arises about intensional operators; that is, if one does not require that if $x \cap D = \{d\}$ then $\pi(x) = d$, but permits arbitrary functions π from $2^{D \cup D^*}$, keeping the clause that $\phi(IxAx) = \pi(\{d: \phi_d^x(A) = T\})$, what formal system characterizes the I operator?

It is not difficult to verify that if one drops (A7) and replaces (R3) by the rule: If x is not free in C and $\vdash C \to [A \leftrightarrow B]$ then $\vdash C \to [IxA = IxB]$ one obtains a complete system. The proof can be obtained by minor modifications of the proof in Section 2.D.

4.C. Other Description Theories

There are minor differences in axioms and in the definition of interpretation between this system and those of Scott, Lambert, and Schock that are due to the requirement that D be nonempty, but we shall ignore these and focus on the interesting ones. Lambert's FD1 (Lambert 1962, and van Fraassen and Lambert 1967) can be obtained by dropping (R3) and adding

(A′8) $s = \iota x(x = s)$

Lambert's FD2 (Lambert 1964, and van Fraassen and Lambert 1967 which is similar to Scott 1967) is obtained by adding (A′9):

(A′9) $(x)[x \neq s \wedge x \neq t] \to s = t$

which renders (R3) redundant.

The definition of interpretation can easily be modified to obtain a characteristic class of models for FD2—one simply requires that D^* be a unit set. This is, in fact, Scott's definition.

FD1 is more difficult to compare. (A′8) can be shown not valid in our system by the interpretation $\langle \phi, D, D^*, \pi \rangle$ such that $D = \{0\}, D^* = \{1, 2\}$, $\phi(z) = 1$, $\pi(X) = 0$ if $0 \in X$, $\pi(\{1\}) = 2$, $\pi(\{2\}) = 1$ and otherwise $\pi(X) = 1$. With this interpretation $\phi(z = \iota x(x = z)) = F$ because $\phi(z) = 1$ and $\phi(\iota x(x = z)) = 2$. On the other hand, our system is not included in FD1, for $(x)Fx \to \iota y Gy = \iota y[(x)Fx \wedge Gy]$, which is a theorem of our system, is not provable in FD1. There is no simple method of proving this last fact, for the van Fraassen–Lambert semantics uses an auxiliary function defined on formulas and terms rather than defining ϕ and π using D^*. The result is that they need not mention 'unreal' objects, but the price is that the semantics is cumbersome. The semantics using D^* and π is closely analogous to ordinary model theory and facilitates comparison between systems.

For example, although it is unclear how to define models for a system with (A′8) without (R3), there is at least one clear condition on π that would render (A′8) valid; if $X = \{d\}$, $\pi(X) = d$.[14]

Schock's system (1968) is rather unlike the one of this paper and FD1 and FD2 in that $s = s$ is true only when 's' denotes, and in general the application of an atomic predicate to a nondenoting term is false. His semantics is in the spirit of Scott rather than van Fraassen and Lambert, however. It can be obtained from our by requiring that identity be treated with other atomic predicates, and that for any atomic predicate $\phi(P^n) \subseteq D^n$. Because of this restriction, the 1-place 0-term 1-formula term-making operators studied in Schock (1964, 1968) are less general than those in Scott (1967) or in Section 4.B.

5. Philosophical Morals

We return now as promised to the questions touched upon in earlier sections concerning the element of intensionality in the treatment of descriptions and the problem of the ontology of the theory. Beginning with the latter, we note first that although the model theory requires a domain D^* of nonentities these objects are unreal only from the point of view of the object language. The statement of the model theory requires only that D^* be nonempty and disjoint from D. When we come to the definition of truth in Section 3, the dubious domain vanishes entirely for no mention of it was made anywhere in that section. One important philosophical point that this system illustrates is that there can be important differences between model theory, the theory of truth for a language in arbitrary structures, and the truth for the language in the sense in which truth is defined in Section 3.

The only aspect of the truth theory that might cause suspicion is the use of sequence expressions that have no denotation, but this seems unproblematic so long as one recalls that we are working within free logic. In the metalanguage, one is free to assert that some sequences assign actual objects to given terms, that other sequences do not assign any object to some terms and in cases of doubt to leave the matter open. The situation seems no different than it is for any other kind of singular term in any other theory, which is probably as it should be.

A point that should be of more interest to philosophers who are concerned with the usefulness of the requirement that an analysis of a portion of natural language should include a definition of truth is that we now have available at least four analyses that meet this demand as it is usually formulated. The only requirements that are usually proposed are that the definition of truth be finitely specified and that it suffice to probe $T(\underline{A}) \leftrightarrow A$ for each $\underline{A}$ of the object language. The theory in this paper, the one obtained by adding (A'8), FD2, and Russell's theory all satisfy these requirements. Since the systems obtained by adding (A'8) and the system FD2 are extensions of the present system, the proof in Section 3 goes through unchanged for those systems. Since Russell's elimination of descriptions reduces the problem to defining truth in ordinary quantification theory, his theory also will satisfy the conditions.

Thus, it appears that we must either look for more criteria of correctness for a definition of truth or admit that the matter is highly relative.

My own view is that we must look to all of the consequences of the theory of truth and attempt to choose among the theories on this wider basis. I have suggested reasons in Section 1 for rejecting the extensional systems, but even if one accepts those reasons as conclusive we have established only that we must look at weaker systems. The system of the present paper may be too weak, but it does provide a starting point since it seems unlikely that any weaker theory will suffice to define truth. Moreover, the semantics are sufficiently transparent to make discussion and comparison of extensions possible.

The semantics also illustrates how one can make sense of intensional operators in terms of extensional semantics within a wider domain. That is, the definite description operator is intensional in the sense that its values do not depend only on evaluations of formulas within the domain, but the operator can be described in straightforwardly set-theoretic terms as depending on the evaluation of the formula with respect to objects not in the domain. This point was already illustrated in one form by possible worlds semantics for modal contexts, but the present example is of some interest in that the "extra" objects are not all possible objects, and issues of cross-identification do not arise.

One could extend the system so as to make it more akin to modal systems. For example, one might add the converse of the principle of extensionality, $N: \iota x A = \iota x B \rightarrow (x)[A \leftrightarrow B]$. Within the system as it stands one can define a sentential operator that bears some resemblance to necessity; that is, one can define $\Box A$ as $\iota x(x = x) = \iota x A$. With this definition one can show that if $\vdash A$ then $\vdash \Box A$, but most of the customary axioms of necessity are missing. For example, $\vdash \Box A \rightarrow A$ is not provable as may be seen by considering one of the extensional models in which A is not satisfied by all objects and $\iota x A$ is not a proper description. Upon the addition of the schema N, however, one would have a system whose theorems included at least those of (S5), including the Barcan formula and its converse. This extension may be of some interest, but it seems unlikely that it is the best approximation to the logic of ordinary language, and so it is outside the scope of this paper.

References

Davidson, Donald, "Theories of meaning and learnable languages," in *Logic, Methodology and Philosophy of Science* (ed. Bar-Hillel, Yoshua), North Holland, Amsterdam (1965) pp. 383–94.

Davidson, Donald, "Truth and meaning," *Synthese*, vol. 17 (1967) pp. 304–23.
Davidson, Donald, "True to the facts," *Journal of Philosophy*, vol. 66 (1969a) pp. 748–64.
Davidson, Donald, "On saying that," in *Words and Objections* (ed. Davidson, D. and Hintikka, J.), Little, Brown & Co., Boston (1969b) pp. 158–74.
Frege, Gottlob, *Philosophical Writings of Gottlob Frege* (ed. Geach, P. and Black, M.), Little, Brown & Co., Boston (1966).
Hintikka, Jaakko, "Existential presuppositions and existential commitments," *Journal of Philosophy*, vol. 56 (1959) pp. 125–37.
Hintikka, Jaakko, "Towards a theory of definite descriptions," *Analysis*, vol. 19 (1959b) pp. 79–85.
Hintikka, Jaakko, "Definite descriptions and self-identity," *Philosophical Studies*, vol. 15 (1964) pp. 4–7.
Lambert, Karel, "Notes on E! III: A theory of descriptions," *Philosophical Studies*, vol. 13 (1962) pp. 51–9.
Lambert, Karel, "Notes on E! IV: A reduction in free quantification theory with identity and descriptions," *Philosophical Studies*, vol. 15 (1964) pp. 85–8.
Leonard, Henry, "The logic of existence," *Philosophical Studies*, vol. 7 (1956) pp. 49–64.
Meyer, Robert K. and Lambert, Karel, "Universally free logic and standard quantification theory," *J. of Symbolic Logic*, vol. 33 (1968) pp. 8–26.
Quine, W. V. O., *Philosophy of Logic*, Prentice-Hall, Englewood Cliffs (1970).
Rescher, Nicholas, "On the logic of existence and denotation," *Philosophical Review*, vol. 69 (1959) pp. 157–80.
Russell, Bertrand, "On denoting," *Mind*, vol. 14 (1905) pp. 479–93.
Schock, Rolf, "On the logic of variable binders," *Archiv für Mathematische Logik und Grundlagenforschung*, vol. 6 (1964) pp. 71–90.
Schock, Rolf, *Logics Without Existence Assumptions*, Almqvist and Wiksell, Stockholm (1968).
Scott, Dana, "Existence and description in formal logic," in *Bertrand Russell: Philosopher of the Century* (ed. Schoenman, R.), Little, Brown & Co., Boston (1967) pp. 181–200.
van Fraassen, Bas C. and Lambert, Karel, "On free description theory," *Zeitschrift für mathematische Logik und Grundlagen der Mathematik*, vol. 13 (1967) pp. 225–40.
Wallace, John, "Equilibrium and intensionality" (1970) unpublished.

Notes

1. Discussions with Tyler Burge and Dana Scott about ancestors of this paper were of great benefit in improving the theory and clarifying its exposition. I am also indebted to David Kaplan for proving that an earlier version was unsound. The research was partly supported by NSF.
2. The system in Schock (1968) is difficult to classify because it uses a free logic and gives a semantics that is not straightforward first-order model theory, but the system is similar to Russell's in that descriptions are eliminable, and thus the frills in the model theory could have been omitted.

3. See Davidson (1965, 1967, 1969a, b), Quine (1970), and Wallace (1970).
4. We shall not go into the matter here, partly because it seems to me too early for a fruitful discussion of the matter and partly because the theory to be presented clearly meets both criteria.
5. One of the systems presented in Lambert (1962) is similar to Scott's system except for the presence of names, but the model theory presented for the Lambert system in van Fraassen and Lambert (1967) is considerably different than Scott's; of this more will be said anon.
6. In the terminology of Meyer and Lambert (1968), the system to be presented is a *free logic* in that no existence assumptions are made with respect to individual constants, but it is not a *universally free logic* because the domain of individuals is assumed to be nonempty.
7. Proper names are assimilated as 0-place functions.
8. Frege (1966, p. 63) can be interpreted as expressing a similar view.
9. Scott (1967, p. 187) is well aware of this consequence and does not consider it a defect for the purposes he is considering; John Wallace persuaded me that it is a defect for the purpose of defining truth and this paper is the result of his challenge to produce a suitable system.
10. The fact that the system lapses into extensionality if one requires free variables to denote was pointed out to me by David Kaplan with respect to an earlier system in which the point was much less obvious. In the present formulation the consequence is immediate, which is one reason for preferring this formulation.
11. This definition permits functions applied to arguments in D^* to have values in D. This may or may not be desirable. I have been unable to find an example where it is clearly plausible, but also know of no general argument against the possibility. Two plausibility examples are 'the father of John's only son', when John has two sons, and 'the square of the first element of $\langle x, y \rangle$' when applied to $\langle 3, \text{Pegasus} \rangle$.

    One could, if desired, alter the definition of interpretation to require that if any element of an n-tuple is in D^*, then the value of any function applied to the n-tuple is in D^*. An axiom schema $(x)[x = f(s_1, \ldots, s_n)] \rightarrow (x_1)(x_1) \cdots (x_n)[x_1 = s_1 \wedge x_2 = s_2 \wedge \cdots \wedge x_n = s_n]$ would have to be added, but with this modification the completeness construction would work without modification.
12. ϕ_d^z is a function identical with ϕ except it assigns d to the variable z.
13. The relativization of A to D is the result of replacing each subformula $\forall xB$ by $(\forall x)[Dx \rightarrow B]$ and each term ιxB by $\iota x(Dx \wedge B)$.
14. It is easy to verify that with the emendation of the semantics the system obtained by adding (A'8) is sound. But it is also complete. Under any interpretation $z = \iota x(x = z)$ must be true, but as we run through different interpretations $\{d: \phi_d^x(x = z) = T\}$ will include all unit sets in $D \cup D^*$ and for each of these $\pi(\{d\}) = d$ must hold if the wff is to be assigned T.

9
Truth and Singular Terms

Tyler Burge

Nondenoting singular terms have been a prime stimulus, or irritant, to students of the use and formal representation of language. Only one other subject (modal logic) has provoked so many differences among logicians over which sentences should be counted valid. It may be that in the case of singular terms these differences are not fully resolvable, especially in light of the various purposes that logical analyses serve. But there are, I think, means of narrowing our options in choosing logical axioms. This paper is devoted to exploring one such means.[1]

I will be viewing a logic of singular terms in the context of a truth-theoretic account of natural-language sentences under their intended interpretations. As a consequence, I shall take native intuitions about truth conditions and truthvalues as evidence in framing a semantical theory for native sentences and for framing a logic underlying that theory. Within this context, I shall argue that some proposed logics for singular terms are unsatisfactory because they lead from true premises to untrue conclusions and that others are faulty because they are too weak to justify transformations needed for an adequate theory of truth. I shall conclude by arguing that one otherwise plausible logical axiom is incompatible with a straightforward means of avoiding these difficulties. I begin by motivating and sketching an account of singular terms that will set the stage for these points.

1. The Motivation for Free Logic

The most clearly semantical problem that nondenoting singular terms raise is that of saying how the truth-conditions of sentences containing them may be determined on the basis of the logical roles of the parts of the sentences. The classical treatments of the problem are, of course, Russell's and Frege's (see Frege 1966; Russell 1905; Whitehead and Russell 1925–1927). Russell's elimination of singular terms pays well-known dividends but fails to account for natural language as it is actually used. Frege's stipulation that intuitively nondenoting singular terms denote the null set forces one to formalize certain negated existence statements in syntactically unnatural ways. Moreover, by thus "identifying" all nondenoting singular terms, the theory counts sentences like 'Pegasus is the smallest unicorn' true. Although some analytical enterprises can perhaps overlook such results, an account of truth in natural languages cannot.

In the last decade or so, a number of free logics have been developed to account for, rather than do away with, nondenoting singular terms (see Grandy 1972; Hailperin and Leblanc 1959; Hintikka 1964; Lambert 1962, 1964; Meyer and Lambert 1968; Schock 1968; Scott 1967; Smiley 1960; van Fraassen and Lambert 1967). These logics seem to me to be on the right track. But there has been little agreement over precisely which inferences such logics should validate or block. The restriction on existential generalization and universal instantiation that is common to all free logics has frequently been justified by reference to various purportedly true singular sentences, like 'Pegasus is winged', from which one cannot existentially generalize. Literally taken, these sentences are, I think, untrue. In so far as they are counted true, they are best seen as involving an implicit intensional context: '(A well-known myth has it that) Pegasus is winged'. The strategy is an extension of Frege's approach to apparent failures of substitution. In our sample sentence (regarded as true), we cannot substitute for 'Pegasus' other singular terms that do not differ in denotation ('the tallest unicorn'), and still preserve truth. So we regard the context as oblique.

One might object that different "nonexistents" are denoted by these singular terms in *all* their occurrences. But as regards the intended interpretation of nondenoting singular terms, this way of speaking is, I think, misleading. Currently, the temptation to speak this way seems to arise only in the face of sentences that are easily seen to be related to one of the standard sorts of intensional contexts (indirect discouse, subjunctives, psychological contexts). The point is clear in our example. When native speakers are asked whether 'Pegasus is winged' is true, they rely on common knowledge and contextual clues to determine what the questioner intends. It is now common knowledge among people who use 'Pegasus' in the relevant contexts that the name is part of, and is meant to be related to, a mythical story. The prefix to the sentence that we supplied above will generally be accepted as producing a paraphrase. But if asked whether the myth itself is true—whether it is a matter of fact (rather than a matter of fiction) that Pegasus is winged—native speakers will reply that the sentence is not literally or 'factually' true. And they will justify this by saying that Pegasus does not exist, "except in the myth." I take this behavior as evidence for embedding the sentence in an intensional context when it is regarded as true. Nondenoting singular terms simply do not have anything as their "ordinary" (nonoblique) denotation. It is just that the oblique reference of some singular terms is the reference they most often have in their everyday uses. Talk of "non-existents" in contexts like the above can perhaps be assimilated to the strategy of finding the oblique reference for singular terms in intensional contexts.[2]

Implementing the strategy and providing an account of oblique contexts is, of course, beyond our present purpose. It is enough to note here that the motivation for free logic may be regarded as independent of issues about apparent substitution failures of singular terms. Consider the sentence '$(x)(x = x)$' from identity theory. By universal instantiation, we derive 'Pegasus = Pegasus'; and by existential generalization we arrive at '$(\exists y)(y = \text{Pegasus})$', which is clearly false. Unless we regress to Russell or Frege, we must either alter identity theory or restrict the operations of instantiation and generalization. Experimentation with the latter two alternatives indicates that the restriction strategy is simpler and more intuitive.

2. A Sketch of a Theory of Singular Terms

We now characterize a logic underlying a formalized metalanguage ML and a theory of truth couched in that language for the sentences of a natural object-language OL. The grammar of ML is that of first-order quantification theory with predicate constants, identity, function signs, and the definite-description operator. The logical axioms and rules underlying ML are as follows:

(A1)   If A is a tautology, $\vdash A$

(A2) $\vdash (x)(A \to B) \to ((x)A \to (x)B)$

(A3) $\vdash (x)(x = x)$

(A4) $\vdash t_1 = t_2 \to (A(x/t_1) \leftrightarrow A(x/t_2))$

(A5) $\vdash (x)A \,\&\, (\exists y)(y = t) \to A(x/t)$

(A6) $\vdash (x)(x = (\iota y)A \leftrightarrow (y)(A \leftrightarrow y = x))$

where variable $x \neq$ variable y, and x is not free in A

(A7) $\vdash (x)(\exists y)(x = y)$

(A8) $\vdash (x)(x = t_1 \leftrightarrow x = t_2) \to (A(y/t_1) \leftrightarrow A(y/t_2))$

where x is not free in t_1 or t_2

(A9) $\vdash At_1 \cdots t_n \to (\exists y_1)(y_1 = t_1) \,\&\, \cdots \,\&\, (\exists y_n)(y_n = t_n)$

where A is any atomic predicate, including identity, and where y_i is not free in t_i

(A10) $\vdash (\exists y)(y = f(t_1, \ldots, t_n)) \to (\exists y_1 = t_1) \,\&\, \cdots \,\&\, (\exists y_n)(y_n = t_n)$

where y is not free in $t_1, \ldots, t_n$, and y_i is not free in t_i

(R1)   If $\vdash A$ and $\vdash A \to B$, then $\vdash B$

(R2)   If $\vdash A \to B$, then $\vdash A \to (x)B$, where x is not free in A

'A' and 'B' range over well-formed formulas of ML; 't', 't_1', ..., 't_n', over terms (including variables); and 'x', 'y', 'y_1', ..., 'y_n', over variables. '$A(x/t_1)$' signifies the result of substituting t_1 for all occurrences of x in A, rewriting bound variables where necessary.

Axioms (A3), (A5), and (A8) are nonindependent. (For the details, see Burge 1971.) They are included for the sake of clarifying our motivation. Alternatively, one might take (A3), (A4), and (A5) as nonindependent and the others as primitive, adding the symmetry and transitivity axioms of identity. The value of this formulation is that it focuses on (A8) instead of (A4). As will be seen, (A8) constitutes the main principle of interchange in the system.

The logic of ML is nearly classical. If (A7) were changed to '$(\exists y)(x = y)$', and if singular terms other than variables were excluded from the language, the logic would revert to classical quantification theory with identity. The chief motivation for (A7) is that, unlike '$(\exists y)(x = y)$', it allows some of the free variables (like some of the other terms) to be uninterpreted. 'Nondenoting' free variables are useful in representing sentence utterances that involve failure of reference with demonstrative constructions.[3]

Axiom (A9) differentiates the syntax of ML from that of Scott (1967). It expresses a deep and widely held intuition that the truth of simple singular sentences (other than those implicitly embedded in intensional contexts) is contingent on the contained singular terms' having a denotation.[4] The pretheoretic notion seems to be that true predications at the most basic level express comments on topics, or attributions of properties or relations to objects: lacking a topic or object, basic predications cannot be true. Given that ML is bivalent, simple singular predications containing nondenoting terms are counted false, and negations of such sentences are true. ('Pegasus is an animal' is false. 'It is not the case that Pegasus is an animal' is true.) Within ML, logical operations such as negation should be intuitively seen as working on simpler sentences as wholes, not as forming complex comments on purported topics or complex attributions to purported objects. This remark would admit of exceptions if we were to provide for singular terms with wide scope ('Pegasus is such that he is not an animal').[5] Then negation operates on an open sentence rather than on a closed one. Nondenoting singular terms with wide scope should cause the sentences they govern, no matter how complex, to be untrue.

Axiom (A9) rests weight on the notion of atomic predicate. As just indicated, I think that the weight has intuitive support, support associated with semantical intuitions about truth and with the pretheoretic notions of property and relation. The axiom should be regarded as a methodological condition on investigations of predication in natural language: count an expression an atomic predicate in natural language only if one is prepared to count simple singular sentences containing it untrue whenever they also contain nondenoting singular terms.[6] Scott's and Lambert's systems show that it is possible to arrange a logically coherent language with atomic predicates that violate our condition. But it is another question whether such predicates have natural-language readings that are best construed as having the logical form of atomic predicates. In numerous cases, intuition backs our condition; the present proposal is that the condition should be used to guide intuition. Needless to say, it must be judged by the quality of its guidance.

Axiom (A9) enables us to derive the Russell equivalence

$B(\iota x)Ax \leftrightarrow (\exists y)((z)(Az \leftrightarrow z = y) \,\&\, By)$

where B is atomic. (This latter restriction amounts to the proviso that the iota operator always takes smallest scope; cf. Note 5.) The present system thus captures Russell's intuitions without using his means of doing so. Whereas we agree with Russell about truth conditions, we disagree with him about logical form. Rather than regarding singular terms on the model of abbreviations for *other* language forms, we take them as primitive in natural language and in formal languages whose purposes include representing natural language. Consequently, rather than give a semantical analysis for singular terms only indirectly, as Russell did, via a semantical analysis for the grammar of quantification theory with identity, we do so directly (cf. Kaplan 1969).

It is worth remarking that in languages where some singular terms fail to denote, (A9) is inconsistent with $[t = t]$. Since some free logics have included this principle, (A9) will be discussed at greater length in Section 4.

Axiom (A10) complements (A9): If n-ary function signs are to be regarded as potentially explicable in terms of $(n + 1)$-ary predicates in the usual way, then where function signs are given primitive status (as they are here), (A10) must be added if (A9) is. Axiom (A10) is Fregean in motivation. The value of a function was, on his view, the result of completing the function with an argument—where "argument" is understood to apply to objects rather than to substituted linguistic items (terms) (see Frege 1966, pp. 24–5; 1967, pp. 33–4, 84).

The model theory for the logic is straightforward. The domain may be empty. Under each interpretation, all sentences are either true or false. Variables, function signs, and complex singular terms are defined by the interpretation function, if at all, on the domain. Only values identical and within the domain satisfy the identity predicate. The clauses for other atomic predicates are as usual. Completeness is provable (for details, see Burge 1971).

We turn now to a theory of truth in ML for a natural object-language (or, better, a canonical reading of a natural object-language) OL. We assume that ML has resources capable of describing the syntax of OL. Further, we assume a general correspondence between the vocabulary of OL and a subvocabulary of ML. This correspondence may be understood in terms of inclusion or in terms of translation. Details of such a translation relation conceived generally are, of course, difficult to state and well beyond the scope of this paper.

I shall first indicate the postulates of the theory of truth and then explain how to read the indications:

(T1) $\vdash (\exists \alpha)(v)(\exists x)(\alpha(v) = x)))$

(T2) $\vdash (\alpha)(v)(x)(\exists \beta)(\alpha \approx^v_x \beta)$

(T3) For each atomic function sign $\bar{f}^n_j$,
$\vdash (x)(x = \alpha(\bar{f}^n_j(\bar{t}_1, \ldots, \bar{t}_n)) \leftrightarrow x = f^n_j(\alpha(\bar{t}_1), \ldots, \alpha(\bar{t}_n)))$

(T4) $\vdash (x)(x = \alpha(\bar{\iota}(v, \bar{A})) \leftrightarrow x = (\iota y)((\exists y)(\alpha \approx^v_y \gamma \ \& \ \gamma \ \text{satisfies} \ \bar{A})))$

(T5) For each atomic predicate $\bar{A}^n_j$,
$\vdash \alpha \ \text{satisfies} \ \bar{A}^n_j(\bar{t}_1, \ldots, \bar{t}_n) \leftrightarrow A^n_j(\alpha(\bar{t}_1), \ldots, \alpha(\bar{t}_n))$

(T6)   ⊢ α satisfies $nega(\bar{A}) \leftrightarrow -(\alpha$ satisfies $\bar{A})$

(T7)   ⊢ α satisfies $condit(\bar{A}, \bar{B}) \leftrightarrow (\alpha$ satisfies $\bar{A} \to \alpha$ satisfies $\bar{B})$

(T8)   ⊢ α satisfies $unquant(v, \bar{A}) \leftrightarrow (\beta)(x)(\alpha \approx_x^v \beta \to \beta$ satisfies $\bar{A})$

Greek letters 'α', 'β', and 'γ' vary over sequences. 'α(v)' is written for 'the assignment of α to v'; analogously for other uses of 'α' in function-sign position. 'v' ranges over variables of OL; '$\bar{A}$' and '$\bar{B}$', over wffs; and '$\bar{t}_1$', ..., '$\bar{t}_n$', over terms. '$\alpha \approx_x^v \beta$' is read 'β agrees with α in all assignments except that it assigns x to v'. Schematically, we use '$\bar{A}_j^n$' as a name of an OL predicate that translates into ML as 'A_j^n'; analogously for the schematic function-sign name '$\bar{f}_j^n$'. '$\bar{A}_j^n(\bar{t}_1, \ldots, \bar{t}_n)$' is read, 'the result of applying the predicate $\bar{A}_j^n$ to any singular terms $t_1, \ldots, t_n$ in the n-place predicative way'. Functional application is analogous. The operation sign '$nega$' is read 'the negation of'. The readings of the other signs will be obvious. The various styles of variables can be eliminated in favor of a single-sorted, first-order quantification theory.

We omit the usual relativization of the quantifiers to the domain of OL. If we were attempting an *explicit* definition of truth for OL, this omission would lead to inconsistency. But since we are content with a finitely axiomatized recursive characterization of truth, the omission may be tolerated. The dividend is that we may intuitively think of the quantifiers of OL (a fragment of our natural language) as ranging over all that there is.

The material adequacy of the theory is shown by proving that ⊢ $Tr(\bar{A}) \leftrightarrow A$, for all closed wffs $\bar{A}$ of OL, where $Tr(\bar{A}) =_{df} (\alpha) (\alpha$ satisfies $\bar{A})$. (Intuitively, 'Tr' is the truth predicate for OL.)[7] The proof of adequacy is reasonably straightforward and can be largely ignored here (for details, see Burge 1971). What is important for our purpose is the treatment of singular terms.

An aim of the proof is to derive biconditionals like:

(1)   α satisfies $\overline{\text{Is-a-number}}$ $(\overline{\text{the successor of}}$ $(\overline{\text{the successor of}}$ $(\bar{0})))$

           ↔ the successor of (the successor of (0)) is a number

If all singular terms denoted something, the steps would be quite ordinary. We would begin by deriving:

(2)   α$(\overline{\text{the successor of}}$ $(\overline{\text{the successor of}}$ $(\bar{0})))$

           = the successor of (the successor of (0))

(The assignment of every sequence α to 'the successor of the successor of 0' is the successor of the successor of 0.) Then we would obtain (1) by using Leibniz's law and (2) to substitute on the right side of this instance of axiom schema (T5):

(T5a) α satisfies $\overline{\text{Is-a-number } \overline{(\text{the successor of } \overline{(\text{the successor of } (\bar{0}))})}}$
↔ α$\overline{(\text{the successor of } \overline{(\text{the successor of } (\bar{0}))})}$ is a number

The derivation of (2) would utilize an axiom for '0' (taken as a 0-place function sign):

(3)   α($\bar{0}$) = 0

(The assignment of every sequence α to '0' is 0.) And it would utilize an axiom for 'the successor of':

(4)   α$\overline{(\text{the successor of } (\bar{t}))}$ = the successor of (α($\bar{t}$))

(The assignment of every sequence α to the result of applying 'the successor of' to any term $\bar{t}$ is the successor of the assignment of α to $\bar{t}$.) These two axioms together with Leibniz's law would suffice to derive (2) and thence (1).

But since some singular terms do not have a denotation (or a sequence assignment), we cannot follow this route. For the use of axioms like (3) and (4) would undermine the truth of the semantical theory in the metalanguage. Axiom (4) is false because, say, 'the successor of α $\overline{\text{(the Moon)}}$' is improper (since there is no successor of the Moon). Though (3) is probably true, other axioms relevantly like it are not. Thus

(5)   the assignment of every sequence α to 'Pegasus' = Pegasus

is intuitively untrue because the terms on both sides of '=' are improper.

Axioms of the form of (T3) and (T4) circumvent this problem. For example, we have instead of (5):

(T3a) (x)(x = α($\overline{\text{Pegasus}}$) ↔ x = Pegasus)

This axiom is true despite the fact that there are nondenoting singular terms in it. Instead of (4), we have

(T3b) (x)(x = α$\overline{(\text{the successor of } (\bar{t}))}$ ↔ x = the successor of α($\bar{t}$)))

Whereas in the case of (3) and (4) we could rely on Leibniz's law to make the substitutions needed to prove sentences like (1), that law is not strong enough to make recursive transformations using (T3a) and (T3b). This is where (A8) is required. It enables us to substitute *different* nondenoting singular terms (e.g., 'Pegasus' and 'α($\overline{\text{Pegasus}}$)') without relying on false identities like (4) or (5).

3. Criticism of Other Accounts

I make no claims of final acceptability for the account set out in the previous section. But I shall put it to normative use in judging other accounts. Roughly speaking, my view is that the published accounts with relatively strong logical axioms yield falsehoods, and that accounts with relatively weak axioms cannot justify substitution of the relevant nondenoting singular terms in the adequacy proof of a truth theory.[8]

The description theories that, I think, are most interesting from a semantical viewpoint are those of Scott (1967), Lambert (1962), van Fraassen and Lambert (1967) (FD2), and Grandy (1972). Given uncontroversial empirical assumptions, each of these logics implies sentences that are uncontroversially untrue under their intended interpretation. Thus, Lambert uses the axiom $[(x)(x \neq t_1 \ \& \ x \neq t_2) \to t_1 = t_2]$, and Scott invokes $[-(\exists y)(y = t) \to t = *]$, where '*' is a constant denoting an object outside the domain of the object-language. Since it is not the case that either the present King of France or the only unicorn on the moon exists, we derive from each axiom

(6)   The present King of France is identical with the only unicorn on the moon

Grandy, who takes these intuitive difficulties with the Lambert and Scott systems seriously, employs in his truth theory two axiom schemas (numbered 'T3' and 'T4') that have untrue instances.[9] For example, they yield the instances:

(7)   What any sequence α assigns to 'the successor of the Moon' is identical with the successor of what α assigns to 'the Moon'

(8)   What any sequence α assigns to 'der Vater von Pegasus' is identical with the father of what α assigns to 'Pegasus'

(9)   What any sequence α assigns to 'the only unicorn on the Moon' is identical with the unique object assigned to the variable v by some sequence $β$ that satisfies 'is a unicorn on the moon'

Sentence (8) is untrue because there is no father of what every sequence α assigns to 'Pegasus', there is no assignment by α to 'Pegasus', and there is nothing that α assigns to 'der Vater von Pegasus'. Analogously for (7) and (9). To give another, slightly oversimplified, example, we can derive in Grandy's truth theory:

(10)   The present King of France is the denotation of 'The present King of France'

and

(11)   The father of Pegasus is the denotation of some expression

I think (10) and (11) are uncontroversially untrue on intuitive grounds. Intuitively, they do not differ in truthvalue from (6) or

(12)   The present King of France is bald

It should be emphasized that the standard defense of consequences like (6) and (7)–(12) (which I think is doubtful in any case) is clearly inappropriate in the present context. It is not sufficient to say that such sentences are unimportant to most cognitive sciences and that a smoother theory is obtained by counting them true. For from the present viewpoint—that of a semantical theory that takes native intuitions as part of its evidence—sentences like the above are not unimportant. And as ordinarily intended, they are untrue. Nor is it evident that significant differences in smoothness of theory are at issue.

The logical principles that lead to untrue conclusions should not be dismissed without taking account of their purpose. Scott notes as reason for 'identifying' all nondenoting singular terms the resulting ability to derive the following principle of extensionality:

(13) $(x)(A \leftrightarrow B) \to (\iota x)A = (\iota x)B$

Together with Leibniz's law, (13) provides substitutivity for nondenoting singular terms as well as for singular terms that denote. As we have seen in the previous section, the availability of substitutivity for different nondenoting singular terms is critical in the recursion steps of a theory of truth. But the logic of Section 2 yields a principle that together with (A8) seems to justify all the reasonable substitutions that (13) and Leibniz's law justify, without leading to untrue sentences like (6), and without depending on axioms like (7)–(10) for their usefulness in a truth theory. This principle is

(14) $(x)(A \leftrightarrow B) \to (x)(x = (\iota x)A \leftrightarrow x = (\iota x)B)$

Several free logics that have been regarded as having the advantage of producing no untrue consequences are too weak to provide the substitutivity needed in a theory of truth. For example, van Fraassen and Lambert's FD (van Fraassen and Lambert 1967, 1972) and most of the very early proposals do not allow for substituting *any* two (different) nondenoting singular terms. Even slightly stronger logics (e.g., Lambert's (1962) FD1) do not appear capable of combining with true semantical axioms such as (T3) and (T4) to yield the needed substitutions. Of course, it is conceivable that one might find semantical axioms other than (T3) and (T4) that are intuitively true and yet strong enough to combine with relatively weak logics to derive the biconditionals, like (1), that are the touchstones of a truth theory. But the prospects for most of the published logics are, I think, dim.

4. Self-identity and Existence

By translating each occurrence of '*' as '$(\iota x)(x \neq x)$' and each occurrence of $\ulcorner t_1 = t_2 \urcorner$ as $\ulcorner (z)(z = t_1 \leftrightarrow z = t_2) \urcorner$, we can prove that a sentence is a theorem of Scott's system if and only if its translation is a theorem of ours. By translating each occurrence of $At_1 \cdots t_n$ as $\ulcorner (\exists y)(y = t_1) \& \cdots \& (\exists y)(y = t_n) \& At_1 \cdots t_n \urcorner$ (where A is atomic), we can prove that a sentence is a theorem of our system if and only if its translation is a theorem of Scott's. (I ignore function signs since Scott's system does not contain them.)

The latter result serves simply to place in a different perspective the view of natural-language predication that I urged in Section 2. The former result indicates that, from the viewpoint of our system, Scott's system is sound (truth-preserving) if and only if his $\ulcorner t_1 = t_2 \urcorner$ is read not as $\ulcorner t_1$ is identical with $t_2 \urcorner$, but as $\ulcorner$anything is identical with t_1 iff it is identical with $t_2 \urcorner$. On our view, Scott's $\ulcorner t_1 = t_2 \urcorner$ says that t_1 and t_2 do not differ in denotation, not that their denotations are the same. In Section 3, I argued from the assumption that Scott's $\ulcorner t_1 = t_2 \urcorner$ was read $\ulcorner t_1$ is identical with $t_2 \urcorner$ to the conclusion that the system was unsound. What can be said for characterizing identity our way?

In the first place, our representation satisfies the minimum restrictions on any identity predicate—the logical laws of identity, (A3) and (A4). The fact that $\ulcorner x = x \urcorner$ is not valid in our system does not show that the self-reflexive law fails, since formulations of the law as $\ulcorner x = x \urcorner$ in classical systems presuppose that variable x always receives a value. That is, the law is standardly interpreted as (A3).

Analogous remarks apply to the Hilbert–Bernays method of simulating identity within a language (cf. Quine 1960, p. 230). Insofar as the method has been regarded as relevant to understanding identity, it has been seen as a means of expressing indiscernibility of *objects* x and y from the viewpoint of the predicates of a given language. Our identity predicate is coextensive with the Hilbert–Bernays open sentence (for any given language)—it is true of just the same objects. If, however, nondenoting singular terms are attached to the identity predicate, one will get a falsehood, whereas if they are substituted into the Hilbert–Bernays open sentence, one will get a truth. Insisting on equivalence of the closed sentences (in addition to the above-mentioned coextensiveness) would commit one to holding that sentences like (6), (10), or (11) are true in the fragments of English that we have been discussing. I know of no good reason for such insistence. It cannot be justified by the claim that the Hilbert–Bernays open sentence *expresses* or *characterizes* identity (as opposed to merely simulating it by being coextensive with it within the given language). For such a claim is quite unintuitive: indiscernibility via a given stock of predicates just does not seem to give the intended interpretation of identity (cf. Quine 1970, p. 63).

The main ground for characterizing identity our way, of course, is intuitive. It avoids the unattractive results of the other systems and accords with those intuitions that are held generally. The intuitions of some, through by no means all, will be crossed by the fact that (A9) contradicts

(15) $t = t$

in a language containing nondenoting singular terms.[10] It would be easy to dismiss acceptance of (15) as the result of misguided applications of universal instantiation to (A3). But a deeper consideration of the matter is worthwhile.

Testing instances of (A9) and (15) on intuition does not resolve the question of which to take as valid. Whereas native speakers are clear and nearly unanimous in their rejection of sentences like (6), they react differently to sentences like

(16)   The present King of France is identical with the present King of France

Some find them clearly untrue. Others take them to be just as clearly acceptable. Hesitations (on both sides) can be elicited by further discussion. Negative reactions seem to become somewhat more widespread in the face of sentences like

(17)   The only square circle is identical with the only square circle

But on the whole, the evidence from intuition as to whether or not all instances of (15) are true is unclear. The decision must rest on more general considerations.

One reason for doubting the validity of (15) is an application of the Fregean considerations raised in Section 1. We cannot preserve the purported truth of (16) if we substitute for one occurrence of 'the present King of France' other singular terms (e.g., 'the only unicorn on the moon')

that do not differ in denotation. Together with (A8) and (8) we can use (15) to derive (6)—a sentence rejected by native speakers in unison. One may, of course, wish to doubt (A8) rather than (15).[11] But (A8) served an important purpose in justifying transformations needed for proving the material adequacy of a theory of truth. It seems fair to ask of anyone who rejects the principle to provide a replacement that effects the relevant transformations without leading to untrue consequences or relying on untrue semantical axioms. Since (A8) is nonindependent in the logic of Section 2 largely because of (A9), we may regard the latter as tentatively preferable to (15).

There is a further consideration against (15)—one that is vaguer and less compelling but nonetheless philosophically interesting. An extremely intuitive feature of Tarski's theory of truth is that it explicates what it is for a sentence to be true in terms of a relation (satisfaction) between language (open sentences) and the world (sequences of objects). The notion of correspondence that had always seemed so integral to truth came clean in Tarski's theory. (This point is forcefully made by Davidson (1969).) It is difficult to see how the purported truth of, say, (16) can be explicated in terms of a correspondence relation.

As mentioned earlier, some may want to give correspondence a toe-hold by assigning 'Pegasus' an unactualized possible. But quite apart from questions about the propriety, clarity, and credibility of the move, it does not apply to (17). Assigning 'Pegasus' itself is not very satisfying either, because it encounters difficulty in explicating '$-(\exists y)(y = $ Pegasus$)$'. The fault is the one we found with Frege's account of singular terms: there is too much correspondence rather than too little.

Loosely speaking, self-identity is a property of objects and all objects have it; sentences expressing identities are true or false by virtue of the relation that the identity predicate and its flanking singular terms bear to the world—never merely by virtue of the identity of the singular terms. Philosophical questions regarding identity seem bound to the notions of existence and object. The point is summed up more austerely by the principle

(18) $(\exists y)(y = t) \leftrightarrow t = t$

which is easily derived in the logic of *ML*.

In their focus on the intended interpretation of the symbols we employ, our truth theory and its underlying logic help clarify how with respect to singular terms we can use the language we use and in the same language believe in the world we believe in. Alternative combinations of theory and logic should be required to do at least as much.

References

Burge, Tyler, "Truth and some referential devices," Dissertation, Princeton University (1971).
Burge, Tyler, "Reference and proper names," *Journal of Philosophy*, vol. 70 (1973) pp. 425–39.
Davidson, Donald, "True to the facts," *Journal of Philosophy*, vol. 66 (1969) pp. 748–64.
Donnellan, Keith S., "Reference and definite descriptions," *Philosophical Review* (1965) pp. 281–304.
Frege, Gottlob, "Function and concept," in *Translations from the Philosophical Writings of Gottlob Frege* (ed. Geach, P. and Black, M.), Blackwell, Oxford (1966); first published 1892.
Frege, Gottlob, "On sense and reference," in *Translations from the Philosophical Writings of Gottlob Frege* (ed. Geach, P. and Black, M.), Blackwell, Oxford (1966); first published 1892.
Frege, Gottlob, *The Basic Laws of Arithmetic* (ed. Furth, M.), University of California Press, Berkeley (1967); first published 1903.
Grandy, Richard, "A definition of truth for theories with intensional definite description operators," *Journal of Philosophical Logic*, vol. 1 (1972) pp. 137–55.
Grice, H. P., "Vacuous names," in *Words and Objections* (ed. Davidson, D. and Hintikka, J.), Reidel, Dordrecht (1969).
Hailperin, T. and Leblanc, H., "Nondesignating singular terms," *Philosophical Review* (1959) pp. 239–43.
Hintikka, Jaakko, "Definite descriptions and self-identity," *Philosophical Studies* (1964) pp. 5–7.
Kaplan, David, "What is Russell's theory of descriptions?" in *Physics, Logic and History* (ed. Yourgrau, Wolfgang), Plenum, New York (1969).
Lambert, Karel, "Notes on E! III: A theory of descriptions," *Philosophical Studies* (1962) pp. 51–9.
Lambert, Karel, "Notes on E! IV: A reduction in free quantification theory with identity and descriptions," *Philosophical Studies* (1964) pp. 85–8.
Meyer, Robert and Lambert, Karel, "Universally free logic and standard quantification theory," *J. of Symbolic Logic* (1968) pp. 8–26.
Quine, W. V., *Word and Object*, MIT Press, Cambridge, Mass. (1960).
Quine, W. V., *Philosophy of Logic*, Prentice Hall, Englewood Cliffs (1970)
Russell, Bertrand, "On denoting," *Mind* (1905) pp. 479–93.
Schock, Rolf, *Logics Without Existence Assumptions*, Almqvist and Wiksell, Stockholm (1968).
Scott, Dana, "Existence and description in formal logic," in *Bertrand Russell: Philosopher of the Century* (ed. Schoenman, Ralph), Allen and Unwin, London (1967).
Smiley, Timothy, "Sense without denotation," *Analysis* (1960) pp. 125–35.
Tarski, Alfred, "The concept of truth in formalized languages," in *Logic, Semantics, Metamathematics*, Clarendon Press, Oxford (1956); first published 1936.
van Fraassen, Bas C. and Lambert, Karel, "On free description theory," *Zeitschrift für mathematische Logik und Grundlagen der Mathematik*, vol. 13 (1967) pp. 225–40.

van Fraassen, Bas C. and Lambert, Karel, *Derivation and Counterexample*, Dickenson Publishing, Encino (1972).
Whitehead, A. N. and Russell, Bertrand, *Principia Mathematica*, vol. I, Cambridge University Press (1925–1927); first published 1910.

Notes

1. I am grateful to Alonzo Church, Donald Kalish, Dana Scott, and especially Richard Grandy for criticisms and suggestions regarding earlier drafts.
2. Some of what Meyer and Lambert (1968) say is congenial with these remarks. However, their distinction between "nominal truth" and "real truth" appears superfluous. "Nominal truth" may be assimilated to "real truth" as applied to sentences involving oblique contexts. One advantage of making explicit the oblique contexts in sentences like the above is that by doing so one uncovers grounds for confirming or disconfirming otherwise puzzling sentences. For example, to confirm or disconfirm fictional sentences, we look at the relevant fiction. Whereas some authors have felt that a sentence like 'Pegasus had fewer than 7 million hairs' should be counted truthvalueless because they could find no plausible reason to count it true or false, our view counts it false (even taken as implicitly oblique) because it is easily disconfirmed.
3. Application of this idea is discussed briefly in Burge (1973). The axioms governing the proper name 'Pegasus' that we discuss later ignore for the sake of brevity the considerations of that paper.
4. Donnellan (1965, esp. pp. 295–304) may seem to be in disagreement with this intuition. But I think that the disagreement is only apparent. It should be noted that the bivalence of ML and the account of negation in the object-language OL (cf. later) are incompatible with some treatments of presupposition in terms of truthvalue gaps. I think that the intuitions backing these accounts can be explicated in other ways. But this issue may be left aside here.
5. Provisions for scope distinctions will be important in a full account of the logical behavior of singular terms in natural languages, especially in treating certain ambiguities that occur with nondenoting terms and in dealing with singular terms in and out of intensional contexts. Such provisions can be added to the present system, but the philosophical issues are complex and will not be discussed here. For a detailed discussion of the problem and an attempt to solve it from a different standpoint from ours, see Grice (1969).
6. It is tempting but mistaken to suppose that the condition prohibits taking both a predicate and another predicate understood as its "contradictory" or "negation" as primitive. In such cases, the condition may be seen as forcing us merely to construe the singular term as having wider scope than the "negative element" attributed to the predicate. As long as the predicate is atomic, it is hard to imagine the situation any other way.
7. Cf. Tarski (1956). In our formulation, relativizations of 'Tr' to a canonical reading, a person, and a time are suppressed for the sake of brevity.
8. An exception is Schock (1968, p. 94). Subsequent to arriving at the theory of Section 2, I found that he uses axioms very like (A9) and (A10). Schock gives

a Frege-type model theory for his logic using the empty set as the denotation of intuitively nondenoting singular terms.
9. It would be a mistake to think that Grandy "identifies" only logically equivalent singular terms. In order to prove the adequacy of his truth theory for the iota case, he must derive an identity sentence containing two nondenoting definite descriptions.
10. Smiley (1960) and Hintikka (1964) are in accord with us on this matter. So, with qualifications, is Russell (Whitehead and Russell 1910, p. 184).
11. (A8) is derivable in both Scott's system and Lambert's FD2. It is not derivable in Grandy's system or in the weaker ones.

10
Presupposition, Implication, and Self-reference*

Bas C. van Fraassen

The two aims of this paper are, first, to explicate the semantic relation of presupposition among sentences, and, second, to employ the distinctions made in this explication in a discussion of certain paradoxes of self-reference. Section 1 will explore informally the distinction between presupposition and implication. In Section 2 we construct a formal paradigm in which this distinction corresponds to a substantial difference. Section 3 explores the relations between presupposition and truth, and, in Section 4, the Liar and some similar paradoxes are finally broached.

1

The best known source for the concept of presupposition is the view (a.o. Strawson's) that a property cannot be either truly or falsely attributed to what does not exist. Thus, the sentence 'The King of France (in 1967) is bald' is neither true nor false, on this view, *because* the King of France does not exist.[1] The explicit characterization of *presupposes* is therefore

---

* An earlier version of this paper was read at Duke University on May 12, 1967, and Sections 1 and 2 were included in a paper presented at a symposium on free logic held at Michigan State University on June 9 and 10, 1967. Acknowledgments and bibliographical references have been collected in a note at the end.
1. The question whether a sentence is true may be said to make sense only relative to an interpretation, or to what is intended by it, and so on. Except when dealing with artificial languages, we shall avoid this issue by assuming that there is a "correct" interpretation known to the reader.

given by

(1)   *A* presupposes *B* if and only if *A* is neither true nor false unless *B* is true

This is equivalent to

(2)   *A* presupposes *B* if and only if:
    (a) if *A* is true then *B* is true
    (b) if *A* is false then *B* is true

From this equivalence it is clear that presupposition is a trivial semantic relation if we hold to the principle of bivalence (that every sentence is, in any possible situation, either true or false). In that case, every sentence presupposes all and only the universally valid sentences.

Why can we not say that, on Strawson's position, 'The King of France is bald' implies 'The King of France exists'? The reason is that presupposition differs in two main ways from implication, on any accepted account of the latter relation. For implication, modus ponens is accepted as valid; from (2a) it is clear that an analog of modus ponens holds also for presupposition. But modus tollens is also generally accepted as valid for implication; what of its analog for presupposition? This question cannot be answered unless we first settle on the meaning of negation. We shall understand the negation of a sentence *A* to be true (respectively, false) if and only if *A* is false (respectively, true). This is not the only possible convention, but it has the virtue of yielding the convenient further characterization of presupposition:

(3)   *A* presupposes *B* if and only if:
    (a) if *A* is true then *B* is true
    (b) if (not-*A*) is true then *B* is true

From (3) it is clear that the analog to modus tollens

(4)   *A* presupposes *B*
      (not-*B*)
      Therefore, (not-*A*)

is not valid; if the premisses are true, the conclusion is not true (and not false). Secondly, (3b) shows that the argument

(5)   *A* presupposes *B*
      (not-*A*)
      Therefore, *B*

is valid: if its premisses are true, so is its conclusion. But of course the analog of (5) for implication does not hold.

Thus, presupposition and implication are not the same, but they have something in common. What they have in common is that, if A either presupposes or implies B, the argument from A to B is valid. This is itself a semantic relation, which we shall here call "necessitation":

(6) A necessitates B if and only if, whenever A is true, B is also true[2]

We see here an obvious possible cause of confusion between implication and presupposition. The standard logic text generally begins with an explanation that an argument is valid if and only if the conjunction of the premises implies the conclusion, if and only if the corresponding conditional is logically true. This explanation then justifies the text's concern with validating arguments (respectively, with proving logical truths) alone. But the standard logic text is concerned only with a case for which the principle of bivalence holds, and we must resist the temptation to extrapolate its teachings to other contexts.

From (3) and (6) we obtain finally the equivalence

(7) A presupposes B if and only if:
    (a) A necessitates B
    (b) (not-A) necessitates B

Thus we have an explication of presupposition in terms of standard semantic notions.

Before turning to the task of making our account precise, we may briefly consider two questions. The first notes that we use 'if . . . then' in, for example, (3) and (6); how is this conditional to be understood? For the sake of clarity, I propose that this be understood as the material conditional. This yields at once the limiting case that, if B is universally valid, every sentence presupposes it. The second question notes that every case of implication or presupposition is a case of necessitation; does the converse hold? This question is not necessarily decidable at this point, because we have not given a complete account of implication (nor presume to be able to do so). But in Section 3 we present a case of necessitation that is not one of presupposition and that, we shall argue, is also not a case of implication.

---

2. This relationship is usually called "semantic entailment," but in the present context that terminology might be confusing. Note that the 'whenever' does not refer to times, but to possible situations.

2

There are two reasons for constructing a formal paradigm at this point. The first is to show the possibility of having distinct semantic relations of implication, necessitation, and presupposition. That is, we wish to show that the distinctions we have drawn are not merely verbal distinctions, but are capable of expressing nonequivalent concepts. The second reason is an obvious one: the notions of implication and presupposition play an important role in certain philosophical arguments. A case in point is the derivation of paradoxes such as the Liar, which we shall examine below. A formal paradigm will give us a means of testing such reasoning.

This may suggest that we mean to construct an axiom system. This is not so. Instead, we intend to serve the two purposes indicated by outlining the construction of a certain kind of artificial language. This kind of language is meant to provide a model for the kind of discourse in which presuppositions are important. Because of our explication of presupposition in terms of necessitation, we shall be able to concentrate on the relations of implication and necessitation alone. Our construction will have to be such that: a case of necessitation need not be a case of implication or of presupposition; a case of presupposition need not be a case of implication; and sentences of which some presupposition is not true are neither true nor false.

An artificial language has two parts: a syntax and a semantics. The syntax comprises a vocabulary and a grammar, which together generate the set of its sentences. The semantics defines what I shall call the admissible valuations; that is, it delineates the possible ways in which these sentences could be true or false together. (In general, the class of admissible valuations is defined in a rather roundabout way, through a (partial) interpretation of the vocabulary.)

The languages whose construction we shall now outline, we call presuppositional languages. By this I mean that in their semantics we make it possible for presuppositions to be made explicit, and we countenance the possibility that, for some of the sentences, presuppositions may fail.

We use 'L' to refer to an arbitrary language of this kind and describe those features that make it a presuppositional language. First, we require that the vocabulary include a negation sign and a disjunction sign (other propositional connectives may be defined in terms of these). I may as well say at once that I will not be satisfied if negation and disjunction do not obey the laws of classical logic. With respect to logic I am conservative: I would resist any imperialism on behalf of classical logic; I would not accept the idea that it is applicable to all contexts or that it is sufficient for all (important) purposes, but on the other hand I have no inclination to change it.

Second, the vocabulary may contain a special sign for implication. The material conditional $(A \supset B)$ may be defined as $(\sim A \vee B)$. But some other sign may be specified as the implication sign. Here we shall require that the semantic assertion that A implies B be so construed that, at least, it is true only if $(A \supset B)$ is logically true. That is, logical truth of the material conditional is a provisionally acceptable explication of implication, and so is any account under which implication entails this logical truth.

In the semantics, we need two preliminary notions. First, the semantics must specify a relation N of necessitation ("nonclassical necessitation") to cover those cases that cannot be cases of implication. How this relation N is specified in the semantics need not concern us here.[3]

The second notion is that of a classical valuation. This is an assignment of truth (T) and falsity (F) to sentences, which disregards the possibility that a sentence might suffer from a failure of presupposition. Given our remarks about *or* and *not*, we know three things about classical valuations:

(8) If v is a classical valuation for the language L and A, B are sentences of L, then
 (a) $v(A) = F$ or $v(A) = T$
 (b) $v(\sim A) = F$ if and only if $v(A) = T$
 (c) $v(A \vee B) = F$ if and only if $v(A) = v(B) = F$

Classical propositional logic is clearly sound with respect to classical valuations, since classical valuations correspond to rows in the ordinary truth tables. But, of course, there may be sentences of L that are assigned T by all classical valuations (classically valid sentences of L) which are not theorems of the propositional calculus. (From here on we shall say that a valuation satisfies A if it assigns T to A, and that it satisfies a set X if it satisfies every member of X.)

In general, no classical valuation will be entirely correct with respect to an actual situation. With respect to an actual situation, the sentences of L are divided into three classes: those that are true, those that are false, and those that are neither. Since a sentence is true if and only if its negation is false, it follows that which sentences are true determines which sentences are false, and hence determines also which sentences are neither true nor

---

3. In this paper, N will be taken to be a relation of sentences to sentences; the general case of such a relation from sets of sentences to sentences will not concern us here.

false. In other words, insofar as the situation can be described in L, it is determined uniquely by the set of sentences of L that are true with respect to it.[4]

Suppose that G is the set of sentences true in a given situation. What do we know about G? First, no sentence in G has a presupposition that fails. Therefore, the sentences of G can all be true together from the point of view of the classical valuations, and what follows from them from this point of view really does follow. (There are sentences about which the classical valuations are radically wrong, since they cannot accommodate a lack of truthvalue; but such sentences do not belong to G.) On the other hand, the classical valuations disregard N; hence there are some consequences of true sentences that they overlook. If A belongs to G, and $N(A, B)$, then B is also true; hence also belongs to G. We may sum this up as follows:

(9)   (a) There is a classical valuation that satisfies G
        (b) If every classical valuation that satisfies G also satisfies B, then B is in G
        (c) if $N(A, B)$ and A belongs to G, then so does B

A set of sentences G for which (9a) holds we call classically satisfiable, and if (9b) and (9c) hold for it, we call G a (necessitation-)saturated set. So our conclusion is that the set of sentences true in an actual situation is a classically satisfiable, saturated set.

It is at this point that we can discuss the admissible valuations. An admissible valuation is to correspond to a possible situation, such that it assigns T to the sentences true in that situation, F to the sentences false in that situation, and does not assign a truthvalue to those sentences that are neither true nor false in that situation. On the basis of our present discussion, we can therefore say the following about admissible valuations:

(10)   An assignment s is an admissible valuation for L only if there is a classically satisfiable, saturated set G such that:
        (a) if A is in G, then $s(A) = T$
        (b) if the negation of A is in G, then $s(A) = F$
        (c) otherwise, $s(A)$ is not defined

We may note that $(\sim \sim A)$ is in G if and only if A is in G; so a sentence has (or lacks) a truthvalue if and only if its negation does. Also, since G is saturated, if A necessitates B and B is not in G, neither is A. From this

---

4. This can be made precise by using the notion of model, but the intuitive notions suffice for our present purpose.

it follows by our characterization of presupposition that if A has a presupposition that is not true, then A is neither true nor false.

There is another important point about admissible valuations, which establishes that whatever logic is sound with respect to the classical valuations is also sound with respect to the admissible valuations. This is the point that an admissible valuation represents what is common to a certain set of classical valuations. The assignment s characterized by (10) is identical with the supervaluation induced by the set G in question—which notion is defined by:

(11)   The supervaluation induced by G is the function that

(a) assigns T to A if all classical valuations that satisfy G assign T to A
(b) assigns F to A if all classical valuations that satisfy G assign F to A and
(c) is not defined for A otherwise

And clearly what is common to classical valuations cannot transgress laws that hold for all classical valuations (in particular, the laws of classical propositional logic).

Specifically, the law of excluded middle continues to hold:

(12)   Any sentence of the form $(A \vee \sim A)$ is valid (assigned T by all admissible valuations)

But the law of bivalence does not hold: some sentences are neither true nor false.

Reasoning concerning presupposition, implication, or necessitation can be tested through our formal paradigm. With a view to this critical function of the formal paradigm, we shall point out certain features of the notion of presupposition that can easily be demonstrated with its help.

First, the reader will have no difficulty in verifying that there can be distinct cases of necessitation, implication, and presupposition, so that these are three distinct semantic relations. Second, we may note that if A presupposes A, then A is never false (it may be true, or neither true nor false). Similarly, if A presupposes $(\sim A)$, then A is never true. If both are the case, if A presupposes a contradiction, then A is always neither true nor false. And if A is presupposed by a valid sentence, then A is valid. (It is clear that the relation N might be such that the language had no admissible valuation: a pathological case that could model only discourse with inconsistent presuppositions.)

3

The first subject to which we shall apply the distinctions drawn above is truth. This is not caprice; the points here made will then play a role in our discussion of the paradoxes of self-reference. The main question before us is: what is the relation between the sentence 'It is true that P' and P itself? (We symbolize the former as $T(P)$.) The answer purports to be given by Tarski's principle:

(13) $T(P)$ if and only if P

But what relationship is this "if and only if" meant to indicate? One obvious answer is that it signifies coimplication. But this is not necessarily so; for example, one might say 'If the King of France is bald, then he exists' to signify that the antecedent necessitates the consequent. (This would be a confusing use of "if...then," to be sure, and would most likely indicate a failure to distinguish between implication and necessitation.)

Do P and $T(P)$ imply each other? Principle (13) is used in the "derivation" of the semantic paradoxes, and has also been used to "derive" the law of bivalence. I wish to consider this latter "derivation" here, and shall present it in its shortest form. Given that (13) must be understood to assert a coimplication, we have in particular

(14)   (a) P implies $T(P)$
        (b) $\sim P$ implies $T(\sim P)$

Suppose now that $T(P)$ is not the case; that its denial is the case.[5] Then, by (a) and modus tollens, the denial of P is the case. But this conclusion and (b) lead by modus ponens to: $T(\sim P)$ is the case. So if $T(P)$ is not the case then $T(\sim P)$ is the case. Since our metalinguistic "if...then" is the material conditional, this means that either P is true or $\sim P$ is true: that is, either P is true or P is false. But if the language is a presuppositional language, this conclusion does not hold.

The first, and obvious, reaction to this argument is that the distinction between use and mention has not been observed. Let us try to observe it. Let L be a presuppositional language, and let us form its metalanguage M as follows:

(15)   (a) Sentences of L are first-level sentences of M[6]
        (b) If P is a sentence of L, then 'P' is a name in M

---

5. This assumption of bivalence for $T(P)$ will be discussed shortly.
6. Strictly speaking, the first level of M should be isomorphic to, rather than identical with L; but this makes no essential difference here.

(c) '*T*' is a predicate of M that is applicable only to names formed by applying quotation marks to sentences of L
(d) If *A* and *B* are sentences of M, so are $\sim A$ and $(A \vee B)$
(e) A sentence that has any well-formed part beginning with '*T*' is a second-level sentence of M

where '$\vee$' and '$\sim$' are also the disjunction and negation signs of L. We continue to let '*T(P)*' stand for '*T*' followed by *P* in quotation marks. We shall use '*P*' and '*Q*' to stand for first-level sentences only.

We do not at this point have the full semantics for M, but at least the semantics of the first level must coincide with the semantics of L. But our argument above is easily restated making use of M (and taking as premiss only the minimal assertion that *P* materially implies *T(P)*). But, however we complete the semantics of M, the argument cannot establish that the principle of bivalence holds for first-level sentences, because that is not so.

We have three possibilities before us. We can reject the premisses, or we can reject the rules whereby the conclusion is derived, or we can accept the conclusion as formulated by the second-level sentence

(16) $T(P) \vee T(\sim P)$

of M, but interpret this sentence differently. Let us consider this last alternative first. We remember that a sentence of the form $P \vee \sim P$, which we had been inclined to interpret as saying that *P* is either true or false, need not be so understood. We could similarly use supervaluations or a many-valued matrix to reinterpret second-level disjunction in such a way that (16) does not say that *P* is either true or false. The most obvious way to do this is to say that when *P* is neither true nor false, then *T(P)* does not have a truthvalue either—as opposed to: then *T(P)* is false. But we shall then have no way of formulating the assertion that a sentence is not true. Nor could we add a third level to *M* in which to formulate this assertion without running into the same problem. That is, assuming that the transition to each higher level obeys the same principle, we would simply get: if *P* is neither true nor false, then neither is *T(P)*, and neither is *T(T(P))*, and so on. Thus we have here a solution, but it means that M is in some important ways not a model of the metalanguage we have actually been using.[7]

The second possibility, of rejecting the validity of the argument, is not a very pleasant one either. The core of the argument, which can be

---

7. This also answers the question of why we assumed bivalence in the formulation of the argument (see fn. 5).

formulated in M, is

(17) $P \supset T(P)$
$\sim P \supset T(\sim P)$
$\sim T(P) \supset \sim P$ contraposition
$\sim T(P) \supset T(\sim P)$ transitivity
$\sim \sim T(P) \vee T(\sim P)$ def. material implication
$T(P) \vee T(\sim P)$ double negation

and is entirely validated by classical propositional logic. Again, I would not be satisfied if this were rejected; such a rejection might throw radical doubt on our own metalogical reasoning. But, instead of rejecting (17), we might reject the original argument on the basis that from 'A implies B' we ought not to conclude that A materially implies B, and that the moves in (17) do not apply to implication proper. I would not care to prevent anyone from introducing a new arrow (this has been a popular and, I would claim, harmless pastime), but I do not really think it would be to the point here.

Rather than turn to such radical departures from the standard logical framework, we may consider the possibility that the argument is not sound. As we originally formulated it, the argument proceeded mainly by modus tollens and modus ponens. Only the latter is valid for necessitation in general. Therefore we will have circumvented the problem very simply by interpreting Tarski's principle not as a coimplication but as a conecessitation.

We can now also complete the semantics of M in a very obvious way:

(18) (a) If the first-level sentence P is true, then $T(P)$ is true, and otherwise $T(P)$ is false
(b) If the second-level sentence A is false, then $\sim A$ is true, and if the second-level sentence A is true, then $\sim A$ is false
(c) If $(A \vee B)$ is a second-level sentence, then if either A is true or B is true, then $(A \vee B)$ is true, and otherwise $(A \vee B)$ is false

Clearly, if A is neither true nor false, then it is a first-level sentence, and so is $\sim A$; if B lacks a truthvalue as well, then $(A \vee B)$ is a first-level sentence. The principle of bivalence holds for second-level sentences. Classical propositional logic remains sound for the language M.[8]

That this semantics satisfies the principles we have adopted is seen as

---

8. This is easily seen by embedding the valuations into a three-valued matrix in which the negation of the middle value gets T and the disjunction of two middle values gets F.

follows: when P is true, so is $T(P)$, and conversely; hence the two necessitate each other. When P is neither true nor false, then $T(P)$ and $T(\sim P)$ are both false, and $T(P) \vee T(\sim P)$ is also false. When P is neither true nor false, then so is $\sim P$, and $T(P)$ is false; hence $(\sim P \vee T(P))$ is false: P does not materially imply $T(P)$. There is, finally, an interesting case of presupposition in M: P does not presuppose $T(P)$, since $\sim P$ does not necessitate $T(P)$. But both P and $\sim P$ necessitate $T(P) \vee T(\sim P)$. Therefore P presupposes its own bivalence; we may call this the ultimate presupposition. Moreover, this is rather a stable result, for to show it we need not appeal to the bivalence of second-level sentences nor to the truth definition for disjunction beyond the unproblematic feature that, when one of the disjuncts is true, so is the disjunction. Not all the results of this section have such stability, as we shall see when we leave the relative security of M.

4

So far we have assumed that there is a neat division between sentences about sentences and the sentences they are about. I do not mean simply that we have observed the distinction between use and mention. I mean that, in addition, this distinction corresponds to a division of the class of sentences with which we have so far been concerned. That is not the same thing: there is certainly a distinction between loving someone and hating someone; yet it is possible to hate someone you love. So we have not only observed the distinction between use and mention, but assumed that no expression is mentioned in the course of its use.

But this assumption is not a necessary one, and this brings us to the subject of self-reference. The famous *Liar* paradox is the obvious point of departure. The Liar says "What I now say is false." Clearly this is an English sentence, perfectly grammatical. Yet it can be construed as mentioning itself: the Liar could equally well have said, "This sentence is false" or "The sentence which I now utter is false." The use of this sentence involves mentioning it; hence the distinction between using sentences and mentioning sentences, while a perfectly good distinction, does not correspond to a neat division among sentences.

The paradoxical element appears when we ask whether what the Liar said was true or false. (In the following argument, we assume principle (13) only in the weakened sense of 'P and $T(P)$ necessitate each other'.) If what he said was true, then what he said was the case; but what he said was that what he said was false. So if what he said was true, then it was false. Similarly we can demonstrate that if what he said was false, then what he said was true. In other words, both the supposition that it was true and the supposition that it was false lead to absurdity.

This conclusion is itself absurd only on the assumption that what he said must be either true or false. But we are now quite used to the failure of bivalence; so we simply say: what he said was neither true nor false. The air of paradox is spurious.

Before we begin to feel too smug about this, however, we must face a second paradox, which I shall call the *Strengthened Liar* and which was designed especially for those enlightened philosophers who are not taken in by bivalence. The Strengthened Liar says "What I say is either false or neither true nor false."

If we now ask whether the sentence is true or false or neither, we find that each of these answers is absurd. For example, suppose that what he says is neither true nor false. Then clearly it is either false or neither true nor false. But then what he said was the case. So what he said was true. And now we seem to be properly caught, our sophistication with respect to bivalence notwithstanding.

One move that one might consider, though unhappily, is to say that there is a fourth possibility. This has first of all the unwelcome consequence of facing us with a Strengthened Strengthened Liar (and so on ad infinitum). Second, it is our desire to conform to classical logic. The principle that has for us replaced bivalence is:

(19) $T(P) \vee T(\sim P) \cdot \vee [\sim T(P) \& \sim T(\sim P)]$

which is a second-level sentence and is valid no matter what P is, for it is a tautology, a theorem of propositional logic. We can deny $T(P) \vee T(\sim P)$, bivalence, for that is not a tautology, but we cannot deny (19).

However, I have not led you through the labyrinthine distinctions among implication, necessitation, and presupposition merely for its own sake. But before I show how these distinctions may be mobilized to help us see our way through the *Strengthened Liar* paradox, I must say something about English sentences and our symbols. We use $\sim P$ to stand for the denial of P, and so when P is an English sentence, say "Tom is tall," we generally read $\sim P$ as "Tom is not tall" or "It is not the case that Tom is tall." But if X is, for example, the *Liar* sentence "What I now say is false," we see that its denial is not expressed by "What I now say is not false." If the Liar were to utter both, he would have to utter them in succession, so the 'now' would refer to two different times. Hence he would not deny his first statement by making the second, the second referring only to itself. Yet X does have a denial—the Liar's audience, or he himself, may respond 'That is not so.' Similarly, in deriving the absurdity we argue, for example, that if what he says is not the case, then what he says is false—using 'what he says is not the case' to express the denial of what he says. The exact English words which may express this denial will depend on who denies it when. This is a clear sign that our symbolism is not nearly adequate to give a complete picture of self-referential language.

But what we can do is to isolate the features of the Liar sentence that play a role in the paradox. In doing so, we may use X to stand for what the Liar says, and $\sim X$ for its denial, however (when, by whom) the denial may be expressed. And then we see that we can take the essential features of X that are appealed to in the derivation of absurdity to be its relations to the sentence $T(\sim X)$ that expresses its falsity. In all other aspects, we shall assume our formal paradigm to govern inferences involving X, for this reduces the problem to the familiar case. And we shall have succeeded if we can then show how the derivation of absurdity is blocked.

First, we might take X to coimply $T(\sim X)$ and also $T(X)$, and $T(\sim X)$ to coimply $\sim X$. This is the first way in which anyone is likely to take the problem, and it leads to absurdity. This shows that this way of taking it is incorrect. (In any case, we have already seen that some of these implications do not hold.) Let us now be cautious and take these relations all to be cases of conecessitation. Then we find that bivalence is needed to derive absurdity; and this dissolves the paradox in the way we indicated informally before.

We could also put this as follows: X necessitates $T(\sim X)$, but so does $\sim X$. Hence, by our definition, X presupposes $T(\sim X)$. We have seen that $T(\sim X)$ cannot be true; therefore, X has a presupposition that is not true. This is why X is neither true nor false. In this way, the distinction between necessitation and presupposition leads to a solution of the paradox.

Thus the *Liar* paradox seems to fit very nicely into our conceptual scheme. But the fact is that is has one feature that does not fit at all. For X is itself an assertion of falsity: the Liar says "What I now say is false." And in the preceding section, when we constructed the metalanguage M, we followed the principle that, although bivalence does not hold generally, it does hold for assertions of truth and falsity. We are here taking the *Liar* sentence to be a first-level sentence, but there are no restrictions on the form of first-level sentences. In particular, this sentence is an assertion of falsity, and it does not have a truthvalue.

This I shall call the basic lesson of the *Liar* paradox: even assertions of truth or falsity do not in general satisfy the law of bivalence.

Turning now to the *Strengthened Liar* paradox, we find a sentence Y that conecessitates its own falsity-or-truthvalueness. That is, Y and $(T(\sim Y) \vee \cdot \sim T(Y) \,\&\, \sim T(\sim Y))$ necessitate each other. Each of the three possible suppositions—that Y is true, that Y is false, that Y has no truthvalue—necessitates a contradiction. We have here three valid arguments with a common conclusion; and this conclusion is a self-contradiction. This conclusion is demonstrated if one of these three valid arguments must have true premisses. This is tantamount to:

(20) $T(Y)$ is true, or $T(\sim Y)$ is true, or $\sim T(Y) \,\&\, \sim T(\sim Y)$ is true

This looks like the tautology (19), but it is not the same unless we persist in the opinion that assertions of truth are themselves always true or false. And this was the basic lesson of the ordinary *Liar* paradox: that opinion is mistaken.

To put it most perspicuously, (20) is related to (19) as bivalence to excluded middle. We cannot conform to logic and also deny (19), or excluded middle. But we can deny bivalence, and we can also deny even that sentences that begin with 'T' are bivalent. The *Strengthened Liar* paradox is averted if we hold that $T(Y)$ and $T(\sim Y)$ are themselves neither true nor false. From this it follows immediately that the sentence $\sim T(Y) \,\&\, \sim T(\sim Y)$ also is neither true nor false.

And we can also give a good reason for holding this. As for every sentence, $T(Y)$ necessitates Y. But, by the tautology (19), $\sim T(Y)$ necessitates $T(\sim Y) \vee \cdot \sim T(Y) \,\&\, \sim T(\sim Y)$. The latter in turn necessitates Y. Hence $T(Y)$ and $\sim T(Y)$ both necessitate Y; in our terminology, $T(Y)$ presupposes Y. As we have seen, Y cannot be true. This is why $T(Y)$ is neither true nor false. (Similarly for $T(\sim Y)$.)

At this point it may be instructive to see how we would extend our formal paradigm to accommodate the deviant sentences X and Y. We must be careful again not to extend it in such a way that classical logic is violated, for then our own reasoning might be drawn in doubt. First we place X and Y among the first-level sentences. They bear no unusual relations to other first-level sentences; hence we need not extend the relation N. But we must now add to the semantics of M a relation N* of nonclassical necessitation, and say that X and $T(\sim X)$ bear N* to each other. Similarly Y and $T(\sim Y) \vee \cdot \sim T(Y) \,\&\, \sim T(\sim Y)$ bear N* to each other. Let us call the thus-extended language "M*." The admissible valuations of M become the classical valuations of M*. We define "saturated set of sentences of M*" as before, except that we use N* instead of N. The admissible valuations for M* are supervaluations generated by saturated set of sentences, as before. Only the sentences belonging to the entourage of X and Y will be affected by this; the others keep their normal truthvalue (if any). That classical logic continues to hold can be demonstrated as before.

The notion of "level" has been made much less sharp; the relation N* imposes translevel semantic relations among the sentences. But there is still a clear and distinct syntactic notion of level. We can also add a third level, in which we can, for example, express the fact that $T(Y)$ is neither true nor false. We would do this by extending M* in just the way that we extended L to produce M. But let us not deceive ourselves: we shall not get to a point where we can say everything relevant and yet not have any presuppositions that could fail. To be presuppositionless may be a regulative ideal in philosophy, but it is not an achievable end.[9]

In conclusion, I should like to describe briefly a further kind of paradox, and attempt to apply my analysis to it. Epimenides the Cretan is reported to have said that all statements by Cretans are false. Clearly, what he said cannot be true. For what he said was said by a Cretan, and hence he has implicitly asserted its falsity. But we can consistently hold that what he said is false. This just means that something said by some Cretan is not false. And this is not as implausible as Epimenides seems to have thought. But, as Church has pointed out, whether or not some statement by a Cretan is not false, is a contingent matter. In particular it entails that the statement of Epimenides that I have just described is not the only one ever made by any Cretan. So the world could have played a neat trick on us: this could have been Epimenides' first and last statement, and all other Cretans could have been entirely dumb. In that case, neither could we have held that what he said was false.

This paradox we shall call the *Weakened Liar*. Epimenides said in effect that all other statements by Cretans are false, and also that his own (this very) statement is false. Let his sentence be Z and let the sentence that expresses that all other statements by Cretans are false be Q. Then Church's point is that the denial of Z necessitates $\sim Q$.[10] But Z necessitates $T(Z)$, which necessitates a contradiction; hence Z also necessitates $\sim Q$. Therefore, $\sim Q$ is a presupposition of Z; and if it fails, Z is neither true nor false. But if $\sim Q$ is true, the case is different: Z and Q are such that $\sim Q$ necessitates $\sim Z$. Therefore if $\sim Q$ is true, then Z is false.

This final example intends to show how our analysis can be applied to

---

9. That we envisage only finitely many levels here is not a necessary limitation, and not essential to this point.
10. In the context of the fact that the statement is made by a Cretan; in the present argument, necessitation and presupposition are relativized to this assumption.

members of this family of paradoxes other than the two for which it was developed explicitly. On the other hand, we have not supplied a general theory of self-referential language. This is the familiar lament that we have no (sufficiently general) formal pragmatics. But the paradoxes of self-reference have been a major obstacle to such a pragmatics, and we shall be satisfied if we have shown that this obstacle at least is not insuperable.

Note

The standard discussions of presupposition are those of P. F. Strawson in his *Introduction to Logical Theory*, Methuen, London (1952), and "On referring," *Mind*, vol. LIX (no. 235) (July 1950) pp. 320–44. Strawson's account has been critically discussed a.o. by W. Sellars, "Presupposing," *Philosophical Review*, vol. LXII (no. 2) (April 1954) pp. 194–215, and G. Nehrlich, "Presupposition and entailment," *American Philosophical Quarterly*, vol. 2 (no. 1) (Jan. 1965) pp. 33–42. The former was answered by Strawson in the same issue of that journal, pp. 216–31, and the latter corrected by Nehrlich himself in "Presupposition and classical logical relations," *Analysis*, vol. XXVII (no. 3) (Jan. 1967)) pp. 104–106. The literature on implication is now too voluminous to be summarized; we refer only to A. R. Anderson and N. D. Belnap, Jr., "The pure calculus of entailment," *J. of Symbolic Logic*, vol. XXVII (no. 1) (March 1962) pp. 19–52. Correspondence with Peter Woodruff, Wayne State University, has helped me to become clearer on the distinction between implication and presupposition.

The basic idea for our treatment of presupposition was suggested by Karel Lambert, University of California at Irvine, to whom I am much indebted, in a discussion of Section VII of this author's "Singular terms, truth-value gaps, and free logic," *Journal of Philosophy*, vol. LXII (no. 17) (Sept. 15, 1966) pp. 481–95. The language of free logic as there described is a presuppositional language in the sense of Section 2 of this chapter. The notion of supervaluation in this chapter is a generalization of that presented in "Singular terms, truth-value gaps, and free logic." Supervaluations have since been used in R. Meyer and K. Lambert, "Universally free logic and standard quantification theory," *J. of Symbolic Logic* (1968) pp. 8–26, and in B. Skyrms' comments on J. Pollock's "The truth about truth" at the APA (Western Division) meetings in Chicago on May 4, 1967 (see below).

The distinction between excluded middle and bivalence, apparently first made by Aristotle, was introduced in this century by the Polish logicians: see S. McCall, "Excluded middle, bivalence, and fatalism," *Inquiry*, vol. 4 (Winter 1966) pp. 384–6. That we are in good company here is witnessed by Quine, who mentions Paul Weiss, Yale University, as having been brought "to the desperate extremity of entertaining Aristotle's fantasy that 'It is true that p or q' is an insufficient condition for 'It is true that p or it is true that q'." (*The Ways of Paradox*, Random House, New York (1966) p. 21.) The nature of validity and the consequence relation in presuppositional languages is explored in my "Presuppositions, supervaluations, and free logic" in the Festschrift in honor of Henry Leonard (ed.

by Karel Lambert)—*The Logical Way of Doing Things*, Yale University Press, New Haven (1969).

Section 3 is an improvement and extension of Section VIII of "Singular terms, truth-value gaps, and free logic." The three-valued matrix mentioned in Section 3 was suggested by footnote 23 of E. Sosa, "Presupposition, the Aristotelian square, and the theory of descriptions" (mimeographed, University of Western Ontario, 1966); the question of the adequacy of such matrices for classical propositional logic is discussed in A. Church, "Non-normal truth-tables for the propositional calculus," *Boletín de la Sociedad Matemática Mexicana*, vol. X (1953) pp. 1–2. I should like to thank John Heintz, University of North Carolina, for helpful suggestions concerning the use of this matrix.

The literature on the paradoxes of self-reference is also huge; this paper has profited most from A. N. Prior, "On a family of paradoxes," *Notre Dame Journal of Formal Logic*, vol. II (no. 1) (Jan. 1961) pp. 16–32. Also suggestive were N. Rescher, "A note on self-referential statements," ibid., vol. V (no. 3) (July 1964) pp. 218–20; and D. Odegard, "On weakening excluded middle," *Dialogue*, vol. V (no. 1) (Sept. 1966) pp. 232–6. Bryan Skyrms, University of Illinois at Chicago Circle, has developed independently a solution to the *Strengthened Liar* paradox through the use of supervaluations, which he presented in the APA paper mentioned earlier. His account is in some ways very similar to ours, but the syntax of the artificial language used is quite different; it may lead to a much finer analysis of the self-referential language used in the paradox than we have provided.

Finally, I should like to acknowledge my debt to stimulating discussions and correspondence on the semantic paradoxes with V. Aldrich, University of North Carolina, and G. Nakhnikian, Wayne State University, and to thank R. Clark, Duke University, and P. F. Strawson, University College, for their encouraging comments on an earlier version of this paper.

11
A Russellian Approach to Truth

Ronald Scales

A primary goal of semantical analysis is to show how the truthvalue of a sentence is determined by the semantical roles of its constituent expressions. Russell's contribution in this area is generally thought to be his elimination rules for singular terms. If we look at the contextual definitions for sentences including definite descriptions from the point of view of truth, however, we find that Russell stands in a long tradition of philosophers whose intuitions have been to count simple subject–predicate sentences false when the subject term fails to denote. In a formal semantics this calls for the use of partial functions in interpreting terms and, following Russell, leads to a highly plausible logic underlying a theory of truth for sentences of a natural language. This truth definition for a predicate logic, including an existence predicate and provision for the semantic dicto–re distinctions available through Russell's scope operators for terms, was carried out in the author's doctoral dissertation (Scales 1969). In later manuscripts, the logic was characterized without reliance on an existence predicate and was extended to modals. Similar semantical investigations have been carried out by Mates (1968) and by Schock (1968). Burge (1974) has proposed an axiomatic development of the semantics as characterizing validity in natural languages. None of these provide for dicto–re distinctions. Mates notes that such a provision (along Russellian lines) is the best way of dealing with paradoxes related to '(9 > 7)'. The provision for wide scope of terms for nonmodal contexts is of course carried out in Russell, and in the author's doctoral dissertation it was put into a syntactic form more conceptually in line with traditional dicto–re distinctions. Class abstracts were used to represent complex predicates of natural language, and the semantic distinctions in Russell were characterized through the use of these abstracts rather than scope indicators for terms. A brief report of this semantics, extended to modals, follows.

According to Russell, the sentence

(1) The present King of France is bald

is true just in case exactly one individual is a present King of France and that individual is bald. Generalizing these truth conditions to arbitrary atomic sentences of a formal logic, we have the rule that such a sentence is true if every term denotes and the *n*-tuple of denotata is a member of the denotation of the predicate, and false otherwise. Introducing classical sentence negation (negation is classical that converts truths to falsehoods and vice versa) yield the interesting feature that $\sim Fa$ is true when the term *a* fails to denote. In natural language, negation provides the conceptually simplest instance of scope ambiguity. The sentence

(2) The present King of France is not bald

Russell held to have alternative semantic representations. In one sense, it is true just in case exactly one individual is a present King of France and this individual fails to be bald. In a second sense, it is true just in case either the condition just stated holds, or there fails to be exactly one present King of France. The first sense entails the second, but not conversely, and the two senses are thus contraries, both true when the first is. Russell's method for indicating this ambiguity syntactically—viz., scope indicators for definite descriptions—is well known and has come in for a good deal of criticism. Representing the alternative semantic readings of such ambiguous sentences through the use of class abstracts undercuts much of this criticism, especially if it turns out that the choice of abstracts rather than scope indicators for terms constitutes no empirical disagreement with Russell. Briefly, the problem of finding an empirical disagreement is presented by the fact that each method imposes just the same entailment relations on sentences of a natural language. (For a proof of the semantic isomorphism of the two methods, cf. Scales 1969.) In any case, abstracts are in line with recent tradition in the treatment of dicto–re distinctions and are conceptually easier to work with than scope indicators for terms. A comparison of the two methods in the treatment of the semantic ambiguity in (2) is as follows: The ambiguity is represented in Russell's method by the alternative forms

(3) $[\iota x K x] \sim B[\iota x K x]$

(4) $\sim [\iota x K x] B(\iota x K x)$

Using class abstracts to represent complex predicates, and for uniformity writing arguments after predicates, we have

(5) $\hat{x}(\sim Bx)(\iota xKx)$

(6) $\sim(B(\iota xKx))$

The abstract $\hat{x}(\sim Bx)$ in (5) represents the complex predicate 'is non-bald'. (5) represents the first and (6) the second sense of (2), the truth conditions being taken from the respective formulas

(7) $(\exists x)((y)(Ky \equiv x = y) \mathbin{\&} \sim Bx)$

(8) $\sim(\exists x)((y)(Ky \equiv x = y) \mathbin{\&} Bx)$

In our formal semantics, the truth conditions for $\hat{x}(\sim Bx)(a)$, where a is any term, are that a denote and its denotation be a member of the class denoted by $\hat{x}(\sim Bx)$ (which will be the complement of the class denoted by B). According terms wider scope, as in contrasting (5) with (6), has the effect of putting such terms in referential position. For nonmodal discourse, this amounts semantically to at most adding existential import to the term in question. Thus, adopting an existence predicate whose range is the domain, (5) is equivalent with $\sim B(\iota xKx) \mathbin{\&} E!(\iota xKx)$. Thus, abstracts are eliminable from nonmodal sentences. The effect of referential position in modal discourse, however, is more complex, requiring the introduction of purely referential denotation functions for the interpretation of de re modalities. We proceed now to a formal development of the semantics.

Models are pairs $\langle D, R \rangle$, D a domain of objects (possibly empty) and R a set of relations on D.

A model M is a frame for a set of formulas Γ iff for each atomic n-place predicate ϕ in formulas of Γ, there is a relation $M(\phi) \in R$ of suitable degree interpreting ϕ in M.

A denotation function $\triangle$ is a partial function from the terms (individual constants, variables, and definite descriptions) of the predicate calculus into the domain of each model. (Denotation functions make assignments to descriptions as a function of the assignments to constants, variables, and atomic predicates in them.) We use the notation $\triangle^M|t|$ to designate the object in the domain of M that $\triangle$ assigns to the term t, and use of this notation indicates that $\triangle$ is defined for t in M. If $\triangle$ assigns no object to t in M, we simply say that $\triangle$ is undefined for t in M. A denotation function $\triangle$ may be defined for t in M but undefined for t in $M' \neq M$, and given $(\triangle, t, M)$, $\triangle$ is defined for t in M or undefined for t in M and not both.

A Russellian Approach to Truth

Important for the semantic interpretation of abstracts containing modal operators (i.e., for the theory of de re modality) is the class of purely referential denotation functions (PRDFs): If $\triangle$ assigns the same object to t in M as it does t in M', or if $\triangle$ is undefined for t in both M and M', we say that $\triangle$ agrees on t for M and M'; otherwise, $\triangle$ disagrees on t for M and M'. Similarly, if $\triangle$ and $\triangle'$ assign the same object in M to t, or if both are undefined for t in M, we say that $\triangle$ and $\triangle'$ agree on t for M; otherwise $\triangle$ and $\triangle'$ disagree on t for M.

A PRDF for $t_1, \ldots, t_n$ is a denotation function that agrees on t_i (for any given t_i, $1 \leq i \leq n$) for any pair of models M, M'. Given $\triangle$ and M, we write $\triangle_{t_1,\ldots,t_n}$ for that PRDF for $t_1, \ldots, t_n$ that agrees on each t_i with $\triangle$ for M.

Thus, PRDFs for t are those that assign the same object, if any, to t in every model in which the object exists (i.e., is in the domain), and PRDFs for $t_1, \ldots, t_n$ are those that satisfy these conditions for each t_i. A PRDF may thus be defined for t in M but undefined for t in $M' \neq M$.

Where M is a frame for Γ, satisfaction is defined only for formulas all predicates of which occur in formulas of Γ. We write '$\triangle^M|A| = 1$' for $\triangle$ satisfies A in M, and '$\triangle^M|A| = 0$' for $\triangle$ fails to satisfy A in M. If M is a frame for Γ and all predicates in A occur in formulas of Γ, then $\triangle^M|A| = 1$ or $\triangle^M|A| = 0$ and not both, for every $\triangle$ on M. All denotation functions agree in M for individual constants; that is, where c is an individual constant and M a model, $\triangle$ agrees with $\triangle'$ on c for M, for every $\triangle$ and $\triangle'$.

(i) $M(=)$ is the identity relation on $D(M)$ (the identity predicate is thus nonreflexive)

where $A(x)$ is a formula with x free,

(ii) If $a \in D(M)$ and $\triangle^M|A(x)| = 1$ only if $\triangle$ is defined for x and $\triangle^M|x|$ is a, then $\triangle^M|\iota x A(x)|$ is a; otherwise, $\triangle$ is undefined for $\iota x A(x)$ in M

where ϕ is an atomic n-place predicate and $t_1, \ldots, t_n$ are terms,

(iii) $\triangle^M|\phi(t_1, \ldots, t_n)| = 1$ iff $\langle \triangle^M|t_1|, \ldots, \triangle^M|t_n| \rangle \in M(\phi)$

where $A(x_1, \ldots, x_n)$ is a formula containing the distinct free variables $x_1, \ldots, x_n$ and $t_1, \ldots, t_n$ are terms, $n > 0$, and $\triangle_{t_1,\ldots,t_n}$ as earlier (on $\triangle, M$),

(iv) $\triangle^M|\hat{x}_1, \ldots, \hat{x}_n(A(x_1, \ldots, x_n))(t_1, \ldots, t_n)| = 1$ iff
    (a) $\triangle_{t_1,\ldots,t_n}$ is defined for each t_i on M, and
    (b) $\triangle^M_{t_1,\ldots,t_n}|A(t_1, \ldots, t_n)| = 1$

where A and B are formulas,

(v) $\triangle^M|\sim A| = 1$ iff $\triangle^M|A| = 0$
(vi) $\triangle^M|A \to B| = 1$ iff $\triangle^M|A| = 0$ or $\triangle^M|B| = 1$ or both
(vii) $\triangle^M|(x)A| = 1$ iff $\triangle_*^M|A| = 1$, for every $\triangle_*$ x-variant of $\triangle$ defined for x in M
(viii) $\triangle^M|\Box A| = 1$ iff $\triangle^M|A| = 1$ for every M

where M is a frame for Γ and every predicate in A occurs in formulas of Γ,

A is true in M iff $\triangle^M|A| = 1$ for every $\triangle$ on M

The semantics is open at this point to allowing closed atomic formulas (and thus a selected subset of nonatomic formulas) to have no truthvalue (i.e., be neither true nor false) if some term fails to denote. However, following Aristotle and Russell, we specify that every closed formula is either true or false.

References

Burge, Tyler, "Truth and singular terms," *Noûs*, vol. 8 (1974) pp. 309–25.
Mates, Benson, "Leibniz on possible worlds," in *Logic, Methodology, and the Philosophy of Science*, vol. 3 (ed. van Ritselaer, G. and Staal, J. F.), North-Holland, Amsterdam (1968) pp. 507–29.
Scales, Ronald, *Attribution and Existence*, University Microfilms, Ann Arbor (1969).
Schock, Rolf, *Logics Without Existence Assumptions*, Almqvist and Wiksell, Stockholm (1968).

PART IV
Metaphysics

McCall's essay, "Abstract individuals," argues that a free logic is needed to represent the notion of an abstract individual—a notion vital to Aristotelian metaphysics—and produces the needed representation. His essay of the same title was published originally in *Dialogue*, vol. 5 (1966) pp. 217–31, and is reprinted here with the permission of the editors. Cocchiarella's essay, "Quantification, time, and necessity," new for this volume, constitutes an application of his own early development of free logic to two of the most important subjects in traditional metaphysics. Mann's essay, "Definite descriptions and the ontological argument," is an assessment of Jan Berg's provocative formulation of the ontological argument, an assessment that relies heavily on a free theory of definite descriptions. Mann's essay is an amended version of the original that appeared in *Theoria*, vol. 30 (1967) pp. 211–29, segments of which are reproduced here with the permission of the editors of *Theoria*. Lambert's study, "Predication and ontological commitment," shows how Quine's preregimented intuitions about predication can be sustained in a bivalent negative free logic while yet avoiding the inconveniences and difficulties Quine believes are inevitably associated with the conception of predication in question. Lambert's essay of the same title was published originally in *Die Aufgaben der Philosophie in der Gegenwart: Aktien des 10. Internationalen Wittgenstein Symposiums* (Hrsg. W. Leinfellner u. F. N. Wuketits), by Hölder-Pichler-Tempsky, Wien (1986) pp. 281–7, and is published here with the permission of the publishers. Finally, Simon's study, new for this volume, constitutes a theory of the part–whole relation formulated in free logic, and an application of that theory to certain notions often regarded as problematic in traditional metaphysics.

12
Abstract Individuals

Storrs McCall

The title of this paper may seem to involve a contradiction: my purpose is to show that it does not.

Individuals fall into two categories; those that depend for their existence upon the existence of other individuals, and those that do not. In the second category are found such things as shoes, ships, cabbages, kings, and discrete bits of sealing wax. These may be called *individual substances*, and the way in which the existence of a cabbage depends upon water and earth, or in which Descartes says the existence of all things depends upon God,[1] will not be in point here. The individuals of the first category are characterized by a much more obvious kind of dependence. They include the sound of an individual shoe falling on the floor, the sinking of the *Bismarck*, the stupidity of George I, the center of gravity of a bit of sealing wax. All these are individuals, though they are not individual substances. They are what I shall call *abstract individuals*, or *abstract particulars*.

There is a natural inclination to say that these so-called abstract individuals are simply properties of individuals—that the stupidity of George I is a property of George I. But this is a mistake. Stupidity, it is true, is a property, but as such it in no way depends for its existence on George I, since there could be stupidity without him. The abstract individual referred to by the words "the stupidity of George I," however, is wholly dependent upon and inseparable from George I. This dependence is built into the very definition of what an abstract individual is.

The distinction between abstract individuals and substances is made by Aristotle in the *Categories*. Aristotle there speaks of what is *in* a subject as opposed to what is *said of* a subject: man for example is said of Socrates, but individual knowledge-of-grammar is not said of Socrates, but is in him. He goes on to explain that what is "in" a subject is firstly not a part of that subject, and secondly cannot exist separately from what it is in. The "individual white," for example, is in a subject, such as a white shirt, though not as a part of it, and cannot exist separately from it. It seems plain that what Aristotle is here describing as *in* a subject is what we have called an abstract individual. What is *said of* a subject, on the other hand, is a species or genus (as distinct from the name of that species or genus). Finally, primary substances are those things that are neither in a subject nor said of a subject, for example, the individual man or the individual horse. They are what we have called individual substances.[2]

From what has been said, it would seem that the general form of reference to an abstract individual would be through the words 'the X of Y'. This is so, but herein lies a danger, for not every phrase of that form designates an abstract individual. Contrast the following list:

> The color of her eyes
> The smell of violets
> The elasticity of rubber
> The existence of God
> The death of Robespierre
> The importance of being earnest
> The end of the affair
> The spin of an electron

with this one:

> The man of the year
> The grain of truth
> The ship of fools
> The sum of 100 shillings
> The author of *Waverley*
> The Omukama of Bunyoro
> The tallest tree of the forest
> The Greek of Aristotle

Plainly there is an important difference between them, which may be brought out in the following way. Of each of the items in the first list, we would say that there was such a thing if and only if the possessor of the thing in question were characterized in a certain specific way. For example, there is such a thing as the color of her eyes if and only if her eyes are colored; such a thing as the smell of violets if and only if violets smell, and so on. But the same does not hold for the second list. It is not the case that there is such a thing as the man of the year if and only if the year is characterized by manhood (or whatever); such a thing as the grain of truth in what you say if and only if truth is granular, and so forth. Roughly and broadly, the first list names abstract particulars, the second individual substances (with the possible exception of Aristotle's Greek, which is a difficult case).[3]

In logic in its present form, there would be no essential difference between the way in which the denoting phrases in the two lists would be treated. Falling under the general form "the so-and-so," they would one and all have to submit to the gentle dismemberment of Russell's theory of descriptions.[4] For example,

(1) The death of Robespierre was unforeseen

would become

(2) $(\exists x)[x$ is a death and x is Robespierre's and x was unforeseen and (y) (if y is a death and y is Robespierre's then y is $x)]$

Here as in all applications of the theory of descriptions, *the* death is analysed in terms of *a* death, *the* color in terms of *a* color, and so on. But in many cases this leads to pure gibberish. Consider analyzing

(3) The importance of being earnest is crucial

into

(4) $(\exists x)[x$ is an importance and x characterizes being earnest and x is crucial and (y) (if y is an importance and y characterizes being earnest then y is $x)]$

The plain answer is, that there is no such thing as an importance. Nor can we cast about for ways of making use of the more familiar predicate "x is important," and transform (3) into something like

(5) Being earnest is important and crucial

since it is not being earnest that is crucial, but its importance. An entirely different approach is called for.

The solution that will be proposed here involves discarding Russell's theory of descriptions as applied to the items of list one above, and, more radically, changing our basic theory of quantification from a logic of predicates to a logic of abstract individuals. As will be seen, nothing need be lost in this process, and the gains will be considerable.

Predicate logic contains among its primitive symbols two kinds of variable, individual variables and predicate or functional variables. But if a new kind of variable ranging over abstract individuals is introduced, then predicate variables can be defined, with the help of the constant identity-function, in terms of abstract individual variables, so that the number of primitive notions is not increased. The definition is made possible by the basic equivalence mentioned above, one of whose concrete instances is the following:

(6) x is colored if and only if there is something that is the color of x

How can we generalize (6)? Let us use small Latin letters in association with the letters w, x, y, and z (which will be reserved for variables ranging over all sorts of individuals, both abstract and substantial) to range over the domain of abstract particulars, and let '$a(x)$' be read 'the a of x'. Then '$c(x)$' will be 'the color of x'. Let 'Cx', in accordance with the conventional style of writing predicates, mean 'x is colored'. Then the formalization of (6) will be

(7) $Cx \equiv (\exists y) y = c(x)$

and in general, where a is any abstract individual variable and A the corresponding predicate, we shall have

(8) $Ax \equiv (\exists y) y = a(x)$

This equivalence makes possible the following means of defining an arbitrary monadic predicate F in terms of the corresponding abstract individual:

(9) $Fx =_{Df} (\exists y) y = f(x)$

and this definition in turn enables us to substitute abstract individual variables for predicate variables among the primitive symbols of logic.[5]

Definition (9), once accepted, commits us to at least one major change in the type of underlying predicate logic we adopt, in addition to the change in primitive notation. Instead of the classical system of *Principia Mathematica*, we shall employ what has come to be known as "free" logic. The reason for this is, that in classical logic the following is a theorem:

(10) $(\exists y) y = x$

so that, by substitution of $f(x)$ for x and application of definition (9), the simple expression Fx would be a theorem. This is unacceptable. Hence a theory of quantification must be found in which (10) is not provable, and free logic presents itself as an obvious candidate. For the details of this logic, whose basic *raison d'être* is to allow for the truth of certain statements containing nondesignating singular terms (e.g., 'Santa Claus lives at the North Pole'), the reader is referred to an article by Karel Lambert, and to the papers mentioned therein.[6]

We shall now attempt to show that a logic that includes abstract individuals in its *Weltanschauung* is an enriched logic, capable of dealing with a more varied class of arguments and inferences than one that does not. Consider the following argument:

> Women are beautiful
> What Considine seeks does not exist
> Therefore, Considine does not seek the beauty of any woman

whose formalization in classical quantification theory would at best be tortured and unnatural. Its formalization using terms for abstract particulars proceeds as follows, where $Wx = x$ is a woman, $Bx = x$ is beautiful, $b(x) =$ the beauty of x, $Sxy = x$ seeks y, and $c =$ Considine:[7]

1.	$(x)(Wx \supset Bx)$	Pr
2.	$(x)[Scx \supset \sim(\exists y)y = x]$	Pr
3.	$E!x \supset (Wx \supset Bx)$	1 UI
→4.	$E!x$	Pr
5.	$Wx \supset Bx$	3, 4 MP
→6.	Wx	Pr
7.	Bx	5, 6 MP
8.	$(\exists y)y = b(x)$	7 Df Bx
9.	$E!b(x)$	8 Df $E!$
10.	$E!b(x) \supset [Scb(x) \supset \sim(\exists y)y = b(x)]$	2 UI
11.	$Scb(x) \supset \sim(\exists y)y = b(x)$	9, 10 MP
12.	$\sim Scb(x)$	8, 11 MT
13.	$Wx \supset \sim Scb(x)$	6, 12 CP
14.	$E!x \supset [Wx \supset \sim Scb(x)]$	4, 13 CP
15.	$(x)[Wx \supset \sim Scb(x)]$	14 UG

The next step in developing the potentialities of the logic of abstract individuals is to introduce singular terms denoting qualities. So far we have dealt with things such as 'the color of x' rather than just 'color'. But expressions denoting the latter may be defined in terms of expressions denoting the former. If '$c(x)$' denotes the color of x, then let the single letter 'c' denote color, simpliciter. Terms such as this, denoting qualities, will be defined contextually as follows, where a is any such term and Fa

any statement containing a:

(11) $\quad Fa =_{Df} (x)[Ax \supset Fa(x)]$

For example, 'color is beautiful' is defined as meaning 'for all x, if x is colored then the color of x is beautiful'. The use of terms denoting qualities enables a whole new range of arguments to be captured in our formal system. For example:

> Only the thickness of the rope saved him
> Strength is not the same as thickness
> Therefore, if the rope existed and was strong, the strength of the rope did not save him

Letting $Sx = x$ is strong, $Vxy = x$ saved y, t = thickness, s = strength, r = the rope, and h = him, the derivation of the conclusion proceeds as follows:

1.	$(x)[x \neq t(r) \supset \sim Vxh]$	Pr
2.	$s \neq t$	Pr
3.	$(x)[s(x) = t(x) \supset s = t]$	Logical truth
4.	$E!r \supset [s(r) = t(r) \supset s = t]$	3 UI
5.	$E!r . Sr$	Pr
6.	$E!r$	5 Simp
7.	$s(r) = t(r) \supset s = t$	4, 6 MP
8.	$s(r) \neq t(r)$	2, 7 MT
9.	$E!s(r) \supset [s(r) \neq t(r) \supset \sim Vs(r)h]$	1 UI
10.	Sr	5 Simp
11.	$E!s(r)$	10 Df S, Df $E!$
12.	$s(r) \neq t(r) \supset \sim Vs(r)h$	9, 11 MP
13.	$\sim Vs(r)h$	8, 12 MP
14.	$(E!r . Sr) \supset \sim Vs(r)h$	5, 13 CP

Note the 'logical truth' assumed at step 3. It is equivalent to the following expression, which will be mentioned again below:

(12) $\quad (x)(y)(z)\{[Ay . Bz . a(x) = b(x)] \supset a(y) = b(z)\}$

In addition to predicate functions of one variable, relational functions of two or more variables may also be introduced. Their definition follows naturally, being analogous to definition (9), once it is perceived that some abstract individuals do not characterize a single individual substance, but instead are shared by two or more substances. For example, the support of a roof by a rafter, the flight of the Israelites out of Egypt, and the bringing of the good news from Aix to Ghent are all abstract individuals, though shared respectively by 2, 2, and 3 individual substances. Just as we denote the color of x by '$c(x)$', let us denote the flight of x out of y by '$f(x, y)$', and the bringing of x from y to z by '$b(x, y, z)$'. Then Fxy, defined as $(\exists z)z = f(x, y)$, will be 'x flees out of y', $Bxyz$ will be 'x is brought from y to z', and so on. The following is an example of an argument using relational individuals.

> The sale of opium by anyone to anyone is forbidden
> Bimbo (who is a real person) is selling opium to everyone
> Therefore, Bimbo's sale of opium to everyone is forbidden

Letting $s(x, y, z) =$ the sale of x by y to z, $Fx = x$ is forbidden, $o =$ opium, and $b =$ Bimbo, the argument may be formalized as follows:

1.	$(x)(y)(z)[x = s(o, y, z) \supset Fx]$	Pr
2.	$(x)Sobx$	Pr
3.	$E!b$	Pr
4.	$E!x \supset Sobx$	2 UI
→5.	$E!x$	Pr
6.	$Sobx$	4, 5 MP
7.	$(\exists y)y = s(o, b, x)$	6 Df S
8.	$E!w . w = s(o, b, x)$	7 EI
9.	$w = s(o, b, x)$	8 Simp
10.	$E!w$	8 Simp
11.	$E!w \supset (y)(z)[w = s(o, y, z) \supset Fw]$	1 UI
12.	$(y)(z)[w = s(o, y, z) \supset Fw]$	10, 11 MP
13.	$E!b \supset (z)[w = s(o, b, z) \supset Fw]$	12 UI
14.	$(z)[w = s(o, b, z) \supset Fw]$	3 13 MP
15.	$E!x \supset [w = s(o, b, x) \supset Fw]$	14 UI
16.	$w = s(o, b, x) \supset Fw$	5, 15 MP
17.	Fw	9, 16 MP
18.	$Fs(o, b, x)$	9, 17 Id
19.	$E!x \supset Fs(o, b, x)$	5, 18 CP
20.	$(x)Fs(o, b, x)$	19 UG

Having given an indication of the range and variety of arguments to which the logic of abstract individuals can be applied, let us now take up the question of the *existence* of these individuals. What inferences presuppose their existence? What statements imply it? Most of the philosophical controversy generated by Russell's theory of descriptions has concerned this matter of existence: does 'The blacksmith of Chicot died yesterday' imply that there is a blacksmith of Chicot? In similar vein, does 'The strength of Hercules was legendary' imply that such strength existed, that is, that Hercules was strong? Not according to the logical system sketched in this paper. For consider the true proposition

(13)   The authenticity of the Seventh Letter is open to doubt

symbolized by $Da(s)$, that is, $(\exists x)x = d(a(s))$. To deduce $(\exists x)x = a(s)$, that is, that the Seventh Letter was authentic, would certainly be fallacious. On the other hand, the truth of (13) presupposes in some very weak sense that there is such a thing as the authenticity of the Seventh Letter. To show this, let us change the example and try to demonstrate the truth of a proposition analogous to (13) using the *dictum de omni*. Can the truth of

(14)   Nero's cruelty is unforgivable

be demonstrated, starting from the general premiss that cruelty is unforgivable? No, the proof of $Uc(n)$ from Uc requires the added premiss Cn, that is, that Nero was cruel. Hence the truth of (14) presupposes that there is such a thing as Nero's cruelty in a weak sense, namely that (14) cannot be proved from 'Cruelty is unforgivable' without that presupposition.

The peculiar status of the existence of such abstract particulars as Nero's cruelty vis-à-vis the proposition (14)—not being implied by it but also in a sense being presupposed—can be put to good use. It sanctions, for example, the validity of the following argument:

Rest-mass is an observable quality
The rest-mass of photons is not an observable quality
Therefore, if they exist, photons have no rest-mass

Letting r = rest-mass, $Ox = x$ is an observable quality, and p = photons, the argument runs as follows:

1. Or — Pr
2. $\sim Or(p)$ — Pr
3. $(x)[Rx \supset Or(x)]$ — 1 Df r
4. $E!p \supset [Rp \supset Or(p)]$ — 3 UI
5. $E!p$ — Pr
6. $Rp \supset Or(p)$ — 4, 5 MP
7. $\sim Rp$ — 2, 6 MT
8. $\sim(\exists x)x = r(p)$ — 7 Df R

9. $E!p \supset \sim(\exists x)x = r(p)$ — 5, 8 CP

Perhaps enough has now been said concerning what *can* be done with the logic of abstract individuals; it remains to say what *cannot*, at least not yet. It would be very nice, for example, to have a way of symbolizing statements like

(15)   Red is a color

(16)   Every one of the criminal's moves was watched

and arguments like

> Harry is honest
> Honesty is a virtue
> Therefore, one of Harry's virtues is honesty

But at the moment it is not at all clear how this may be done. For example, how should (15) be symbolized? Certainly not as Cr: the predicates 'is a color' and 'is colored' are different. A possibility is $(x)[(\exists y)y = r(x) \supset (Ey)y = c(x)]$, but while this would permit the inference from 'x is red' to 'x is colored', it is not certain that its logical powers would be such as to enable the conclusion of the following argument to be derived:

> The rainbow is colored with every color
> Red is a color
> Therefore, the rainbow is red[8]

Here it seems that we need a way of symbolizing, not only 'red is a color', but the more general 'x is a color', and perhaps also 'x is a color of y'. No doubt these predicates could be defined with the help of more *recherché* abstract individuals—for example, 'x is a color' could be defined as '$(\exists y)y$ = the chromaticity of x'—but this may be the wrong direction to move in.

The situation is complicated even more by the fact that the proper logical force of such statements as (15) cannot be determined without investigating the various types of valid inference into which they can enter. And the validity of these inferences depends upon what sort of world, with what sort of semantic rules, our logical system is to apply to. For example, confining ourselves to color worlds and color rules (these yield problems enough), is the following inference valid?

(17) X is red
Red is a color
Therefore, the color of X is red

Well, yes, in some color worlds it is, even if we interpret the conclusion as being false if X is some other color as well as red. For example, the world might be such as to consist entirely of amorphous color patches, each of the same pure red, pure orange, pure yellow, pure green, or pure blue[9] (or alternatively, and assuming hypersensitive eyes, the number of such distinct colors might be infinite). Furthermore, to each color might be assigned one and only one color name. In this 'speech-situation' (named S_0) not only would argument (17) be valid, but the following formulas would all be true, where a and b are any colours and $c(x)$ is the color of x:

(18) $\quad (x)(y)[x = y \supset a(x) = a(y)]$

(19) $\quad (x)(y)[a(x) = a(y) \supset c(x) = c(y)]$

(20) $\quad (x)(y)\{[Ax \,.\, c(x) = c(y)] \supset Ay\}$

(21) $\quad (x)(y)[(Ax \,.\, Ay) \supset a(x) = a(y)]$

(22) $\quad (x)[(Ax \,.\, Bx) \supset a = b]$

(23) $\quad (x)(y)(z)\{[Ay \,.\, Bz \,.\, a(x) = b(x)] \supset a(y) = b(z)\}$ (see (12))

Hence, a logical system tailored to S_0 would sanction (17), and include (18)–(23) as logical truths. But let us not be impatient to include (17)–(23) within a general logic of abstract individuals. No sooner is S_0 imagined than more complicated alternatives spring to mind; in fact there are three different ways in which complexity may be introduced.

(i) The basic S_0-principle of 'one color, one name' may be retained, and the colors remain the same, but the colored objects, instead of being amorphous, may be piebald. In this world (18), (20), (21), and (23) continue to hold, but (17) and (22) do not, and (19) is doubtful.

(ii) As in S_0 each object may be of one color only, and the colors be distinct, but the color terminology may become ambiguous. Thus 'red', for example, might be correctly applicable to objects that in S_0 would be either red or orange, 'orange', to orange objects only, 'yellow' to orange, yellow, and green objects, and so on. In this world (18), (19), and (20) would hold, but (17), (21), (22), and (23) would not.

(iii) Again each object may be of one color only, and the paradigm color-samples used for naming be all distinct and of different names, but there may be many colors, and the color of some of the objects in the world may fall between two paradigms, in such a way as to be indistinguishable from both. So again there is ambiguity, not built into the language, but arising from the limitations of sense-perception, and (22) and (23) fail. Furthermore, two objects may be indistinguishable in color, although, being each indistinguishable from different paradigms, different color names are assigned to them. Hence (20) fails. Finally, the same color name may be assigned to two objects that are of distinguishable colors. Hence (21) fails. Of the rest, (17) fails, (18) holds, and (19) is doubtful.

The plethora of different logical principles, which hold or fail in different situations, ought to discourage any lighthearted attempt to axiomatize the logic of abstract individuals. Of course, it is easy to translate Lambert's axioms for free logic into abstract individual notation (see the Appendix), and so arrive at *a* calculus of abstract individuals, but this simply avoids the main problems. Nevertheless I believe that the subject is an important one, and is worth further study. Is it logic, or is it metaphysics? I think that here there is no sharp distinction between the two. Metaphysics has surely always involved the search for certain basic or ultimate categories, in terms of which the anatomy of the universe can be fixed. The notion of an abstract individual may turn out to be one such category. Instead of the protocol 'Red here now', perhaps we should say 'The red of X here now'! But whether or not the concept of an abstract particular proves fruitful is going to be determined, in this day and age, by the richness and usefulness of the logical system that embodies it. The edibility of the metaphysical pudding lies in its powers of proof.

APPENDIX

Axiomatic Basis of Free Logic as a Calculus of Abstract Individuals

I. *Primitive symbols*
   Individual variables (IV): $w, x, y, z, w_1, x_1, \ldots$
   Abstract individual variables (AIV), divided into monadic, dyadic,

   ... n-adic:

   $a, b, c, \ldots$

   Logical constants: $\sim, \supset, =$.
   Brackets.

II. *Rules of formation*
   We first define a term as follows:
   1. If α is an IV then α is a term.
   2. If α is an n-adic AIV and $\beta_1 \cdots \beta_n$ are terms, then $\alpha(\beta_1, \ldots, \beta_n)$ is a term.

3. Nothing else is a term.

Next follows the definition of a well-formed formula (wff):
1. If α and β are terms then $\alpha = \beta$ is a wff.
2. If α and β are wffs then $\sim\alpha$ and $\alpha \supset \beta$ are wffs.
3. If α is an IV and β is a wff then $(\alpha)\beta$ is a wff.
4. Nothing else is a wff.

III. *Definitions*
1. The usual definitions of '.', '$\vee$', '$\equiv$', and '$\exists$'.
2. Definition of an arbitrary n-adic predicate variable F, where f is the corresponding n-adic AIV, and $\alpha_1 \cdots \alpha_n$ are IV distinct from y:

$$F\alpha_1 \cdots \alpha_n =_{Df} (\exists y) y = f(\alpha_1, \ldots, \alpha_n)$$

3. Contextual definition of an arbitrary quality variable a, that is, a monadic AIV standing alone as a term within a wff:

$$\text{---}a\text{---} =_{Df} (x)(Ax \supset \cdots a(x)\text{---})$$

4. Contextual definition of an arbitrary abstract relational variable a, that is, an n-adic AIV ($n > 1$) standing alone as a term within a wff:

$$\text{---}a\text{---} =_{Df} (x_1)\cdots(x_n)(Ax_1\cdots x_n \supset \cdots a(x_1, \ldots, x_n)\text{---})$$

IV. *Axioms*[10]
1. $(y)\{[(x)(\exists z)z = f(x)] \supset (\exists z)z = f(y)\}$
2. $(x)[(\exists y)y = f(x) \supset (\exists y)y = g(x)] \supset [(x)(\exists y)y = f(x)$
   $\supset (x)(\exists y)y = g(x)]$
3. $x = x$
4. $x = y \supset [(\exists z)z = f(x) \supset (\exists z)z = f(y)]$

V. *Rules of inference*
1. Substitution for IV and AIV.
2. Modus ponens.
3. From $\alpha \supset \beta$, if γ is an IV not free in α, infer $\alpha \supset (\gamma)\beta$.

Notes

1. "For it is clear, when one considers the nature of time, that just the same power and agency is needed to preserve any object at the various moments of its duration, as would be needed to create it anew if it did not yet exist." (Meditation III, translated by Anscombe and Geach)
2. The relevant section of the *Categories* is 1a20–2a34. The author was greatly helped by J. L. Ackrill's translation and notes (*Aristotle's 'Categories' and*

'De Interpretatione', Oxford University Press, 1963), and by Miss Anscombe's commentary (*Three Philosophers*, Oxford University Press, 1963).
3. This is not to say that there are not significant differences among the items of list one, as was pointed out to the author by Bas van Fraassen. For example, the death of Robespierre and the end of the affair are (concrete, not abstract) events, which can be dated, while the other terms are not. But for the purposes of this paper these differences, compared with those between lists one and two, are unimportant.
4. Or alternatively (in logical systems richer than the theory of quantification alone) to Frege's theory. See for example Quine, *Mathematical Logic*, Cambridge University Press (1951) pp. 146–52.
5. See the appendix for fuller details. Note also that definition (9) facilitates a sharper distinction between such pairs of negations as 'x is not trustworthy' and 'x is untrustworthy' than is usually made in predicate logic. The former is translated as $\sim(\exists y)y = t(x)$, while the latter is $(\exists y)y = u(x)$. A formal means of sanctioning the entailment of the first by the second lies beyond the scope of this paper.
6. Lambert, K., "Existential import revisited," *Notre Dame Journal of Formal Logic*, vol. 4 (1963) pp. 288–92. See also Lambert's paper "Explaining away singular non-existence statements," *Dialogue*, vol. 1 (1963) pp. 381–9.
7. The natural deduction rules used for deriving conclusions are those of I. Copi, *Symbolic Logic*, 2d edition, Macmillan, New York (1965) pp. 102–14, modified for free logic as follows. First '$E!x$' is defined as '$(\exists y)y = x$' (note that only in free logic could 'x exists' be translated in this way, because of (10) above). Copi's rules, while retaining all their present restrictions, then become:

UI $\quad \dfrac{(\mu)\Phi\mu}{E!v \supset \Phi v}$

EG $\quad \dfrac{E!v \,.\, \Phi v}{(\exists\mu)\Phi\mu}$

EI $\quad \dfrac{(\exists\mu)\Phi\mu}{E!v \,.\, \Phi v}$

UG $\quad \dfrac{E!\mu \supset \Phi\mu}{(v)\Phi v}$

Id $\quad$ (same as in Copi)

8. Compare also:

> God is the Perfect Being
> Existence is a perfection
> Therefore, God exists

9. Compare 'Speech-situation S' in J. L. Austin's "How to talk," reprinted in his *Philosophical Papers*, Oxford University Press (1961). Austin's 'Speech-situation S_1' is a combination of the complications (ii) and (iii) below.
10. For the axioms and rules of inference of free logic with identity, valid for every domain, see the first paper of Lambert's mentioned in Note 6.

13
Quantification, Time, and Necessity

Nino B. Cocchiarella

The fundamental assumption of a logic of actual and possible objects is that the concept of *existence* is not the same as the concept of *being*.[1] Thus, even though necessarily whatever exists has being, it is not necessary in such a logic that whatever has being exists; that is, it can be the case that there *be* something that does not exist. No occult doctrine is needed to explain the distinction between existence and being, for an obvious explanation is already at hand in a framework of tense logic in which being encompasses past, present, and future objects (or even just past and present objects) while existence encompasses only those objects that presently exist. We can interpret modality in such a framework, in other words, whereby it *can* be true to say that some things do not exist. Indeed, as indicated in Section 3, infinitely many different modal logics can be interpreted in the framework of tense logic. In this regard, we maintain, tense logic provides a paradigmatic framework in which *possibilism* (i.e., the view that existence is not the same as being, and that therefore there can *be* some things that do not exist) can be given a logically perspicuous representation.

Tense logic also provides a paradigmatic framework for *actualism* as the view that is opposed to possibilism; that is, the view that denies that the concept of existence is different from the concept of being. Indeed, as we understand it here, actualism does not deny that there can be names that have had denotations in the past but that are now denotationless, and hence that the statement that some things do not exist can be true in a semantic metalinguistic sense (as a statement about the denotations, or lack of denotations, of singular terms). What is needed, according to actualism, is not that we should distinguish the concept of existence from the concept of being, but only that we should modify the way that the concept of existence (being) is represented in *standard* first-order predicate logic (with identity). A first-order logic of existence should allow for the possibility that some of our singular terms might fail to denote an existent object, which, according to actualism, is only to say that those singular terms are denotationless rather than that what they denote are objects (beings) that do not exist. Such a logic for actualism amounts to what nowadays is called *free logic*.

In what follows we shall first formulate a logic of actual and possible objects in which existence and being are assumed to be distinct second-level concepts represented by the different quantifiers $\forall^a$ and $\forall$, respectively (with $\exists^a$ and $\exists$ defined in terms of $\forall^a$ and $\forall$ in the usual way).[2] The free logic of actual objects, where existence is not distinguished from being—but also where it is not assumed that all singular terms denote—is then described as a certain subsystem of the logic of actual and possible objects. Of course, it is only from the perspective of possibilism that the logic of actual objects is to be viewed as a *proper* subsystem of the logic of being, since, according to possibilism, the logic of being includes the logic of possible objects as well. From the perspective of actualism, the logic of actual objects is all there is to the logic of being.

Both the free logic of actual objects and the logic of actual and possible objects are formulated in Section 1 in such a way as not to presuppose any further encompassing framework, such as tense or modal logic, where these logics find their most natural applications. We describe a framework for tense logic in Section 2, where we distinguish an application of the logic of actual and possible objects from an application of the free logic of actual objects *simpliciter*. In Section 3 we indicate how different modal logics can be interpreted in terms of tense logic, and, in that regard, how an application of the logic of actual and possible objects in modal logic can be distinguished from an application of the free logic of actual objects *simpliciter*. In Section 4 we indicate the kinds of qualifications that are required in the statement of the laws involving the interplay of quantifiers, tenses, and modal operators. The three tense-logical frameworks for which these laws are stated provide logically perspicuous representations of the differences between actualism and possibilism (including a restricted version of temporal possibilism where determinate being includes only what did or does exist, leaving the future as an indeterminate realm of nonbeing). Tense logic, as these developments indicate, is indeed a paradigmatic framework in which to formally represent the differences between actualism and possibilism.

1. A Logic of Actual and Possible Objects

We take a *language* to be a set of predicate and operation expressions of arbitrary (finite) degrees. The set of *terms* and *atomic formulas* of a language are understood to be in accordance with their usual definitions within standard first-order logic with identity. We use two quantifiers—though only one style of individual variable—one for quantification over possible objects, or *possibilia*, and the other for quantification over actual objects. The *formulas* of a language L are those objects that belong to every set K containing the atomic formulas of L and such that $\sim\phi$, $(\phi \to \psi)$, $\forall x\phi$, $\forall^a x\phi \in K$ whenever $\phi, \psi \in K$ and x is a variable.

By a *model* suited to a language L we understand a 3-tuple $\langle A, B, R \rangle$, where (1) A, called the *universe* of the model, is included in B, the domain or set of possibilia of the model, (2) B is nonempty, and (3) R is a function with L as domain and such that for all n, π, δ, (i) if n is a natural number and π is an n-place predicate expression in L, then $R(\pi) \subseteq B^n$, and (ii) if n is a natural number and δ is an n-place operation expression in L, then $R(\delta) \in B^{B^n}$. Satisfaction and truth are defined in the usual Tarski manner, except that the satisfaction clause for the actual quantifier applies only to the universe of the model in question, whereas the satisfaction clause for the possible quantifier covers the entire domain of discourse, that is, the set of *possibilia* of the model. Finally, ϕ is said to be *logically true* if for some language L of which ϕ is a formula, ϕ is true in every model suited to L.

Where ϕ, ψ, χ are formulas, x, y are variables, and ζ, η are terms, universal generalizations of all instances of the following axiom schemas, together with modus ponens as the only inference rule, yield all and only the logical truths:

(A1) $\phi \to (\psi \to \phi)$

(A2) $[\phi \to (\psi \to \chi)] \to [(\phi \to \psi) \to (\phi \to \chi)]$

(A3) $(\sim\phi \to \sim\psi) \to (\psi \to \phi)$

(A4) $\forall x(\phi \to \psi) \to (\forall x\phi \to \forall x\psi)$

(A5) $\forall^a x(\phi \to \psi) \to (\forall^a x\phi \to \forall^a x\psi)$

(A6) $\phi \to \forall x\phi$, where x is not free in ϕ

(A7) $\forall x\phi \to \forall^a x\phi$

(A8) $\exists x(\zeta = x)$, where x does not occur in ζ

(A9) $\forall^a x \exists^a y (x = y)$

(A10) $\zeta = \eta \to (\phi \to \psi)$, where ϕ, ψ are atomic formulas and ψ is obtained from ϕ by replacing an occurrence of η by ζ

This sound and complete axiom set, it should be noted, does not involve the notion of the proper substitution of a term for a variable (and even the notion of bondage and freedom in (A6) can be replaced by the notion of occurrence *simpliciter*). If we restrict ourselves to *standard formulas* (i.e., those in which the actual quantifier does not occur), then (A1)–(A4), (A6), (A8), and (A10) yield all and only the standard logical truths. (The completeness of the latter system is due to R. Montague and D. Kalish, their result being obtained by a modification of an original formulation by A. Tarski.) If we restrict ourselves to *E-formulas* (i.e., those in which the possible quantifier does not occur), then (A1)–(A3), (A5), (A9), (A10), together with the schemas $\phi \to \forall^a x \phi$, where x is not free in ϕ, and $(\zeta = \zeta)$, where ζ is an arbitrary term, yield all and only those logical truths that are E-formulas. Thus, whereas the standard formulas that are logically true constitute the *logic of possible objects simpliciter*, the E-formulas that are logically true constitute the *logic of actual objects simpliciter*. All of the formulas together (i.e., the standard formulas, the E-formulas, and the formulas that contain both the possible and the actual quantifiers) that are logically true constitute *the logic of actual and possible objects*.

Note that whereas by (A8) every term denotes a possible object (i.e., that there *be* an object denoted by that term), it is not true that every term denotes an actual (existent) object. In the logic of actual objects *simpliciter* (i.e., as interpreted from the perspective of actualism), this means that some terms may be denotationless. Also, whereas the law of universal instantiation,

$\forall x \phi \to \phi(\zeta/x)$

is logically true for the possible quantifier without any qualification, only the following qualified version is logically true for the actual quantifier:

$\exists^a y (\zeta = y) \to [\forall^a x \phi \to \phi(\zeta/x)]$

where y does not occur in ζ and ζ is free for x in ϕ. In addition to the second-level concept of existence represented by $\forall^a$ as a primitive logical constant, the first-level concept of existence is definable as follows (where x and y are distinct variables):

$E!(x) =_{df} \exists^a y (x = y)$

2. A Completeness Theorem for Tense Logic

In applying the logic of actual and possible objects to the framework of tense logic we add to our logical constants the unary formula operators P and F (read, respectively, as 'It was the case that' and 'It will be the case that') and define the *tensed formulas* of a language L to be those objects that belong to every set K containing the atomic formulas of L and such that $\sim\phi$, $(\phi \to \psi)$, $P\phi$, $F\phi$, $\forall x\phi$, $\forall^a x\phi \in K$ whenever $\phi, \psi \in K$ and x is a variable. (We assume & and $\vee$ to be defined in terms of $\sim$ and $\to$ in the usual way.)

Where $\mathfrak{B}$ is a model (as defined earlier), we set $\mathfrak{U}_\mathfrak{B}$ as the universe of $\mathfrak{B}$ and $\mathfrak{P}_\mathfrak{B}$ as the set of possibilia of $\mathfrak{B}$. If L is a language and R is a relation (set of ordered pairs), then we say that $\mathfrak{A}$ is an R-history with respect to L if there are a nonempty index set I included in the field of R and an I-termed sequence $\mathfrak{B}$ of models suited to L such that (i) $\mathfrak{A} = \langle R, \mathfrak{B} \rangle$; (ii) I is identical with the field of R if I has more than one element; (iii) $\bigcup_{j \in I} \mathfrak{U}_{\mathfrak{B}_j} \subseteq \mathfrak{P}_{\mathfrak{B}_i}$, for all $i \in I$; and (iv) $\mathfrak{P}_{\mathfrak{B}_i} = \mathfrak{P}_{\mathfrak{B}_j}$, for all $i, j \in I$.

Where $\langle R, \mathfrak{B} \rangle$ is such a history, we take the members of the set I indexing $\mathfrak{B}$ to be the *moments* of the history and R to be the *earlier-than* relation ordering those moments. The structure of R is the temporal structure of the history; for example, it may have a beginning, or an end, both, or neither, and it may be discrete, dense, or continuous, and so on. Condition (iii) stipulates that whatever is actual at one time or another in a history is a possible object of that history. Condition (iv) states the requirement that whatever is a possible object at one moment of a history is a possible object at any other moment of that history.

Satisfaction and *truth in* a history $\langle R, \mathfrak{B} \rangle$ *at* a given moment i of $\langle R, \mathfrak{B} \rangle$ is understood, except where the tense operators are involved, as satisfaction and truth in the model $\mathfrak{B}_i$. The satisfaction clauses for the tense operators have the obvious references to the models associated with the moments before and after the moment i. *Validity in* a history is defined as truth at all times in that history.

If R is a relation, then ϕ is said to be R-*valid* if ϕ is a tensed formula of some language L such that for each R-history $\mathfrak{A}$ with respect to L, ϕ is valid in $\mathfrak{A}$. A tensed schematic formula ϕ is understood to *characterize* a class K of relations if for each relation R, ϕ is R-valid if, and only if, $R \in K$. Special schematic formulas can be shown to characterize various classes of relations.

In regard to the characterization of logical truth as extended to all tensed formulas, we restrict our considerations—in deference to this fundamental feature of time (or rather of *local* time)—to *serial* histories (i.e., histories whose temporal ordering is a series) and say that ϕ is *tense-logically* true if for some language L of which ϕ is a tensed formula, ϕ is valid in every serial history suited to L. Equivalently, a tensed formula ϕ is tense-logically true if, and only if, for every series R, ϕ is R-valid. Given as inference rules modus ponens, universal generalization (for $\forall$), and

(i) if $\vdash_t \phi$, then $\vdash_t \sim P \sim \phi$
(ii) if $\vdash_t \phi$, then $\vdash_t \sim F \sim \phi$

then all instances of the following schemas, together with all instances of (A1)–(A10) of Section 1 (applied now to tensed formulas), yield all and only the tense-logical truths:[3]

(A11) $\sim P \sim (\phi \to \psi) \to (P\phi \to P\psi)$

(A12) $\sim F \sim (\phi \to \psi) \to (F\phi \to F\psi)$

(A13) $\phi \to \sim P \sim F\phi$

(A14) $\phi \to \sim F \sim P\phi$

(A15) $PP\phi \to P\phi$

(A16) $FF\phi \to F\phi$

(A17) $P\phi \,\&\, P\psi \to P(\phi \,\&\, \psi) \lor P(\phi \,\&\, P\psi) \lor P(\psi \,\&\, P\phi)$

(A18) $F\phi \,\&\, F\psi \to F(\phi \,\&\, \psi) \lor F(\phi \,\&\, F\psi) \lor F(\psi \,\&\, F\phi)$

(A19) $P(\psi \,\&\, F\phi) \to P(\phi \,\&\, P\psi) \lor (\phi \,\&\, P\psi) \lor (F\phi \,\&\, P\psi)$

(A20) $F(\psi \,\&\, P\phi) \to F(\phi \,\&\, F\psi) \lor (\phi \,\&\, F\psi) \lor (P\phi \,\&\, F\psi)$

(A21) $(x = y) \to \sim P \sim (x = y) \,\&\, \sim F \sim (x = y)$, where x, y are variables

If we restrict ourselves to *tensed standard formulas* (i.e., those tensed formulas in which the actual quantifier does not occur), then (A1)–(A4), (A6), (A8), (A10)–(A21) yield all and only those tense-logical truths that are tensed standard formulas. These tense-logical truths constitute the logic of possible objects *simpliciter* as applied in the framework of tense logic (for local time). On the other hand, if we restrict ourselves to *tensed E-formulas* (i.e., those in which the possible quantifier does not occur), then (A1)–(A3), (A5), (A9), (A10)–(A21), together with the schemas $\phi \to \forall^a x \phi$, where x is not free in ϕ, and $(\zeta = \zeta)$, where ζ is an arbitrary term, as well as the following inference rules (for each natural number n)

added to those already noted above (but with universal generalization for $\forall^a$ instead of $\forall$):

(iii) if $\vdash_{t^e} \sim P \sim (\phi_0 \to \sim P \sim [\phi_1 \to \cdots \to \sim P \sim (\phi_{n-2} \to \sim P \sim \phi_{n-1}) \cdots])$

and x is not free in $\phi_0, \ldots, \phi_{n-2}$, then

$$\vdash_{t^e} \sim P \sim (\phi_0 \to \sim P \sim [\phi_1 \to \cdots \to\\ \sim P \sim (\phi_{n-2} \to \sim P \sim \forall^a x \phi_{n-1}) \cdots])$$

(iv) if $\vdash_{t^e} \sim F \sim (\phi_0 \to \sim F \sim [\phi_1 \to \cdots \to \sim F \sim (\phi_{n-2} \to \sim F \sim \phi_{n-1}) \cdots])$

and x is not free in $\phi_0, \ldots, \phi_{n-2}$, then

$$\vdash_{t^e} \sim F \sim (\phi_0 \to \sim F \sim [\phi_1 \to \cdots \to\\ \sim F \sim (\phi_{n-2} \to \sim F \sim \forall^a x \phi_{n-1}) \cdots])$$

yield all and only those tense-logical truths that are tensed *E*-formulas. These tense-logical truths constitute the logic of actual objects *simpliciter* (i.e., the logic of actualism), as applied to the framework of tense logic (for local time).

In assuming that being and existence are not the same concept, possibilism does not also assume that whatever *is* (i.e., whatever has being) either did exist, does exist, or will exist, a thesis we shall call *temporal possibilism*. Formally, this thesis is stated as follows: $\forall x[PE!(x) \lor E!(x) \lor FE!(x)]$. If we add this formula as a new axiom, then to render it tense-logically true we need only require that the condition stated in clause (iii) of the definition of an *R*-history be an identity rather than just an inclusion. A somewhat stronger assumption than the thesis of temporal possibilism—but one that still falls short of actualism as a claim about existence and being—can be made to the effect that whatever *is* either did exist or does exist (i.e., that being covers only past or present existence), leaving future existence to the realm of nonbeing (apparently because the future is indeterminate metaphysically and not just epistemically, whereas the past and the present are at least metaphysically determinate). Quantification over past objects, as well as quantification over past and present objects, can be defined in the tense logic of actual and possible objects as follows (where $\exists^p$ and $\exists^p_p$ are defined in the usual way as the duals of $\forall^p$ and $\forall^p_p$, respectively):

$$\forall^p x \phi =_{df} \forall x[PE!(x) \to \phi]$$
$$\forall^p_p x \phi =_{df} \forall x[PE!(x) \lor E!(x) \to \phi]$$

Thus, the metaphysical thesis in question, namely that what *is* either did exist or does exist, can be stated as follows: $\forall x \exists_P^P y(x = y)$. Alternatively, instead of having the concept of being in such a framework represented by the possibilist quantifier $\forall$, we can take it to be represented directly by $\forall_P^P$ as a primitive quantifier together with $\forall^a$ for the concept of existence. A sound and complete axiom set for this system is then given by (A1)–(A3), (A5), (A9), (A10), together with the schemas

$\forall_P^P x(\phi \to \psi) \to (\forall_P^P x\phi \to \forall_P^P x\psi)$

$\phi \to \forall_P^P x\phi$, where x is not free in ϕ

$\forall_P^P x\phi \to \forall^a x\phi$

$\forall^a x \exists^a y(x = y)$, where x and y are distinct variables

$\forall_P^P x[\exists_P^P y(x = y)\ \&\ \sim F \sim \exists_P^P y(x = y)]$

$\forall_P^P x \sim P \sim \phi \to\ \sim P \sim \forall_P^P x\phi$

$\sim F \sim \forall_P^P x\phi \to \forall_P^P x \sim F \sim \phi$

$(\zeta = \zeta)$, where ζ is an arbitrary term

and the inference rules modus ponens, universal generalization (for $\forall_P^P$), rules (i), (ii) as described above, and the counterpart of rule (iv) above using $\forall_P^P$ in place of $\forall^a$.[4]

3. Modality Within Tense Logic

It is significant that the first modal concepts to be discussed and analyzed in the history of philosophy are concepts based on the distinction between the past, the present, and the future, that is, concepts that can be analyzed in terms of the temporal modalities that are represented by the standard tense operators. Diodorus, for example, is reported as having argued that the possible is that which either is or will be the case, and that the necessary is that which is and always will be the case. Formally, the Diodorean modalities can be defined as follows:

$\diamond^f \phi =_{df} (\phi \vee F\phi)$

$\square^f \phi =_{df}\ \sim \diamond^f \sim \phi$

Aristotle, on the other hand, included the past as part of what is possible; that is, for Aristotle the possible is that which either was, is, or will be the case (in what he assumed to be the infinity of time), and therefore the necessary is what is always the case:

$\Diamond \phi =_{df} (P\phi \vee \phi \vee F\phi)$

$\Box \phi =_{df} \sim \Diamond \sim \phi$

Where L is a language, let $\Diamond FM_L$ be the intersection of all sets K containing the atomic formulas of L and such that $\sim \phi$, $\Diamond \phi$, $(\phi \to \psi)$, $\forall x \phi$, $\forall^a x \phi \in K$ whenever $\phi, \psi \in K$ and x is a variable. We call ϕ an *S5-formula* if $\phi \in \Diamond FM_L$, for some language L. We also say that ϕ is *S5-valid* if ϕ is an S5-formula that is tense-logically true. We obtain the system we call S5^t if to the axioms (A1)–(A10) of the logic of actual and possible objects we add all instances of schemas of the following forms:

(S5^t-1) $\Box \phi \to \phi$

(S5^t-2) $\Box(\phi \to \psi) \to (\Box \phi \to \Box \psi)$

(S5^t-3) $\Diamond \phi \to \Box \Diamond \phi$

(S5^t-4) $(x = y) \to \Box(x = y)$, where x, y are variables

and take in addition to modus ponens and universal generalization the following inference rule:

if $\vdash_{S5^t} \phi$, then $\vdash_{S5^t} \Box \phi$

It can be shown that for each S5-formula ϕ, ϕ is a theorem of S5^t if, and only if, ϕ is S5-valid—which is our completeness theorem for S5^t. For the logic of actual objects as applied to S5-formulas, we need only restrict the latter to those that are *E-formulas* (i.e., formulas in which the possible quantifier does not occur), and use only the logic of actual objects as described in Section 1 together with the axiom schemas (S5^t-1)–(S5^t-4) and one new inference rule added to those of S5^t. That is, where S5^t_e is that subsystem of S5^t that is the result of replacing (A1)–(A10) of the logic of actual and possible objects by the axioms for the logic of actual objects *simpliciter* and adding to the inference rules of S5^t the following,

if $\vdash_{S5^t} \Box(\phi_0 \to \Box[\phi_1 \to \cdots \to \Box(\phi_{n-2} \to \Box \phi_{n-1}) \cdots])$

and x is not free in $\phi_0, \ldots, \phi_{n-2}$, then

$\vdash_{S5^t} \Box(\phi_0 \to \Box[\phi_1 \to \cdots \to \Box(\phi_{n-2} \to \Box \forall^a x \phi_{n-1}) \cdots])$

then for each S5-formula ϕ that is also an E-formula, ϕ is a theorem of S5^t_e if, and only if, ϕ is S5-valid, which is our completeness theorem for S5^t_e.

For the Diodorean modalities, let $\Diamond^f FM_L$, where L is a language, be the intersection of all sets K containing the atomic formulas of L and such that $\sim\phi$, $\lozenge^f\phi$, $(\phi \to \psi)$, $\forall x\phi$, $\forall^a x\phi \in K$ whenever $\phi, \psi \in K$ and x is a variable. We say that ϕ is an S4.3-formula if $\phi \in \lozenge^f FM_L$, for some language L. Also, we say that ϕ is *S4.3-valid* if ϕ is an S4.3-formula that is tense-logically true. We obtain the system we call S4.3^t if we add to the axioms (A1)–(A10) of the logic of actual and possible objects all instances of schemas of the following forms:

(S4.3^t-1) $\square^f\phi \to \phi$

(S4.3^t-2) $\square^f(\phi \to \psi) \to (\square^f\phi \to \square^f\psi)$

(S4.3^t-3) $\square^f\phi \to \square^f\square^f\phi$

(S4.3^t-4) $\lozenge^f\phi \,\&\, \lozenge^f\psi \to \lozenge^f(\phi \,\&\, \psi) \vee \lozenge^f(\phi \,\&\, \lozenge^f\psi) \vee \lozenge^f(\psi \,\&\, \lozenge^f\phi)$

(S4.3^t-5) $\lozenge^f(x = y) \to \square^f(x = y)$, where x, y are variables

(S4.3^t-6) $\forall x \square^f\phi \to \square^f \forall x\phi$

and take in addition to modus ponens and universal generalization the same modal inference rules already described for S5^t except for having $\square^f$ where $\square$ occurs in those rules. It can be shown that for each S4.3-formula ϕ, ϕ is a theorem of S4.3^t if, and only if, ϕ is S4.3-valid, which is our completeness theorem for S4.3^t.

For the logic of actual objects as applied to S4.3-formulas (i.e., to obtain the subsystem S4.3^t_e of S4.3^t when the latter is restricted to E-formulas), we must first delete the axiom schema (S4.3^t-6), which is not an E-formula, and then replace (A1)–(A10) of the logic of actual and possible objects by the axioms of the logic of actual objects simpliciter and adopt the same modal inference rules as already described for S5^t_e except for using $\square^f$ instead of $\square$ in those rules. Then, it can be shown that for each S4.3-formula ϕ that is also an E-formula, ϕ is a theorem of S4.3^t_e if, and only if, ϕ is S4.3-valid, which is our completeness theorem for S4.3-formulas when the latter are restricted to E-formulas.

Infinitely many other modal logics can be generated in ways similar to the above by various combination of tenses (e.g., such as merely iterating new occurrences of F in the definition of the Diodorean modalities). In addition to these temporal notions of modality, the semantics for yet another can be given corresponding roughly to the idea that a formula is (conditionally) necessary (in a given history at a given moment of that history) because of the way the past has been. The semantics for this notion also yields a completeness theorem for an S5 type modal structure, and it may be used for a partial or full explication of the notions of causal modality and counterfactuals.

Finally, the semantics for yet another temporal notion of modality is available parallel to the Diodorean approach, except that the role the future plays in the Diodorean concept is altered to that of the causal future as determined by a *relativistic system* of R-histories, where R is a series. This temporal notion of modality results in an S4 type structure.[5]

4. Some Observations on Quantifiers in Tense and Modal Logic

In describing some of the theorem schemas involving quantifiers, tenses and modal operators in these different logics, we shall use $\vdash_t$, $\vdash_{t^e}$, and $\vdash_{t^P}$ to stand for *being a theorem of tense logic (for local time) with quantification over* (a) *both actual and possible objects*, (b) *over just actual objects*, and (c) *over past and present objects*, respectively. As already indicated,

$$\{\phi: \vdash_{t^e}\phi\} \subseteq \{\phi: \vdash_{t^P}\phi\} \subseteq \{\phi: \vdash_t\phi\}$$

and, therefore, we may use $\vdash_{t^e}$ to state what is provable in all three systems (a)–(c), and $\vdash_{t^P}$ for what is provable in (a) and (c). In stating some of these theorem schemas we shall also use the following counterparts of notions already defined:

$$\forall^f x \phi =_{df} \forall x [FE!(x) \to \phi]$$

$$\diamond^P \phi =_{df} (\phi \vee P\phi)$$

$$\square^P \phi =_{df} \sim \diamond^P \sim \phi$$

Leibniz's Law. We assume that ζ, η are terms, ϕ is a formula, and that ψ is obtained from ϕ by replacing one or more free occurrences of ζ by free occurrences of η. Then,

(1) $\vdash_{t^e} \square(\zeta = \eta) \to (\phi \leftrightarrow \psi)$

(2) $\vdash_{t^e} (\zeta = \eta) \to (\phi \leftrightarrow \psi)$ if ζ, η are variables

(3) $\vdash_{t^e} (\zeta = \eta) \to (\phi \leftrightarrow \psi)$ if ζ does not occur in ϕ within the scope of a past or future tense operator

(4) $\vdash_{t^e} \square^P(\zeta = \eta) \to (\phi \leftrightarrow \psi)$ if ζ does not occur in ϕ within the scope of a future tense operator

(5) $\vdash_{t^e} \square^f(\zeta = \eta) \to (\phi \leftrightarrow \psi)$ if ζ does not occur in ϕ within the scope of a past tense operator

Identity and Nonidentity. Although identity and nonidentity as expressed

Quantification, Time, and Necessity

in terms of individual variables is always necessary, that is,

$$\vdash_{te}(x = y) \to \Box(x = y), \qquad \vdash_{te}(x \neq y) \to \Box(x \neq y)$$

the same is not true for other singular terms. The relevant qualifications are as follows, where it is assumed that x and y do not occur in ζ and η, respectively:

(6) $\quad \vdash_t \exists x \Box(x = \zeta) \,\&\, \exists y \Box(y = \eta) \to$
$$[(\zeta = \eta) \leftrightarrow \Box(\zeta = \eta)] \,\&\, [(\zeta \neq \eta) \leftrightarrow \Box(\zeta \neq \eta)]$$

$\vdash_{te} \exists^a x \Box(x = \zeta) \,\&\, \exists^a y \Box(y = \eta) \to$
$$[(\zeta = \eta) \leftrightarrow \Box(\zeta = \eta)] \,\&\, [(\zeta \neq \eta) \leftrightarrow \Box(\zeta \neq \eta)]$$

$\vdash_{tP} \exists^P_P x \Box(x = \zeta) \,\&\, \exists^P_P y \Box(y = \eta) \to$
$$[(\zeta = \eta) \leftrightarrow \Box(\zeta = \eta)] \,\&\, [(\zeta \neq \eta) \leftrightarrow \Box(\zeta \neq \eta)]$$

Similar theorems hold when $\Box$ is uniformly replaced throughout (6) by $\Box^P$ or $\Box^f$, respectively.

Universal Instantiation. The law of universal instantiation does not hold in general in these logics without qualification. The different qualifications are as follows, where x and y are variables, ζ is a term in which y does not occur, and ζ is free for x in ϕ:

(7) $\quad \vdash_t \exists y \Box(y = \zeta) \to [\forall x \phi \to \phi(\zeta/x)]$

$\vdash_{te} \exists^a y \Box(y = \zeta) \to [\forall^a x \phi \to \phi(\zeta/x)]$

$\vdash_{tP} \exists^P_P y \Box(y = \zeta) \to [\forall^P_P x \phi \to \phi(\zeta/x)]$

(8) If either ζ is a variable or x does not occur in ϕ within the scope of a past or future tense operator, then

$\vdash_t \forall x \phi \to \phi(\zeta/x)$

$\vdash_{te} \exists^a y(y = \zeta) \to [\forall^a x \phi \to \phi(\zeta/x)]$

$\vdash_{tP} \exists^P_P y(y = \zeta) \to [\forall^P_P x \phi \to \phi(\zeta/x)]$

(9) If x does not occur in ϕ within the scope of a future tense operator, then

$\vdash_t \exists y \Box^P(y = \zeta) \to [\forall x \phi \to \phi(\zeta/x)]$

$\vdash_{te} \exists^a y \Box^P(y = \zeta) \to [\forall^a x \phi \to \phi(\zeta/x)]$

$\vdash_{tP} \exists^P_P y \Box^P(y = \zeta) \to [\forall^P_P x \phi \to \phi(\zeta/x)]$

(10) If x does not occur in ϕ within the scope of a past tense operator, then

$$\vdash_t \exists y \Box^f(y = \zeta) \to [\forall x \phi \to \phi(\zeta/x)]$$
$$\vdash_{te} \exists^a y \Box^f(y = \zeta) \to [\forall^a x \phi \to \phi(\zeta/x)]$$
$$\vdash_{t^P} \exists^P_P y \Box^f(y = \zeta) \to [\forall^P_P x \phi \to \phi(\zeta/x)]$$

Laws of Commutation. The possible quantifier $\exists$ commutes with both the past and future tense operators and therefore with $\Diamond^f$, $\Diamond^P$, and $\Diamond$ as well. Dually, $\forall$ commutes with $\sim P \sim$ and $\sim F \sim$ and therefore with $\Box^f$, $\Box^P$, and $\Box$ as well:

(11) $\vdash_t P \exists x \phi \leftrightarrow \exists x P \phi,\qquad \vdash_t \sim P \sim \forall x \phi \leftrightarrow \forall x \sim P \sim \phi$

$\vdash_t F \exists x \phi \leftrightarrow \exists x F \phi,\qquad \vdash_t \sim F \sim \forall x \phi \leftrightarrow \forall x \sim F \sim \phi$

$\vdash_t \Diamond^f \exists x \phi \leftrightarrow \Diamond^f \exists x \phi,\qquad \vdash_t \Box^f \forall x \phi \leftrightarrow \forall x \Box^f \phi$

$\vdash_t \Diamond^P \exists x \phi \leftrightarrow \Diamond^P \exists x \phi,\qquad \vdash_t \Box^P \forall x \phi \leftrightarrow \forall x \Box^P \phi$

$\vdash_t \Diamond \exists x \phi \leftrightarrow \exists x \Diamond \phi,\qquad \vdash_t \Box \forall x \phi \leftrightarrow \forall x \Box \phi$

The actual quantifier $\exists^a$ does not commute with the past or future tense operators except under special conditions, and even then different conditions are required for each direction—unless it is assumed that nothing ever comes to exist or ceases to exist (in symbols, $\Box \forall^a x \Box E!(x)$), in which case $\exists^a$ commutes with $\Diamond^f$, $\Diamond^P$, and $\Diamond$ as well (and therefore $\forall^a$ commutes with $\sim P \sim$, $\sim F \sim$, $\Box^f$, $\Box^P$, and $\Box$):

(12) $\vdash_{te} \forall^a x \sim P \sim E!(x) \to (\exists^a x P \phi \to P \exists^a x \phi)$

$\vdash_{te} \sim P \sim \forall^a x \sim F \sim E!(x) \to (P \exists^a x \phi \to \exists^a x P \phi)$

$\vdash_{te} \forall^a x \sim F \sim E!(x) \to (\exists^a x F \phi \to F \exists^a x \phi)$

$\vdash_{te} \sim F \sim \forall^a x \sim P \sim E!(x) \to (F \exists^a x \phi \to \exists^a x F \phi)$

$\vdash_{te} \Box \forall^a x \Box E!(x) \to (\exists^a x P \phi \leftrightarrow P \exists^a x \phi) \,\&\, (\exists^a x F \phi \leftrightarrow F \exists^a x \phi)$

$\qquad \&\, (\exists^a x \Diamond^f \phi \leftrightarrow \Diamond^f \exists^a x \phi) \,\&\, (\exists^a x \Diamond^P \phi \leftrightarrow \Diamond^P \exists^a x \phi)$

$\qquad \&\, (\exists^a x \Diamond \phi \leftrightarrow \Diamond \exists^a x \phi)$

Assumptions weaker than the condition that nothing ever comes into or goes out of existence—such as that everything presently existing always has existed and always will exist, or that everything now existing will never cease to exist, or that everything now existing always has existed—yield commutations in only one direction:

$$\vdash_{te} \forall^a x \Box E!(x) \rightarrow (\Box \forall^a x \phi \rightarrow \forall^a x \Box \phi)$$

$$\vdash_{te} \forall^a x \Box^f E!(x) \rightarrow (\Box^f \forall^a_x \phi \rightarrow \forall^a_x \Box^f \phi)$$

$$\vdash_{te} \forall^a x \Box^P E!(x) \rightarrow (\Box^P \forall^a x \phi \rightarrow \forall^a_x \Box^P \phi)$$

The quantifier $\exists^P_P$ commutes with the past and future tense operators in only one direction, each the converse to the other, and therefore it commutes with $\Diamond^P$ and $\Diamond^f$ in only one direction as well. Similarly, $\forall^P_P$ commutes with $\sim P \sim$ and $\sim F \sim$, and therefore with $\Box^P$ and $\Box^f$, in only one direction:

(13) $\quad \vdash_{tP} P \exists^P_P x \phi \rightarrow \exists^P_P x P \phi, \qquad \vdash_{tP} \forall^P_P x \sim P \sim \phi \rightarrow \sim P \sim \forall^P_P x \phi$

$\quad \vdash_{tP} \exists^P_P x F \phi \rightarrow F \exists^P_P x \phi, \qquad \vdash_{tP} \sim F \sim \forall^P_P x \phi \rightarrow \forall^P_P x \sim F \sim \phi$

$\quad \vdash_{tP} \exists^P_P x \Diamond^f \phi \rightarrow \Diamond^f \exists^P_P x \phi, \qquad \vdash_{tP} \Box^f \forall^P_P x \phi \rightarrow \forall^P_P x \Box^f \phi$

$\quad \vdash_{tP} \Diamond^P \exists^P_P x \phi \rightarrow \exists^P_P x \delta^P \phi, \qquad \vdash_{tP} \forall^P_P x \Box^P \phi \rightarrow \Box^P \forall^P_P x \phi$

$\forall^P_P$ commutes with $\Box^P$ in both directions if every past and present object always was a past or present object:

$$\vdash_{tP} \forall^P_P x \Box^P [E!(x) \lor PE!(x)] \rightarrow (\Box^P \forall^P_P x \phi \leftrightarrow \forall^P_P x \Box^P \phi)$$

Strong conditions are needed in order to commute $\forall^P_P$ with $\Box$, and in fact only a very strong condition suffices for commutation in both directions:

$$\vdash_{tP} \forall^P_P x \Box[E!(x) \lor PE!(x)] \rightarrow (\Box \forall^P_P x \phi \rightarrow \forall^P_P x \Box \phi)$$

$$\vdash_{tP} \Box \forall^P_P x \Box[E!(x) \lor PE!(x)] \rightarrow (\forall^P_P x \Box \phi \leftrightarrow \Box \forall^P_P x \phi)$$

Concluding Remarks

Tense logic is not the only framework in which both the logic of actual and possible objects and the logic of actual objects *simpliciter* have natural applications and in which the differences between possibilism and actualism can be made perspicuous. There is also, for example, the logic of intentional discourse and the differences between the possible quantifier and the actual quantifier binding variables otherwise occurring free within the scope of operators for propositional attitudes. Still, even these other frameworks, it would seem, must presuppose some account of the logic of tenses, in which case the differences between possibilism and actualism within tense logic becomes paradigmatic. Indeed, as we have indicated, this is certainly the case for the differences between possibilism and actualism in modal logic, since some of the very first modal concepts ever to be discussed in the history of philosophy have been modal concepts that can be analyzed in the framework of tense logic.

Notes

1. Sections 1 through 3 of this paper are slightly revised versions of three abstracts (with the same titles as those sections) from *The Journal of Symbolic Logic*, vol. 31 (no. 4) (1966) pp. 688–91. The abstracts are summaries of lectures given at the December, 1965 meetings of the Association for Symbolic Logic. (A preliminary version of those lectures was given at UCLA in 1963, and a final version was given at UCLA in the spring of 1965 at a public lecture constituting the defense of my doctoral dissertation.) Section 4 is from a handout that was part of a lecture given in 1966 at the Berkeley campus of the University of California.

    The revisions made here have mainly to do with symbolic notation, replacing, for example, the Tarski quantifiers $\bigwedge$ and $\bigvee$ with $\forall$ and $\exists$, and making several other similar changes. Other revisions have to do with maintaining continuity of text, correcting typographical errors, and making explicit certain points that were originally left as implicit. The observation at the end of Section 2 about using a primitive quantifier over past and present objects is from the handout of the 1966 lecture at Berkeley. It is added to Section 2 where its content is more appropriate rather than left in Section 4.
2. We could use $\forall^e$ and $\exists^e$ instead of $\forall^a$ and $\exists^a$, especially in an applied framework in which we do not want existence to have the connotation that actuality has—namely, of existing in time and being a component of causal nexuses—which is exactly the connotation that we want in a framework such as tense logic. In the pure logic of actual and possible objects that we describe in Section 1 we ignore the differences, if any, between existence and actuality and consider them to be the same concept.
3. (A19) and (A20) were subsequently shown by E. J. Lemmon to be redundant.
4. The counterpart of rule (iii) with $\forall^p_P$ in the place of $\forall^a$ is provable, and so it does not need to be assumed as a primitive rule.
5. This notion was later described in more detail in Section 15 of my essay, "Philosophical perspectives on quantification in tense and modal logic," in *Handbook of Philosophical Logic*, vol. II (ed. Gabbay, D. and Guenthner, F.), Reidel, Dordrecht (1984) pp. 309–53.

14
Definite Descriptions and the Ontological Argument*

William E. Mann

Jan Berg has presented, in a painstaking and highly compressed paper, St. Anselm's Ontological Argument, dressed in the garb of a formal language, L.[1] L is the first-order predicate calculus with identity and, particularly, with Russell's theory of descriptions.[2] It is the task of Berg's paper to reconstruct Anselm's argument in L with an eye towards (1) preserving historical accuracy and (2) preventing the argument from begging the question. In this paper I will argue that he has not completely satisfied either objective, and further, that the reason he has not is that Russell's theory of descriptions is particularly unsuitable for Anselm's argument. I will then investigate the implications of reconstructing the argument in L, supplemented not with Russell's theory, but with some alternative theories.

1

Berg presents four major arguments as candidates, only the fourth of which he thinks is successful in not begging the question. I will discuss only the first, third, and fourth; the second falls as the first does.[3]

Let 'G' be an interpreted predicate such that 'Gx' means that nothing greater than x can be conceived. 'E' will be interpreted as 'exists *in re*';

---

* This paper has benefited greatly from the criticism and encouragement of Professor Gareth B. Matthews. [The present version is somewhat revised, incorporating new translations of the passages from Anselm's works.]

thus, 'Ex' means that x exists in re. Then we have the following arguments in L:

(A1)

(P1)   God $=_{df} (\iota x)(Gx)$ Presupposition
(P2) $-(E(\iota x)(Gx)) \supset -(G(\iota x)(Gx))$ Presupposition
(P3) $G(\iota x)(Gx)$ Presupposition
(1) $E(\iota x)(Gx)$ (P2), (P3), Modus tollens
(2) $E(\text{God})$ (P1), (1), Substitution

We can eliminate 'E' as a predicate in favor of the existential quantifier and an identity sign: 'Ex' will then be expressed by '$(\exists y)(y = x)$'. (A1) will then become:

(A1')

(P1)   God $=_{df} (\iota x)(Gx)$ Presupposition
(P2') $-(\exists y)(y = (\iota x)(Gx)) \supset$
        $-(G(\iota x)(Gx))$ Presupposition
(P3) $G(\iota x)(Gx)$ Presupposition
(1') $(\exists y)(y = (\iota x)(Gx))$ (P2'), (P3), Modus tollens
(2') $(\exists y)(y = \text{God})$ (P1), (1'), Substitution

One problem with (A1) and (A1') lies in (P3). As an instance of (*Principia Mathematica*) *PM* *14.22, we have

(A) $E!(\iota x)(Gx) \equiv G(\iota x)(Gx)$

From (P3) and (A) we get

(B) $E!(\iota x)(Gx)$

which is equivalent by definition (*PM* *14.02) to

(C) $(\exists y)(x)(Gx \equiv x = y)$

However, (C) is equivalent (by *PM* *14.202 and substitution) to

$(\exists y)(y = (\iota x)(Gx))$

which is (1'). Of course, we need not have gone further than (B): with (P3) as a presupposition, we establish not only the existence of God, but of a monotheistic one at that. Thus (A1) and (A1') beg the question.[4]

Dropping (P3), then, Berg considers another possible argument, (A3).

(A3)

(P1)	God $=_{df} (\iota x)(Gx)$	Presupposition
(P2b)	$-(\exists y)(y = (\iota x)(Gx)) \supset$ $-((\iota x)(Gx) = (\iota x)(Gx))$	Presupposition
(P3b)	$(\iota x)(Gx) = (\iota x)(Gx)$	Presupposition
(1′)	$(\exists y)(y = (\iota x)(Gx))$	(P2b), (P3b), Modus tollens
(2′)	$(\exists y)(y = \text{God})$	(P1), (1′), Substitution

(A3) offers little solace for someone of Anselm's persuasion, for we have trouble with (P3b). As an instance of *PM* *14.28, we have

(D) $\quad E!\,(\iota x)(Gx) \equiv ((\iota x)(Gx) = (\iota x)(Gx))$

which, along with (P3b), yields (B) again, and thus eventually, (1′) again.

Berg finally offers a reformulation of the argument that he thinks is free from the circularity that plagues (A1)–(A3). There are three alternatives tendered to patch up our troublesome "(P3)" position: it will be sufficient for my purposes to consider any one of them.

(A4)

(P1)	God $=_{df} (\iota x)(Gx)$	Presupposition
(P2c)	$-(\exists y)(y = (\iota x)(Gx)) \supset$ $(-G(\iota x)(Gx))$	Presupposition
(P3c′)	$-(-G(\iota x)(Gx))$	Presupposition
(1′)	$(\exists y)(y = (\iota x)(Gx))$	(P2c), (P3c′), Modus tollens
(2′)	$(\exists y)(y = \text{God})$	(P1), (1′), Substitution

2

Throughout all his reformulations of the ontological argument, there is one presupposition that Berg leaves constant and unaltered—(P1). One ought to be chary about maintaining that Anselm would hold 'the unique x such that nothing greater than x can be conceived' to be a definition of God. In Anselm's *De Grammatico* it is made clear that to define a thing is to specify its essence, for ". . . the essence of any individual thing is fixed in a definition."[5] Furthermore, in Chapter X of the *Monologion* we get the picture that we know or understand a thing, if and only if we know its essence, or its definition.

> For I express *man* in one way, when I signify him by this name, that is, "man"; in another, when I think the same name silently; in another, when the mind

considers the man himself, either through an image of the body or through the reason. Through an image of the body, when it [the mind] imagines his sensible form; through the reason, when it thinks of his universal essence, which is *rational mortal animal*.

These three distinct varieties of speaking constitute the words of one's nation. But the words of that [kind of] speaking that I have put third and last, when they are of things not unknown, are natural and are the same among all nations. And since all other words are grounded on the basis of these, where these are, no other word is necessary for recognizing a thing, and where these cannot be, no other [word] is helpful for making a thing known.[6]

Is it so clear that, when we consider God as the being than whom nothing greater can be conceived, we now know all the essential properties of God? And, therefore, that 'the unique x such that nothing greater than x can be conceived' is a definition of God? It seems that Anselm did not think so, for after using the notion of a being than whom nothing greater can be conceived to show *that* God is, Anselm devotes the rest of the *Proslogion* to uncovering *what* God is. In the beginning of Chapter V, he asks: "What then are you, Lord God, than whom nothing greater can be conceived?"[7] The very asking of this question would be superfluous, as indeed would most of the remainder of the *Proslogion*, if the answer were simply 'the being than whom nothing greater can be conceived'. Moreover, if one took 'the unique x such that nothing greater than x can be conceived' as a definition of God, then it would seem that one would be hard-pressed to make much sense of the enigmatic Chapter XV of the *Proslogion*: "Therefore, Lord, not only are you that than which a greater cannot be conceived, but you are also something greater than can be conceived."[8] This amounts to saying that God is so great that he cannot be entirely conceived. I take it that this would imply that his essence cannot be fully understood. But if his essence were to be the being than which nothing greater can be conceived, then it would follow that 'the being than which nothing greater can be conceived' cannot be fully understood. Yet Anselm maintained in *Proslogion* II that even the fool understands 'the being than which none greater can be conceived'.[9] Following through on this line of reasoning, we would be forced to conclude that Anselm has contradicted himself between Chapters II and XV. To absolve Anselm of the charge of inconsistency, it is sufficient to distinguish between God's essence and God's characterization as the being than which nothing greater can be conceived.

Of course, it may be replied that the phrase 'the being than which nothing greater can be conceived' plays a dual role for Anselm. There is the low-grade role where just enough of its meaning is employed early in the *Proslogion* to reduce the fool to absurdity, and there is the high-grade role which is such that if it were possible to explicate fully this phrase, we would have God's essence or definition. So the phrase can be understood sufficiently to establish with certainty that God exists, yet not even Anselm himself understands it completely.[10] However, to admit this distinction is to admit that as it stands, the phrase 'the being than which nothing greater can be conceived' is not a specification of God's essence, simply because explication of the phrase is necessary. If so, it would be the *explicans* that gives us God's essence (and would thus be a definition of God), and not the explicated phrase.

If the relation between the term 'God' and the phrase 'the being than which nothing greater can be conceived' is not one of definiendum to definiens, what then is it? Whatever else might be involved in this relation, this much seems clear: 'the being than which nothing greater can be conceived' is a definite description and 'God' is a name, both of which purport to pick out one and the same being. On the basis of this, we might try to modify (P1) into an identity statement:

(P1′)   God = $(\iota x)(Gx)$

Some unwelcome results are attendant upon (P1′), however. As an instance of *PM* *14.13, we have

(E) $(\text{God} = (\iota x)(Gx)) \equiv ((\iota x)(Gx) = \text{God})$

(P1′) and (E) give us

(F) $(\iota x)(Gx) = \text{God}$

From *PM* *14.15 we get

(G) $((\iota x)(Gx) = \text{God}) \supset (F(\iota x)(Gx) \equiv F(\text{God}))$

where 'F' is any predicate whatsoever. Applying modus ponens to (F) and (G), we get

(H) $F(\iota x)(Gx) \equiv F(\text{God})$

By *PM* *14.01, '$F(\iota x)(Gx)$' is definitionally equivalent to '$(\exists y)((x)(Gx \equiv x = y) \& Fy)$', and applying a definitional interchange to (H) yields

(I) $(\exists y)((x)(Gx \equiv x = y) \& Fy) \equiv F(\text{God})$

Inasmuch as 'F' stands for any property, we can let it be Berg's property G and get

(I′) $(\exists y)((x)(Gx \equiv x = y) \& Gy) \equiv G(\text{God})$

But look what has happened now. One-half of the biconditional (I′) asserts that if God is such that nothing greater than he can be conceived, then there exists exactly one entity such that nothing greater than that entity can be conceived. We get an even more general result from (I). If God is claimed to have any property at all, then he exists. Yet there certainly are properties for which it is difficult to see how they could be denied of God. For example, let 'H' be the predicate 'is identical with God'. Then, from (I), we have

(I″) $(\exists y)((x)(Gx \equiv x = y) \& Hy) \equiv H(\text{God})$

If we say, then, that God is identical with himself, then on Russell's theory of descriptions, we obtain the left-hand side of the biconditional (I″).[11] Similarly, let 'J' be the predicate 'is identical with the unique x such that nothing greater than x can be conceived'. Substituting into (I) gives us

(I‴) $(\exists y)((x)(Gx \equiv x = y) \& Jy) \equiv J(\text{God})$

But '$J(\text{God})$' just is (P1′), and once again we obtain the left-hand side of our biconditional.

(I) puts Anselm's fool in quite an embarrassing position: either he cannot say anything about God at all, or he already agrees that God exists. Not only is the fool embarrassed; so is Anselm, for now the ontological argument has been trivialized, and Anselm is indeed guilty of the charge that God has been presupposed into existence.

Well and good, it might be argued, but so far it has only been shown that (P1′) has this undesirable consequence. It may be that if we reinstate (P1), even though it does not seem to be something that Anselm would have accepted, then (A4) will go through. After all, the kind of definition that (P1) is need not be taken to be what Anselm would have called a definition—something that gives the essence or nature of some being. We can just treat (P1) as a stipulation or an abbreviation, with none of the metaphysical trappings of an Aristotelian real definition.

This will not help matters either. For one of the cardinal features of a definition in a system (such as L) is that definiens and definiendum be mutually substitutable, at least in all extensional contexts, without change of truth value.[12] Consider any extensional predicate 'F', and suppose it is predicated, either truly or falsely, of God. Thus we have '$F(\text{God})$'. But by our principle of substitutivity, this gives us '$F(\iota x)(Gx)$'. On Russell's theory, this is definitionally equivalent to '$(\exists y)((x)(Gx \equiv x = y) \& Fy)$', and again the fool (and Anselm) is in a box. It was the task of Berg's paper to show that formulations (A1)–(A3) ran into difficulties at (P2) and (P3) and their variants, whereas (A4) would save Anselm. However,

Definite Descriptions and the Ontological Argument 263

(A4) still contains (P1) or my variant (P1'), in which case even it presupposes the existence of God.[13]

3

All the difficulties encountered above are engendered, I suggest, by the utilization of Russell's theory of descriptions. The theory has come under attack from many quarters, for many reasons, and the reason why it is an unfortunate candidate for giving a sympathetic rendition of Anselm's argument in L can be given on an intuitive level. Consider these two sets of sentences:

(A)

(1) The winged horse ridden by Bellerophon was born, full-grown, from Medusa's slain body.
(2) The winged horse ridden by Bellerophon lives an immortal life on Olympus.
(3) The author of "On denoting" wrote *Introduction to Mathematical Philosophy* while in prison.
(4) The author of "On denoting" once thought the ontological argument was sound.

(B)

(1) The winged horse ridden by Bellerophon was sired by Bellerophon.
(2) The winged horse ridden by Bellerophon won the Triple Crown in 1959.
(3) The author of "On denoting" has never changed his mind in philosophy.
(4) The author of "On denoting" is an arch-conservative.

We should like to say that the members of (A) have at least one thing in common—they are all true. Similarly, all the members of (B) are false. There is a difference between those members numbered (1) and (2), and those numbered (3) and (4). In the former, the subject does not exist; in the latter, the subject does. On Russell's view, (A1) and (A2) are false, and this is where his theory parts company from ordinary life. On Russell's theory, (A2), for example, will be true just in case (a) there is a winged horse ridden by Bellerophon, (b) there is only one such horse, and (c) he lives an immortal life on Olympus. (A2) is false, then, because condition

(a) is not met. (B1) will be false for exactly the same reason. But in ordinary life we would like to say that (B1) is false, all right, but not simply because Pegasus does not exist—rather, that his father was Poseidon, not Bellerophon.

The trouble with Russell's theory, with respect to the ontological argument, is that ordinarily we sometimes want to affirm some properties, and deny others, to things we know not to exist, or whose existence is putative. On Russell's view, the privilege of ascribing properties is reserved only for those things we know to exist, or, to put it differently, predication presupposes the existence of the subject of predication. Russell's theory is inappropriate because Anselm's argument depends essentially upon a notion of predication more akin to what we might call the ordinary sense of predication. In order to justify this claim, I want to present a way of looking at Anselm's motivation and strategy behind the ontological argument.[14]

It is illuminating to consider Anselm as being sensitive to the problems of negative existential statements—that is, statements denying the existence of some individual or species. Traditionally, one such problem might be put as follows: The statement, 'Pegasus does not exist', is apparently true, yet on closer inspection, it seems to be either false or meaningless (assuming we do not equate, as has often been done, a false statement with a meaningless one). For if Pegasus in no sense exists, then the noun of the sentence 'Pegasus does not exist' has no referent, or is not about anything, and thus the sentence is meaningless. On the other hand, if 'Pegasus' does have a referent, then the statement is false. There has been a host of solutions proposed to deal with this, and other kindred problems, and Russell's theory itself (especially as it is employed by Quine) is one of the most brilliant of them all.

Anselm has a solution to this sort of problem, although it is an implicit one: there is no explicit discussion that I know of in the Anselmian corpus. The reader is referred to the previously mentioned Chapter X of the *Monologion*, and this passage from Chapter II of Anselm's *Response*:

> For just as what is conceived is conceived by conception, and what is conceived by conception, as it is conceived so it is in conception; so what is understood is understood by the understanding, and what is understood by the understanding, as it is understood so it is in the understanding. What is clearer than that?[15]

Given this background, it would seem natural enough for Anselm to maintain that in the sentence 'Pegasus does not exist', we understand the term 'Pegasus' because and only because Pegasus is in our understanding.

Pegasus thus has a mode of existence, *in intellectu*. What we are denying when we deny the existence of Pegasus is that he exists *in re*, in the real world outside our understanding or imagination. So the statement 'Pegasus does not exist' can be counted as meaningful and true, if it is taken in the sense of 'Pegasus does not exist *in re*'. The statement is false if it is uttered significantly and intended in the sense of 'Pegasus does not exist *in intellectu*'.[16]

If we accept this account of what Anselm would say about negative existential statements, we can now come to grips with a negative existential statement that does interest him very much—'The being than which nothing greater can be conceived does not exist'. It is Anselm's claim that when the fool says this, he certainly does not mean that the being than which nothing greater can be conceived does not exist *in intellectu*.[17] Rather, the fool means to say that no such being exists *in re*. From this assertion, Anselm proceeds to derive a contradiction, thus establishing the existence of the being than which nothing greater can be conceived.

The relevant feature that emerges from this is that Anselm has a way of making sense of negative existential statements, and moreover, has a way of ascribing properties to things that do not exist *in re*, or whose existence in re is being argued for. In the case of, say, 'Pegasus is swift', the predicate 'swift' is predicated of the Pegasus which exists *in intellectu*; the statement is about something that exists in our understanding. There is no commitment made to Pegasus' existence *in re* from the mere fact that we assign some properties to him. Similarly, there is no apparent commitment to the real existence of God if one assents to the statement 'The being than which nothing greater can be conceived is greater than all other beings'. As it stands, we need say no more than that this statement is about the being than which nothing greater can be conceived which exists in *intellectu*.

It is important to notice that Anselm is quite explicit in maintaining that when one understands something, that very thing exists in the understanding. We have seen one passage already from Chapter II of the *Response* bearing this out. Consider now the very next paragraph, paying attention to the occurrences of the word 'it':

> Next I said that if *it* [the being than which nothing greater can be conceived] is even in the understanding alone, *it* can be conceived to exist in reality also, which is greater. Therefore if *it* is in the understanding alone, *it* itself, namely 'that than which a greater cannot be conceived', is that than which a greater can be conceived. What, I ask, is more logical? For if *it* is even in the understanding alone, can *it* not be conceived to exist also in reality? And if *it* can be, does not he who conceives of this conceive of something greater than that being, if *it* is in the understanding alone? Thus what is more logical than

that if 'that than which a greater cannot be conceived' is in the understanding alone, that is the same as to be that than which a greater can be conceived?[18]

Finally, consider these passages concerning the claim that there is no commitment to a thing's existing *in re* from the simple fact that it exists *in intellectu*:

> For it is one thing for a thing to be in the understanding, and another to understand a thing to exist. For when a painter conceives beforehand that which he is to make, he certainly has it in the understanding, but he does not yet understand to exist that which he has not yet made. However, when he has painted it, he both has it in the understanding and understands that that which he has now made exists.[19]
>
> ... I was trying to prove what was doubtful, to whom at first it was enough to show that being to be understood and to be in the understanding in some way, so that it might be considered subsequently whether it was in the understanding alone, as false things are, or also in reality, as true things are. For if false things and doubtful things are understood and are in the understanding in this way, that when they are spoken of, the hearer understands what the speaker signifies, nothing prohibits what I have spoken of from being understood and being in the understanding.[20]

There is one all-important exception to this latter doctrine, according to Anselm, and that is in the case of God. What establishes God's existence is not that we admit that the being than which nothing greater can be conceived has some properties or other—Pegasus has properties, but does not, for all of that, exist. God's existence is established by the logically unique properties involved with the being than which nothing greater can be conceived. Such a being cannot exist only *in intellectu*, for then a greater being can be conceived—one who in addition, exists *in re*. This case is the exceptional one. For any other being, real or imaginary, it is never self-contradictory to deny its *in re* existence,[21] but in the case of the being than which nothing greater can be conceived, it is self-contradictory.

Now we are in a position, hopefully, to see why Russell's theory of descriptions is unsuitable for Anselm's purposes. For Russell, to assign a predicate to a description-term is to presume that the description-term designates an entity *in re*.[22] For Anselm, this presumption has an important qualification. To assign a predicate to a description-term is to presume that the term designates, but not necessarily anything *in re*. In cases where a thing is known not to exist *in re*, or where its existence is moot, the receptacle of predicates is something *in intellectu*. What keeps Anselm's argument from being blatantly circular is his distinction between modes of existence. Russell's theory fails in fairly reproducing Anselm's argument not only because it intentionally ignores the notion of modes of existence, but, more basically, because it proceeds upon a rather stringent view concerning the connection between predication and existence.

4

A formal language like L, supplemented with Russell's theory of descriptions, does not produce a very happy result for Anselm's argument. Can we supplement L with another theory of descriptions and hope to do any better? If we remember that for Anselm, to talk about something, and to ascribe properties to it, is not to assume that the thing exists *in re*, it would seem natural to turn to a theory of descriptions that is free of such existential presuppositions. A discussion of such theories arises from an interest in developing a "free" logic, which is basically a language like the classical first-order predicate calculus, but with an important modification. The requirement that every singular term (i.e., either individual constants or definite descriptions) have a denotation is waived. Any singular term, from a purely formal standpoint, may or may not have a denotation in a free logic.

One such theory of descriptions has been formulated by Jaakko Hintikka.[23] As the basis for his theory, he puts forth the following contextual definition schemata:

(SA) $a = (\iota x)(Fx) \equiv (Fa \,\&\, (x)(Fx \supset x = a))$

and

(SB) $y = (\iota x)(Fx) \equiv (Fy \,\&\, (x)(Fx \supset x = y))$

Consider these examples:

(1) Russell = the author of "On denoting" if and only if Russell authored "On denoting" and nobody else did.
(2) Homer = the author of *The Iliad* if and only if Homer authored *The Iliad* and nobody else did.
(3) Pegasus = the winged horse ridden by Bellerophon if and only if Pegasus was a winged horse ridden by Bellerophon and nothing else was a winged horse ridden by Bellerophon.
(4) God = the being than which nothing greater can be conceived if and only if God is a being than which nothing greater can be conceived and nothing else is a being than which nothing greater can be conceived.

Clearly, (1)–(4) are all translations of substitution instances of (SB). Moreover, in Hintikka's system, there are no existential presuppositions involved in the singular terms; an individual constant or definite description may or may not have a denotation. Thus, if we transcribe (4) into a statement of the form of (SA), symbolizing 'God' by an individual constant of L and 'the unique x such that nothing greater than x can be conceived' by a definite description, it does not follow that we have presumed the existence of anything at all. It is because of this that Hintikka's theory is more amenable to Anselm's argument than Russell's.

It is natural to ask how one does indicate existence in Hintikka's system. That is, how do we show that some particular individual constant, 'a', has a denotation? Hintikka argues that the formal counterpart to 'a exists' is '$(\exists x)(x = a)$'.[24] Given that this serves to show that a exists, we can use the same method to claim that '$(\iota x)(Fx)$' has a denotation— '$(\exists y)(y = (\iota x)(Fx))$'. From (SB), by substitution, we now have a schema that explicates what it means for the unique x such that x Fs to exist:

(SC) $(\exists y)(y = (\iota x)(Fx)) \equiv (\exists y)(Fy \,\&\, (x)(Fx \supset x = y))$

Employing Hintikka's theory, we turn back now to Berg's (A1'), and modify it to produce the following argument:

(A1'')

(P1')	God $= (\iota x)(Gx)$	Presupposition
(P2')	$-(\exists y)(y = (\iota x)(Gx)) \supset$	
	$-(G(\iota x)(Gx))$	Presupposition
(P3)	$G(\iota x)(Gx)$	Presupposition
(1')	$(\exists y)(y = (\iota x)(Gx))$	(P2'), (P3), Modus tollens
(2')	$(\exists y)(y = $ God$)$	(P1'), (1'), Substitution

A curious feature emerges from using Hintikka's theory in this argument. It turns out that presupposing (P3) is otiose, and that it can be derived from (P1'). The derivation is as follows. As an instance of (SB) we have

(K)   God $= (\iota x)(Gx) \equiv (G(God) \,\&\, (x)(Gx \supset x = $ God$))$

(P1') and (K) yield

(L) $G($God$) \,\&\, (x)(Gx \supset x = $ God$)$

from which, by simplification, we get

(M) $G($God$)$

By a substitution licensed by (P1′) we get, from (M),

$G(\iota x)(Gx)$

which is (P3).

At any rate, it appears that we now have a valid, noncircular argument for the existence of God. No longer are we bothered with the presuppositions (P1) or (P1′), which, in the framework of Russell's theory, shipwreck Anselm's aspirations. For in Hintikka's theory, there is no presupposition that the terms in (P1′) designate anything *in re*.

Our hopes are short-lived, however. Lambert has shown that, if we accept the unconditional validity of '$(\iota x)(Fx) = (\iota x)(Fx)$', where '$F$' is any predicate whatsoever, then Hintikka's system for free description theory is inconsistent. Schema (SB) yields

(SD) $y = (\iota x)(Fx) \supset (Fy \ \& \ (x)(Fx \supset x = y))$

Truth-functionally, (SD) gives us

(SE) $y = (\iota x)(Fx) \supset Fy$

If we allow that '$(\iota x)(Fx) = (\iota x)(Fx)$' is true whether '$(\iota x)(Fx)$' designates or not, then by substituting into (SE), we obtain

(SF) $(\iota x)(Fx) = (\iota x)(Fx) \supset F(\iota x)(Fx)$

and by modus ponens,

(SG) $F(\iota x)(Fx)$

But as indicated in Note 4, having (SG) as a theorem results in a contradiction when 'F' is interpreted as the predicate '$(\lambda x)(Fx \ \& \ -Fx)$'.[25]

Lambert has investigated the possibility of remedying this theory, while still cleaving to a free logic, and has offered the following alternatives. We can avoid the difficulty engendered by Hintikka's theory by basing a free description theory on the two axiom-schemata,

(SH) $(y)(y = (\iota x)(Fx) \equiv ((x)(Fx \supset x = y) \ \& \ Fy))$

and

(SI) $y = (\iota x)(x = y)$

in conjunction with a modified rule of universal specification. We replace 'From $(x)(Fx)$ to infer Fy' with 'From $(x)(Fx)$ and $(\exists y)(y = z)$ to infer Fz'.[26] We can effect a reduction in this system by introducing one contextual definition schema, viz.,

(SJ) $(\iota x)(Fx) = y \equiv (z)(y = z \equiv (Fz \ \& \ (x)(Fx \supset x = z)))$

from which (SH) and (SI) are derivable.[27]

If we inspect (A1″) in terms of Lambert's theory, we will find that (P3) is no longer deducible from (P1′), for we are not allowed to treat

God $= (\iota x)(Gx) \equiv ((x)(Gx \supset x =$ God$) \ \& \ G($God$))$

as an instance of (SH) unless we know that

$(\exists y)(y =$ God$)$

is true, which is just what (A1″) is attempting to prove. Of course, (P1′) carries no existential presuppositions for Lambert, and it appears now that we have secured a valid, noncircular argument for Anselm. In fact, we have more than one, for (A4) also seems to meet the test, as does (A3). Concerning (P3b) in (A3), Lambert is willing to allow the unconditional assertability of a statement of its form,[28] although, as we have seen, Russell and Hintikka are not.

If these arguments are valid, it is up to a detractor of them to challenge their soundness. But certainly that task is more than I promised in this paper.

Notes

1. Jan Berg, "An examination of the ontological proof," *Theoria*, vol. 27 (1961) pp. 99–106.
2. The theory is presented formally in Alfred North Whitehead and Bertrand Russell, *Principia Mathematica*, 2d edn., vol. 1, Cambridge University Press (1960) pp. 30–1, 66–71, 173–86 (*14).
3. See Berg, *op. cit.*, pp. 104–5.
4. Berg shows that (P3) cannot be taken as an instance of a theorem of the form '⊢ $F(\iota x)(Fx)$', where 'F' is any predicate whatsoever, for if this were a theorem, then, by *PM* *14.01, we could derive '⊢ $(\exists y)((x)(Fx \equiv x = y) \ \& \ Fy)$', and by distribution of the existential quantifier, obtain '⊢ $(\exists y)(x)(Fx \equiv x = y)$' and '⊢ $(\exists y)(Fy)$'. Of these, the former asserts that every property is exemplified by exactly one thing, while the latter says that every property is exemplified. (See Note 1, pp. 104–5.) Karel Lambert has shown that a contradiction can be derived from our purported theorem. If 'F' can be any predicate at all, then it can be the predicate '$(\lambda x)(Fx \ \& \ -Fx)$', from which we get, by concretion, '$F(\iota x)(Fx \ \& \ -Fx) \ \& \ -F(\iota x)(Fx \ \& \ -Fx)$'. See Lambert, "Notes on E! III: A theory of descriptions," *Philosophical Studies*, vol. 13 (1962) pp. 51–9, especially p. 54.
5. St. Anselm, *De Grammatico*, in *S. Anselmi Cantuariensis Archiepiscopi Opera*

Omnia (ed. Franciscus Salesius Schmitt), Friedrich Frommann Verlag [Günther Holzboog], Stuttgart-Bad Cannstatt (1968), Tome I, Volume I, p. 152. All translations are my own.

6. St. Anselm, *Monologion*, X, in Schmitt (Note 5), p. 25.
7. *Proslogion*, V, in Schmitt (Note 5), p. 104.
8. *Proslogion*, XV, in Schmitt (Note 5), p. 112.
9. See Schmitt (Note 5), p. 101.
10. "So when 'that than which nothing greater can be conceived' is said, without doubt what is heard can be conceived and understood, even if that being, than which a greater cannot be conceived, cannot be conceived or understood." St. Anselm, *Responsio Editoris*, IX, in Schmitt (Note 5), p. 138.
11. We might attempt to block this consequence by allowing identity ascriptions to hold only when the term in question is known to designate. Such a restriction strikes me as too ad hoc. Consider this elegant and succinct argument of LeBlanc and Hailperin: "We feel indeed that a statement of the form $W = X$ is true if and only if X designates whatever W designates. But W designates whatever W designates, whether or not W designates anything. Hence $W = W$ should be true, whether or not W designates anything." (Hugues LeBlanc and Theodore Hailperin, "Nondesignating singular terms," *Philosophical Review*, vol. 68 (1959) pp. 239–43, especially p. 242.)
12. See, for example, Whitehead and Russell (Note 2), p. 11, where the stronger claim, "without change of meaning," is made.
13. It is perhaps significant to note that textually, Anselm himself does not invoke the doctrine of (P1) until after he has proved the existence of the being than which nothing greater can be conceived. (See Berg's (Q4) and (Q5), p. 101.) Thus, perhaps a more historically accurate rendition of Anselm's argument would proceed only with, say, (P2c) and (P3c') to get (1'). The transition from (1') to (2') might then be licensed by the following "meaning postulate": 'God' means the same as 'the being than which nothing greater can be conceived'. It would seem that the difficulties of (P1) and (P1') are circumvented, because the crucial terms are being mentioned, and not used.
14. This way of appreciating Anselm was pointed out to me by Professor Matthews. This does not imply that he would agree with my presentation of it, or with the conclusions that I draw from it.
15. *Responsio Editoris*, II, in Schmitt (Note 5), p. 132.
16. Not only false but self-defeating: On the theory that I am attributing to Anselm, if a person uttered this sentence, understanding its terms, he would be making a pragmatic contradiction.
17. "Therefore, even the fool is convinced that something than which nothing greater can be conceived is at least in the understanding, since when he hears this, he understands it, and whatever is understood is in the understanding." *Proslogion*, II, in Schmitt (Note 5), p. 101.
18. *Responsio Editoris*, II, in Schmitt (Note 5), p. 132.
19. *Proslogion*, II, in Schmitt (Note 5), p. 101.
20. *Responsio Editoris*, VI, in Schmitt (Note 5), p. 136.
21. Although it may be "existentially self-defeating" to deny one's own existence. See Jaakko Hintikka, "*Cogito, Ergo Sum*: inference or performance?" *Philosophical Review*, vol. 71 (1962) pp. 3–32. (Chapter 7 in this volume.)
22. See especially *PM* *14.01. The language of *Principia Mathematica* does not

recognize modes of existence; in particular, the existential quantifier is univocal.
23. Jaakko Hintikka, "Towards a theory of definite descriptions," *Analysis*, vol. 19 (1958–1959) pp. 79–85. See also his "Existential presuppositions and existential commitments," *Journal of Philosophy*, vol. 56 (1959) pp. 125–37; "Definite descriptions and self-identity," *Philosophical Studies*, vol. 15 (1964) pp. 5–7; "Studies in the logic of existence and necessity," *The Monist*, vol. 50 (no. 1) (1966) pp. 55–76.
24. See especially "Existential presuppositions and existential commitments," pp. 133–4; "Studies in the logic of existence and necessity," pp. 63–4, 70–3. (See Note 23.)
25. See Lambert (Note 4), pp. 53–4. In his reply to Lambert, "Definite descriptions and self-identity," Hintikka suggests that we ought not to accept the universal applicability of '$(\iota x)(Fx) = (\iota x)(Fx)$'. For a critique of Hintikka's proposal, see Karel Lambert, "Definite descriptions and self-identity: II," *Philosophical Studies*, vol. 17 (1966) pp. 35–43.
26. See Lambert (Note 4), pp. 52, 57–8.
27. See Karel Lambert, "Notes on E! IV: A reduction in free quantification theory with identity and descriptions," *Philosophical Studies*, vol. 15 (1964) pp. 85–8.
28. See Lambert, "Definite descriptions and self-identity: II" (Note 25), pp. 39–40.

15
Predication and Ontological Commitment

Karel Lambert

1. Introduction

Current theory about what there is is awash in new and resurrected kinds of objects. Perry and Barwise have discovered situations; Fine has resurrected arbitrary objects; a few years ago Scott was championing virtual objects; Simons, Mulligan, and Smith, the first of whom professes formal ontology almost in the suburbs of Kirchberg, have sought lasting status for moments; and, finally, Routley, Parsons, and many others, have rediscovered nonexistent objects. The remarks in the rest of this chapter bear directly only on nonexistent objects, an enduringly fascinating if highly dubious kind of entity.

Now it is an historical fact that one of Russell's greatest philosophical contributions was to highlight the role that premisses about logical form play in ontological arguments. A pair of quotations will introduce his point that great metaphysical systems are often not only based on, but are debased by, the belief that certain statements of philosophical discourse are logically subject–predicate in form.

Speaking of Hegel's Absolute Idealism—Russell wrote in *Our Knowledge of the External World*:

> Mr. Bradley has worked out a theory according to which, in all judgment, we are ascribing a predicate to Reality as a whole; and this theory is derived from Hegel. Now the traditional logic holds that every proposition ascribes a predicate to a subject, and from this it easily follows that there can be only one subject, the Absolute, for if there were two, the proposition that there were two would not ascribe a predicate to either. Thus Hegel's doctrine, that

philosophical propositions must be of the form, "the Absolute is such and such," depends on the traditional belief in the universality of the subject–predicate form. This belief, being traditional, scarcely self-conscious, and not supposed to be important, operates underground, and is assumed in arguments which, like the refutation of relations, appear at first sight to establish its truth.[1]

Speaking of Meinong's theory of objects, Russell wrote in the *Introduction to Mathematical Philosophy*:

> The question of "unreality," which confronts us ... is a very important one. Misled by grammar, the great majority of those logicians who have dealt with this question have dealt with it on mistaken lines. They have regarded grammatical form as a surer guide in analysis than, in fact, it is ... [So] many logicians have been driven to the conclusion that there are unreal objects. It is argued, e.g., by Meinong [*Untersuchungen zur Gegenstandtheorie und Psychologie* (1904)], that we can speak about "the golden mountain," "the round square" and so on; [and] we can make true propositions of which these are the subjects ...[2]

And even earlier, in the famous proof in *Principia Mathematica* that definite descriptions are incomplete symbols, Russell urged (in effect) that if the false statement

(1)   The round square exists

were regarded as having the same logical form as the statement

(2)   Socrates is mortal

a statement which is logically subject–predicate in form, it would commit one to the round square, a commitment he, Russell, argued to be paradoxical.[3] The important point is that it is not the fact that a statement is a true predication but the mere fact that it is a predication that commits one to objects.

There is one prestigious contemporary philosopher who has long believed that Russell and Meinong were wrong in their common view that statements such as the statements

(3)   Vulcan rotates on its axis

and

(4)   The round square is round

would commit one to the nonexistent objects Vulcan and the round square in virtue of their being logically subject predicate in form, were that indeed their form. That philosopher is W. V. Quine. Yet, in a way, Quine ultimately reneges on that opinion. I, in contrast, shall persevere in it.

And that is the major purpose of this chapter; I shall argue that Quine should have held fast, and that his reasons for not doing so are unconvincing.

2. Predication and Ontological Commitment: Meinong and Russell vs. Quine

In what follows, the statements of concern are parts of unregimented philosophical discourse. By 'unregimented' I mean that the statements have not yet been paraphrased into a technical idiom of one's choice. For instance, the technical idiom preferred by Quine is an extensional idiom consisting of truthfunctions, quantifiers, general terms and variables. An unregimented statement such as (3), that is, the statement

Vulcan rotates on its axis

gets paraphrased into the regimented idiom as

(5)   There exists something that vulcanizes and rotates on its axis

where the expression 'vulcanizes' is the general term correlate in the regimented idiom of the singular term 'Vulcan' in the unregimented idiom. The distinction between regimented and unregimented idioms is important here because Quine's own account of predication concerns, in the primary instances, statements in unregimented philosophical discourse.

What, then, does the Meinong–Russell conception of predication look like? A fair reconstruction, I believe, is this: a statement is logically subject–predicate in form if it consists of a general term juxtaposed to a singular term whose truthvalue, true or false, is a function of whether the general term is true (or false) of what the singular term refers to.

Two things are important about this way of expressing the Meinong–Russell conception of predication. First, the language is Quineian in character in order to facilitate later comparison with Quine's own conception of predication. Second, though a departure from the language of Meinong and Russell—I do not know that either ever used the expression 'singular term', despite the fact that it was in use in philosophy during the time they flourished—it captures their intentions well. To see this, assume the principle of bivalence, and assume also the principle that if a general term is true (or false) of what a singular term refers to, then there is an object that is the referent of that singular term. (These are principles both Meinong and Russell espoused though they probably would have expressed them differently.) Then the inference

(6)   '*Gs*' is a predication
      Therefore, there is an object that is *s*

is valid, where *G* is a general term placeholder and *s* a singular term placeholder. And indeed as urged earlier, this is a hallmark of the Meinong–Russell conception of predication. Of course, the two philosophers differed greatly in which statements are predictions. Meinong, who believed that the statements

   Vulcan rotates on its axis

and

   The round square is round

are predications, reasoned downward, on the basis of (6), that there is an object that is Vulcan and that there is an object that is the round square. But Russell, unable to ingest either nonexistent possible or nonexistent impossible objects, reasoned upward, on the basis of (6), that the example statements are not predications, and asserted that Meinong was driven to believe otherwise only because, as documented earlier, he was "misled by [presumably German] grammar."

Let me digress a moment to mention the obvious. If one or the other of the above assumptions is abandoned, then the conception of predication in question is not ontically committing, that is, (6) no longer holds. For instance, I can think of two contemporary philosophers who, at least during their nonmodal or nonepistemic moments, apparently accept the present conception of predication (or something very much like it), but reject (6) because they reject the principle of Bivalence. One is Bas van Fraassen in his essay "Singular terms, truthvalue gaps and free logic,"[4] (Chapter 4 in this volume) though he makes an exception for existence statements such as 'Vulcan exists' which he regards as false in virtue of 'Vulcan' being irreferential. The other is Kit Fine who does not even rule out statements of the form such as '*s* exists' as truthvalueless where '*s*' is genuinely irreferential.[5]

To return to the topic of concern, Quine, who believes both in bivalence and in the other principle expressed above, nevertheless rejects (6). The implication is that where Meinong and Russell go wrong is in the very conception of predication itself. And indeed this is borne out by comparing Quine's own statement of predication with the reconstruction of the Meinong–Russell view presented earlier.

Here is Quine's statement of predication: "Predication," he says, "joins a general term and a singular term to form a sentence that is true or false according as the general term is true or false of the object, if any, to which the singular term refers."[6] The only real difference between this statement of predication and the reconstruction of the Meinong–Russell conception mentioned earlier is the occurrence of the words "if any" in Quine's characterization; but it is a very significant difference. For those words allow a statement to be a predication even when containing only irreferential singular terms. Contra Russell, for instance, the statement

Vulcan rotates on its axis

qualifies as a predication for Quine even though, contra Meinong, the singular term 'Vulcan' may specify no object at all. It follows on Quine's view that the inference (6), that is, the inference

'Gs' is a predication
Therefore, there is an object that is s

is invalid; predication is not ontically committing. An indeed is not this what one would expect from the philosopher who has emphasized that it is only certain quantificational statements, and perhaps what implies them, that are ontically committing?

Nevertheless, when one considers what Quine calls "the canonical language," statements such as

Vulcan rotates on its axis

do not even occur let alone survive as predications. The only sentences that could conceivably qualify as predications in the canonical language are sentences with free variables, sentences that, on assignment, yield true or false statements in accordance with his conception of predication. Indeed, they are even predications in the stronger sense of Meinong and Russell. Why is this so? Why does Quine support an ontically noncommitting sense of predication yet disallow any statements in the canonical idiom to so qualify? The answer lies in his concerns about irreferential singular terms and in his subsequent policy for the elimination of all singular terms.

Among the reasons cited by Quine for disallowing irreferential singular terms in the canonical language, and hence statements such as

Vulcan rotates on its axis

are these. First, predications, he holds, are supposed to be contexts par-excellence in which singular terms enjoy purely referential position. Now "the intuitive idea behind 'purely referential position' [is] supposed to be that the term is used purely to specify the object, for the rest of the sentence to say something about."[7] But the singular term 'Vulcan' in the predication 'Vulcan rotates on its axis' clearly fails in its allotted task. So it is anomalous, though not contradictory, to hold that 'Vulcan' has purely referential position in the predication

　　Vulcan rotates on its axis

So in the interests of a nonanomalous concept of having purely referential position, predications with irreferential singular terms should be disbarred from the canonical language. Note that if predications in the canonical language are confined to sentences with free variables, the notion of having purely referential position is no longer anomalous because, for Quine, free variables always refer in predications.

Second, Quine believes that many, if not most, predications of the 'Vulcan rotates on its axis' variety introduce truthvalue gaps; that is, they are truthvalueless. But this does not harmonize with the two-valued character of the canonical language, a language of reform supposed to be adequate for the purposes of doing science and philosophy. So in the interests of eliminating truthvalue gaps, predications containing irreferential singular terms cannot be allowed to survive in the canonical language.

There is another matter, a matter of principle rather than of convenience, that Quine does not notice but that might be used to support the banishment of predications such as

　　Vulcan rotates on its axis

from the canonical language, namely, that their inclusion threatens the extensionality of the canonical language in the sense that counting them as predications does not imply that coextensive predicates (or general terms) substitute *salva veritate*.

The argument, which has been detailed elsewhere,[8] is this. Assume that the quantifiers have existential import, that general terms are coextensional (à la Quine) just in case they are true of the same (existent) objects, that the singular term 'Vulcan' specifies no existent object, and that the statement 'Vulcan exists' is false. Then the general terms 'rotates', 'object such that it exists and rotates', and 'object such that it rotates if existent' are coextensive. But the predication

　　Vulcan rotates

if true, yields the false statement

　　Vulcan exists and Vulcan rotates

upon appropriate substitution of the complex general term 'object such that it exists and rotates' for the general term 'rotates'. On the other hand, if interpreted as false, the predication

   Vulcan rotates

yields the true statement

   If Vulcan exists, then it rotates

upon appropriate substitution of the complex general term 'object such that it rotates if existent' for the general term 'rotates'. (This argument assumes that the principle of general term abstraction holds. That principle says, roughly, that where s is a singular term, s is an object such that it is so and so if and only if s is so and so.) So, if predications such as

   Vulcan rotates

were allowed in the canonical idiom, its extensional character would prima facie be threatened.

The importance of Quine's proposed grand elimination of all singular terms from the canonical language is now clear; truthvalue gaps, the anomalous concept of having purely referential position and the threat of extensionality evaporate in the full sweep of Quineian paraphrase. So, though Meinong and Russell are wrong about the ontically committing character of predication, a reformed idiom in which such a concept does not survive is on the whole to be preferred for various reasons of convenience and principle. This is what I meant when I said earlier that Quine, in a sense, reneges on his nonontically committing sense of predication.

3. Predication and Ontological Commitment: Quine's Reasons Undone

The position to be argued from here on is this: there is no need to exclude irreferential singular terms from the Quineian canonical language nor, therefore, a nonontically committing sense of predication. My procedure will be to outline such an idiom showing in each case how it treats Quine's worries about truthvalue gaps and an anomalous concept of having purely referential position, and the Quine-like concern about the nonextensionality of such an idiom. I shall end by making some general remarks about the wisdom of trying to eliminate all singular terms from an idiom alleged to be sufficient for the purposes of doing philosophy and science.

In 1969, Ronald Scales devised a version of negative free logic.[9] It is a regimented language exactly like Quine's canonical idiom except that it also contains singular terms and a term forming operator 'λ' that makes complex general terms out of open sentences. (This operator is not to be confused with Church's use of the lambda notation for creating names of functions.) The idiom is called "negative" because all simple statements—and indeed all predications—containing at least one irreferential singular term are counted false. Most importantly, the inference from

'Gs' is a predication

to

There is an object that is s

does not hold in this idiom; predication, in other words, is not ontically committing. For instance, the statement

Vulcan is the same as Vulcan

is a two-place predication for Scales, but the statement

There is an object that is Vulcan

is false.

With this picture in mind, let us return to Quine's worries and look, first, at the anomalous character of having purely referential position.

A position p in a context C is purely referential just in case for any singular term t occupying p replacement of t by a codesignative singular term t' preserves truthvalue. The concept of purely referential position is important in the Quineian scheme of things as a device for determining when quantification into a context is acceptable. It is because the position of the numeral '9' in the statement 'Necessarily 9 is the number of the planets' is not purely referential that the quantificational statement 'There exists something x such that x is the number of the planets' is not acceptable.

Concerning the alleged anomalous character of having purely referential position, I have two remarks. First, today it is believed in many quarters that it is not the position of a singular term in a context that is so important vis-à-vis quantifying into that context, but rather the kind of singular term occurring in that context. If the singular term is a "genuine name," a "rigid designator," or what have you, quantifying in is OK. But, if it is a garden variety definite description—or at least a "nonrigid designator"—then quantifying in is perilous. The moral, of course, is that since the concept of purely referential position is not so important after all, there is no need to worry about the alleged anomalous character of certain expressions having it.

Second, and more importantly, since in the canonical idiom devised by Scales the positions of singular terms outside the abstracts in predications are purely referential in Quine's sense, the alleged anomaly associated with irreferential singular terms like 'Vulcan' in predications such as 'Vulcan rotates' is easily subverted by inserting the words "if any" at the appropriate place in the statement of motivation underlying the concept of purely referential position. Such motivation is now reexpressed thus: the idea behind the concept of purely referential position is that the singular terms occupying it specify the object, if any, for the rest of the statement to talk about. As far as I can see, this minor alteration does no violence to any of the uses Quine wishes to make of the concept of purely referential position.

Turning now to Quine's concern about the truthvalue gaps introduced by preregimented sentences containing at least one irreferential singular term, it should be clear, even from the sparse description of Scales' canonical idiom given earlier, that failure of the principle of bivalence is not inevitable when irreferential singular terms are allowed to occur in the statements of the regimented language. In fact, the sentential fragment of Scales' formal idiom is fully classical. To be sure the quantificational fragment with identity in Scales' formal idiom is nonclassical in the sense that the inference from a statement of the form

As/x

to a statement of the form

$(\exists x)A$

and the inference from a statement of the form

$(\forall x)A$

to a statement of the form

As/x

holds only on the condition that

$(\exists x)(x = s)$

But, to paraphrase an influential Harvard philosopher, this supporting premiss is less to be deplored as uneconomical than prized as an unfolding of a tacit assumption, an articulation of the inarticulate.[10]

Turning now to the worry that admission of predications such as

> Vulcan rotates on its axis

threatens the extensionality of the canonical language in the sense that their inclusion does not imply that coextensive general terms substitute salva veritate, that worry is eliminated in the idiom invented by Scales. To see this note, first, that the principle of abstraction that founds the nonextensionality argument outlined earlier does not hold in Scales' idiom. What does hold is the following variant of abstraction: s is an object such that it is so and so just in case s is so and so and s exists, where s is a singular term. The immediate result is that, second, the predication

> Vulcan rotates

which is false in Scales' treatment, does not yield the true

> If Vulcan rotates, then it rotates

but rather the false

> (If Vulcan rotates, then it rotates) and Vulcan exists

Third, and finally, it can be proved in general in Scales' language that where 'F' and 'G' are coextensive general terms, they substitute salva veritate in any context expressible in that idiom. Moreover, the same holds for coextensive singular terms and coextensive statements. In short, Scales' language is completely extensional. So, a nonontically committing conception of predication can be introduced into that idiom without threatening its extensionality.

Turning finally to Quine's elimination program for singular terms, the principal motivation for which was the avoidance of the alleged difficulties associated with irreferential singular terms outlined earlier, I would like to raise an additional point or two beyond the obvious suggestion that it seems not to be needed—even in the Quineian scheme of things. I will be brief because these points have been made in greater detail elsewhere.[11]

My first remark is that Quine's elimination procedure for singular terms—as laid out in *Word and Object*—is not exhaustive because there are singular terms and sentences containing terms that are not acceptably paraphrasable into the canonical language. 'Acceptably paraphrasable' means that the paraphrase does not disagree with the truthvalue Quine himself associates with, and thinks important to, the statement being paraphrased.

Consider, for example, the pair of virtual class truths in his *Set Theory and Its Logic*:[1,2]

$$\{x: x \notin x\} = \{x: x \notin x\}$$

and

$$\{x: x \notin x\} = \{y: y \in \{x: x \notin x\}\}$$

The paraphrases of these identities containing virtual class names, a kind of irreferential class name, turn out false by the method of *Word and Object*, and hence they are unacceptably paraphrasable. To be sure, Quine applies another kind of procedure in *Set Theory and Its Logic* to eliminate virtual class names that yields acceptable paraphrases. But this raises the really important questions: Which elimination procedure applies to which kinds of singular term? If they do not conflict, where is the proof that there does not lie somewhere some stubborn, recalcitrant species? I conclude that Quine's elimination program is essentially an article of faith.

My second, and final, complaint is that Quine's canonical language sans singular terms may lack explanatory power. In his 1980 Presidential Address to the Pacific Division of the American Philosophical Association, David Kaplan proposed a new explanation of epistemic discourse and quantification thereunto based on "the thesis of direct reference," that is, the thesis that there are sentences containing indexical expressions whose contents—the singular propositions expressed by those sentences—contain individuals directly referred to by those indexical expressions. The question arises whether such an explanation is possible in a language eschewing singular terms. It seems unlikely, because even admitting propositions—a very unQuineian thing to do—there are no such sentences in the Quineian canonical language (or cousins of same) having individuals in the propositions expressed by those sentences. For Quine's elimination procedure yields general sentences as counterparts to the original singular sentences. But the propositions associated with these general sentences would not have individuals as constituents of those propositions, a key part of the thesis of direct reference. So the Quineian singulartermless, idiom bars the door to an interesting explanation of epistemic discourse and thus lacks explanatory power.

To sum up, I have tried to argue that Quine's nonontic committing conception of predication can be sustained in a language like his own canonical language except for containing singular terms and complex general terms. By that I mean that it resolves certain worries thought to be associated with such a conception of predication, worries thought to require an idiom in which predication, if it occurs at all, is ontically committing. And, lastly, I have tried to argue that the program for eliminating singular terms on which the avoidance of the worries just mentioned relies has serious difficulties of its own.

Notes

1. Russell, Bertrand, *Our Knowledge of the External World*, 2d edn., W. W. Norton, New York (1929) p. 41.
2. Russell, Bertrand, *Introduction to Mathematical Philosophy*, 2d edn., George Allen & Unwin, London (1919) pp. 168–9.
3. Whitehead, A. and Russell, B., *Principia Mathematica*, Cambridge University Press (1950) p. 66.
4. van Fraassen, B., *Journal of Philosophy*, vol. 63 (1966) pp. 481–95.
5. He does not regard 'Vulcan' in the usual contexts as irreferential. But he thinks there are occasions where a singular term can be irreferential. See his essay "The problem of nonexistent objects," *Topoi*, vol. 1 (1982) pp. 97–140; and especially footnote 6.
6. Quine, W. V., *Word and Object*, Wiley, New York (1960) p. 96. My italics.
7. Ibid. (Note 6), p. 177.
8. See my essay "Predication and extensionality," *Journal of Philosophical Logic*, vol. 3 (1974) pp. 255–64.
9. Scales, Ronald, *Reference and Attribution*, University of Michigan Microfilms (1969).
10. Ibid. (Note 6), p. 187.
11. Lambert, Karel, "On the elimination of singular terms," *Logique et Analyse*, vol. 108 (1984) pp. 378–92.
12. Revised edition, Cambridge University Press (1969).

16
Free Part–Whole Theory*

Peter M. Simons

1. Classical Mereology

Although philosophers since antiquity have concerned themselves with the concepts of part and whole, formal theories of these and cognate concepts are quite recent. An adequate account had to wait for a proper logical treatment of relations. In his *Logical Investigations* of 1901[1], Husserl envisaged such a formal theory of part and whole but did not work it out logically beyond a few obvious theorems. Properly formal theories were developed around 1915, independently, by Whitehead, for whom it was the theory of "extension,"[2] and by Leśniewski, who first called it "theory of sets (manifolds)" but later coined the less misleading term "mereology," from the Greek *meros*, part.[3] When Leonard and Goodman redeveloped the same ideas somewhat later they used the title "calculus of individuals." I find this label a little unfortunate and shall not use it. Hereafter, I shall use Leśniewski's term *mereology*.

Whitehead's treatment, in the surviving writings at least,[4] was formally less worked out (and less adequate)[5] than that of Leśniewski, but the latter couched his theory in terms of his own logical system, called (again, somewhat unhappily) "ontology." This was, and remains, unfamiliar to most potential users.[6] Tarski adapted some of Leśniewski's concepts for

---

* My thanks go to Karel Lambert for his helpful comments on an earlier version of this paper. It was written while I was a guest of the University of California at Irvine, and I wish to thank here the members of the Philosophy Department who helped to make my stay an enjoyable one.

a classically formulated (*Principia Mathematica*) logic,[7] and his system does not differ essentially from the one that made formal part–whole theory best known, Leonard's and Goodman's "The calculus of individuals and its uses" of 1940.[8]

Despite the differences in the logical bases they used, Leśniewski, Tarski, Leonard and Goodman did not differ substantially in the principles they thought a mereology should satisfy.[9] One can express these using a classical first-order predicate logic with identity and a new primitive predicate, written '<' and read 'part' (short for 'proper or improper part'). To the logic add the following proper axioms:

PART $\forall x \forall y(x < y \equiv \forall z(\exists v(v < z \ \& \ v < x) \supset \exists w(w < z \ \& \ w < y)))$
GSUM $\exists x(Fx) \supset \exists y(\forall z(\exists w(w < z \ \& \ w < y)$
            $\equiv \exists v(Fv \ \& \ \exists u(u < v \ \& \ u < z))))$
ASPT $\forall x \forall y((x < y \ \& \ y < x) \supset x = y)$

Apart from ASPT, which asserts the antisymmetry of '<', these are not particularly transparent. They become clearer if we define a predicate 'o', read 'overlaps', to mean 'has a common part with':[10]

Df o $x \text{ o } y \equiv \exists z(z < x \ \& \ z < y)$

Then PART and GSUM are respectively equivalent to

CMT1 $\forall x \forall y(x < y \equiv \forall z(z \text{ o } x \supset z \text{ o } y))$
CMT2 $\exists x(Fx) \supset \exists y(\forall z(z \text{ o } y \equiv \exists v(Fv \ \& \ z \text{ o } v)))$

While we are defining, it is useful to add two new predicates: '≪' (proper part) and '|' (disjoint), whose meanings are obvious from their definitions:

Df ≪ $x \ll y \equiv x < y \ \& \ \sim(y < x)$
Df | $x \mid y \equiv \ \sim \exists z(z < x \ \& \ z < y)$ [i.e., $\sim(x \text{ o } y)$]

From PART and logic alone it follows that '<' is totally reflexive and transitive, which together with ASPT make it a partial ordering. But to be a part concept a relation must fulfill a special further condition, a supplementation principle:[11]

WSP $\forall x \forall y(x \ll y \supset \exists z(z \ll y \ \& \ z \mid x))$

(an object that has a proper part has a second proper part disjoint from the first). This follows by logic from the definition of '≪' and PART alone, so PART does most of the work required to determine that '<' is a part predicate. GSUM, the general sum principle, ensures that for any fulfilled predicate *F* there is an object, the mereological sum of *F*s, which consists of just the *F*s and nothing outside them. ASPT ensures that objects with the same parts are identical:

MEXT $\forall x \forall y (\forall z(z < x \equiv z < y) \supset x = y)$

Because MEXT is analogous to the extensionality principle for sets, I call it the *principle of mereological extensionality* and a mereology that satisfies MEXT an *extensional mereology*. One that also satisfies GSUM I call a *classical extensional mereology* (CEM), "classical" not just because it is based on classical logic, but because these conditions are fulfilled by the "classical" systems of Leśniewski, Tarski, Leonard, and Goodman (modulo the underlying logic).

Not all extensional mereologies are classical in this sense: that of Whitehead, for instance, is not, because it fails to satisfy GSUM.

In the presence of asymmetry ASPT, GSUM tells us not only that for any fulfilled predicate there is a sum of objects satisfying it, but further that this sum is unique. So it is natural to define a general sum operator σ using definite descriptions:

Df σ $\sigma x(Fx) := \mathrm{I}x(\forall y(y \circ x \equiv \exists z(Fz \ \& \ y \circ z))$

and then to use this to define certain other mereological constants: a general product operator

Df π $\pi x(Fx) := \sigma x(\forall y(Fy \supset y < x))$

a binary sum and product

Df $+$ $x + y := \sigma z(z = x \vee z = y)$
Df $\cdot$ $x \cdot y := \sigma z(z < x \ \& \ z < y)$ [i.e., $\pi z(z = x \vee z = y)$]

a mereological difference

Df $\setminus$ $x \setminus y := \sigma z(z < x \ \& \ z \mid y)$

the universe

Df U $U := \sigma x(x = x)$

and complement

Df $-$ $-x := \sigma y(y \mid x)$

[i.e., $U \setminus x$], all of which are pretty familiar from Boolean algebra (or, in the cases of general product and general sum, Boolean algebra extended with infinitary operators). Indeed, as Tarski first pointed out,[12] models for CEM differ from those for extended Boolean algebras solely by lacking a Boolean zero or null element. A large number of theorems, similar to those of extended Boolean algebra but made more complicated by the lack of a zero element, can be proved on the basis of the axioms PART, GSUM, and ASPT. This minor inelegance and inconvenience has prompted a few people to add a null individual in analogy with the null class. Most mereologists, who tend to employ mereology because of their ontological views, regard this as offensive to their robust sense of reality, or, to be more blunt, as fantasy and falsification that algebraic convenience cannot excuse.

2. Motivations for Reforming Mereology

I have introduced o and the other operators in terms of a definite description operator 'I', and mereology would be much impaired without the complex terms thereby made possible. Complex terms which in a Boolean algebra would denote the zero element are empty in mereology. The following are all complex terms that may turn out empty:

$\sigma x(Fx)$,   if nothing is F
$x \cdot y$,   if x and y do not overlap
$\pi x(Fx)$,   if the Fs do not have any common part
$x \setminus y$,   if $x < y$

The term '$-U$' is unconditionally empty. Clearly such terms should not be substituted for bound variables if they do not denote anything. So even within the confines of CEM there are at least reasons of convenience for wanting to be able to manipulate empty complex terms just like any others.

Enter free logic. It is natural to consider formal theories of part and whole based not on classical but on free logic: free mereologies. Historically speaking, the first fully worked out mereology, and to this day still the best-researched, that of Leśniewski, was based on a logic that has in common with free logic that its terms need not denote. But Leśniewski's logic has features that set it apart from free logic, which many consider just classically formulated predicate logic with the existence assumptions for its singular terms made explicit. Leśniewski's terms may be not only (semantically) singular or empty, they may also be plural, that is, denote many individuals. For another thing, neither of Leśniewski's quantifiers carries existential import. That is, whether a term b denotes or not, from a sentence of the form $\forall a(\ldots a \ldots)$, the sentence $(\ldots b \ldots)$ may be validly inferred, and from the latter $\exists a(\ldots a \ldots)$ may be validly inferred. So a reason for developing a mereology based on a more "standard" free logic, where no terms are plural and where the standard quantifier inferences are not valid, is that it is then easier to compare the mereology of

Leśniewski with those of Tarski, Leonard, Goodman, and others, and to determine that their mereologies indeed "assert the same things" despite the differences in underlying logics.[13]

Now CEM is seriously defective as a theory of part and whole, no matter on which logic it is based.[14] It is too strong because it requires the existence of mereological sums where it seems there are none, and imposes mereological extensionality. Sums have quite often been criticized on the grounds that there are no good reasons to believe they always exist: algebraic simplicity is not a good reason for adding entities unnecessarily to one's ontology. Many past mereologists were nominalists and wanted merology to provide a nominalistically acceptable alternative to set theory, so it was natural for them to make the instrument as powerful as possible. But divorced from such programs, the sum principle loses most of its ontological attraction. Indeed, there have been mereologies, including Whitehead's, in which the sum principle does not hold.[15] If, as seems correct, we nevertheless admit that sums sometimes exist, then we still need a sum notation like that using σ, and the cases where σ-terms are empty will be more numerous than in CEM. Once again, free logic provides the most suitable vehicle for handling these.

The rejection of the extensionality principle is a more radical move and not many have considered making it. However, extensionality of parts is not analytically contained in the concept part, and is to be rejected in general. The argument is detailed elsewhere.[16] In the light of these objections, only PART remains untouched as a part–whole principle. It is, of course, possible to weaken mereology in this way without leaving classical logic. But other defects of CEM make the adoption of an underlying free logic more natural. CEM omits both the temporal and modal dimensions of mereology. The temporal aspect appears when one considers objects that may have different parts at different times, something that holds of organisms (such as ourselves), and of artefacts such as ships. Classical mereology as such does not have the resources to recognize a tensed or temporally indexed part–whole relation. If one takes a tense-logical approach to tensed mereology, it is natural to relativize the domain to times, and regard a sentence 's exists' as true now only if s exists now, or—on another meaning of 'exists'—if s exists now or existed in the past. In either case, terms may denote at some times and not at others, and this will go for complex terms as well. For instance if patient P receives on Tuesday a heart transplant from victim V who died on Monday, then on Sunday it is false that H, the heart which was part of V on Sunday, overlaps with P, whereas by Wednesday this is true. So the product of P and H does not exist on Sunday but does exist on Wednesday.[17]

The modal aspect of mereology arises from the need to distinguish in many objects the parts that they could not have failed to have (essential parts) from those that they might not have had (accidental parts). Such concepts are involved, for example, in discussion of the Leibniz–Chisholm thesis of mereological essentialism,[18] according to which any object properly so called has only essential parts. An adequate vocabulary for discussing this controversial thesis clearly requires a modal predicate logic whose nonmodal predicate basis is free.

The customary way of dealing with modal discourse in semantics is by relativizing the domain and the extensions of predicates to possible worlds. In such circumstances the natural nonmodal logic on which to build one's modal predicate logic is a free logic, in which some terms are empty in some worlds and not in others. To recall the heart transplant example, it is clearly contingent that P received V's heart H, so while the product P·H exists at some time in the actual world, in other worlds that product does not exist at any time. The modal mereology I have given elsewhere is based on a version of quantified S5 in which individuals need not exist in all worlds.[19]

In what follows I shall consider only nonmodal and nontemporal free mereologies, in which mereological principles of varying strength may be formulated. These can form a basis for the mereologies needed for the more complex tasks outlined earlier.

3. Free Logical Preliminaries

Once classical predicate logic (with identity) is abandoned, the number of ways in which empty singular terms, especially empty definite descriptions, can be added while the quantifiers retain their standard meaning becomes very large. Since free logicians have explored many alternatives in some detail, I can be fairly brief in developing the free logic on which to base free mereology.[20]

The free logic in question may be described metalinguistically without concern for the exact form of the object language. The following runs of metavariables are used, where all letters may be numerically subscripted as necessary:

    For singular variables: u v w x y z
    For singular parameters: a b c
    For wffs: A B C
    For singular terms: r s t

Singular terms comprise singular variables, parameters, and definite descriptions, and therefore also include terms introduced by definition using descriptions. The following symbols stand for the logical constants regularly associated with them:

$\sim$ & $\vee$ $\supset$ $\equiv$ $\forall$ $\exists$ = I

Parentheses will be used in a standard way and detailed rules are not given. There are the following additional conventions:

If A is a wff, then $A[s/t]$ denotes the unique wff obtained from A by uniformly substituting s for t, with the proviso that there is no such wff if s is a bound variable that already occurs bound in A: it is stipulated that in no wff is there any occurrence of a variable within the scope of more than one operator binding that variable (so, e.g., $\exists x(Fx \,\&\, \forall x(Fx \supset Gx))$ is not a wff).

If A is a wff, then $(\forall)A$ denotes the unique wff that results from A by universally binding all its free variables in alphabetical order (its bound universal closure).

If A is a closed formula, then $(\forall)A$ is just A. Only closed formulas (sentences) will be admitted in the system as theorems, but the effect of free individual variables may be obtained using parameters.

The meta-axioms and rule schemata of the underlying free logic are as follows:

FA0   If A is a substitution instance of a tautology of propositional calculus, $(\forall)A$ is a theorem
FA1 $(\forall)(\forall x(A \supset B) \supset (\forall xA \supset \forall xB))$
FA3 $(\forall)(A \supset \forall xA)$, where x does not occur in A
FA4 $(\forall)(\forall y(\forall xA \supset A[y/x])$
FA5 $a = a$
FA6 $(\forall)(s = t \supset (A \supset A[s/t]))$

with the rules

FR1   If A and $A \supset B$ are theorems, B is a theorem
FR2   If A is a theorem and a occurs in A and x does not, then $\forall xA[x/a]$ is a theorem

It remains to consider 'I'. There are numerous free description theories. Elsewhere I made use of one known as FD2, which has among its theorems the following two:

> FD2T1 $(\sim\exists xA \ \& \ \sim\exists xB) \supset IxA = IxB$
> FD2T2 $s = Ix(x = s)$

The former in particular facilitates comparison with Leśniewski's systems. However, when modal operators are added to the free mereology, both principles are questionable. If $Ix(x = s)$ is regarded—as I think it should be—as meaning 'the (actually existing) object that is identical with s' then '$s = Ix(x = s)$' is false if s does not exist. If A and B are two conditions that independently merely happen not to be fulfilled, such as being a greatest common part of a and b (which happen to be disjoint) and being a greatest common part of c and d (which likewise happen to be disjoint), there is no reason to think that $a \cdot b = c \cdot d$ in general.

In these circumstances it is advisable to adopt the weakest reasonable principles for definite descriptions. Such a minimal theory may be obtained by adding the axiom scheme known as Lambert's Law:[21]

> FA7 $\forall x(x = IyA \equiv (A[x/y] \ \& \ \forall z(A[z/y] \supset z = x)))$

This is a principle that expresses the basic property of 'I' in free description theory. I am here, of course, skating rapidly over difficult questions about the meaning of definite descriptions and identity, especially in nonextensional contexts. Being minimalistic, I hope to avoid error, possibly at the cost of not revealing the whole truth.

4. Free Mereology

The same caution can be applied to free mereology. As with definite descriptions, once the theory is freed from the constraints of classical logic, the alternatives multiply. Consider for example the classical theorem $s < s$. What if s does not exist? Is it true, false, or neither, or sometimes one, sometimes another? One intuition, that behind supervaluational semantics for free logic, is that it is always true on the grounds that in all cases where s does exist, the statement is true. Another line of thought, one that I have espoused, regards it as false if s does not exist. One way of expressing this thought is to say that < is (or appears to be) an existence-entailing attribute (at both ends).[22] It is hard to find motivation for this policy that is not circular, but one inclining reason is that on the definition Df o of 'overlaps', s o s is false if s does not exist, since there is no common part of s and s. But if $s < s$ is also asserted, the intuitively attractive link between having a part and having something that is a part is broken. In the end, perhaps this is just another way of stating that 'part' is existence-entailing, a principle I have elsewhere, following Kit Fine, called the *falsehood principle* for $<$:[23]

FPPT $s < t \supset (\exists x(x = s) \& \exists y(y = t))$

The more radical position that $s < s$ is truthvalueless if s does not exist violates bivalence, of course. A stance on this issue can be avoided simply by saying nothing about $s < s$. Adopting PART as the first principle of free mereology does not commit us to a position on such an issue, since PART employs bound variables and enables us to derive only

FMT1 $\forall x(x < x)$
FMT2 $\exists x(x = s) \supset s < s$

both of which are desirable and unobjectionable.

So a minimal free mereology, containing nothing to which one could reasonably object, consists of free logic FA0–7, FR1–2, together with PART. For stronger free mereologies add either or both of GSUM, ASPT. The free logic allows us to handle in many contexts definite descriptions whose matrices are not uniquely satisfied, and the basic theorem for free mereology with definite descriptions is then

FMT3 $\exists x(x = \sigma y A) \equiv \exists x(\forall y(y \circ x \equiv \exists z(A[z/y] \& y \circ z)) \&$
$\forall w(\forall y(y \circ w \equiv \exists z(A[z/y] \& y \circ z)) \supset w = x)$

This is a rather timid mereological theory. Though one's strong feelings about the mereology of nonexistents may be few, one would not normally be inclined to believe that an existent has a nonexistent as its part or is part of a nonexistent. So it is plausible to add to any of the four free mereologies so far outlined the further principle of nonexistent parts:

NEPT $(\exists x(x = s) \& (s < t \lor t < s)) \supset \exists y(y = t)$

There are nonetheless applications of free part–whole theory that require rejection of this principle. Three such applications will be considered below. Adding the falsehood principle FPPT renders NEPT derivable, but without FPPT one may add either or both of unrestricted reflexivity and transitivity of parts:

REFL $s < s$
TRAN $(r < s \& s < t) \supset r < t$

There are many other variations in mereological theory based on the minimal free logic with descriptions, but it is difficult to motivate preference for one rather than another without considering particular applications. Another course is to alter the underlying free description theory. By supplementing it with the principle FD2T1 (given earlier) the much stronger free description theory FD2 is obtained; FD2 approximates the Fregean "chosen object" account of definite descriptions in classical logic. This logic, which allows the intersubstitution *salva veritate* of empty terms, is extensional, and as such is particularly apt when comparing free mereology with Leśniewski's version. This stronger system of free mereology has been described elsewhere in greater detail.[24]

It pays to be flexible when considering the principles a free mereology may satisfy. In the final sections I outline three possible philosophical applications for free mereology. The first considers Meinong's notion of implexion of objects in other objects. The second is an Aristotelian theory of potential parts of continua. The third is an account of the parts of states of affairs. Each of these involves rejecting the principle NEPT. The suggestion in each case is that free mereology allows the doctrines in question to be represented in a way that is better than formulations involving some alternative to mereology, such as set theory. Being better means representing the questions in a natural vocabulary (that of part and whole) that comes already loaded with some principles governing what is acceptable and what is not (and so does not allow us to make just any constructions we like) but is still neutral enough not to prejudge the issue of whether the doctrine is correct or not.

5. Implexive Containment

Alexius Meinong is undoubtedly the greatest exponent of the view that there are nonexistent objects. Meinong does not discuss the mereology of nonexistents, but he does consider a relation, which he calls implexion, that has all the formal properties of the usual part–whole relation and that relates nonexistents and existents. Meinong embraced a theory of incomplete objects:[25] an object is incomplete for Meinong if there is some property P such that the object has neither P nor its contradictory P'.[26] Assuming that 'P' ranges over only nuclear properties,[27]

Df cpt $\text{cpt}(a) \equiv \forall P(\sim(a \text{ is } P) \supset a \text{ is } P')$

I shall not go into what is required beyond completeness for it to be possible for an object to exist; the question is somewhat involved and not directly relevant. Let us suppose only that some objects are possible, that is, could exist. Now according to Meinong,[28] an incomplete object has implexive being if there is at least one possible object in which it is implexively contained. What then is implexive containment? An object a is implexively contained in an object b iff every nuclear property of a is a nuclear property of b:

Df imp $a \text{ imp } b \equiv \forall P(a \text{ is } P \supset b \text{ is } P)$

Notice that among possible objects implexive containment reduces simply to identity, assuming Leibniz's principle of the indiscernibility of identicals for nuclear properties.[29] It is only when we consider incomplete objects that it becomes an interesting notion. Clearly implexive containment is reflexive and transitive, and (again, assuming that Leibnizian principle) it is also antisymmetric. So it partially orders objects. Does it then satisfy the formal requirements for being a part–whole relation, that is, does it additionally satisfy WSP? What we have so far is that

$$a \text{ imp } b \ \& \sim (b \text{ imp } a) \supset (\forall P(a \text{ is } P \supset b \text{ is } P) \ \&$$
$$\exists P(b \text{ is } P \ \& \sim (a \text{ is } P)))$$

So far this does not give us an object helping to make up the difference between a and b, but we get this from an additional assumption of Meinongian object theory, a kind of Meinongian comprehension principle:

MCOMP   For every set of nuclear properties there is an object that has just these nuclear properties[30]

So suppose a imp b and not b imp a, and consider the object whose nuclear properties are just those had by b but not by a: we might call it the *implexive remainder* of b after a: b is the disjoint sum of a and the remainder, and the remainder object satisfies the condition in the consequent of WSP. Hence implexive containment is, as far as its formal properties go, a part–whole relation. This is reflected in Findlay's translation of 'implektieren' as 'embed',[31] a term that is now standard in the literature on Meinong. But clearly it is not the usual part–whole relation, as Meinong explicitly says.[32] Take an actual individual, such as George Washington. The parts of George Washington, in the usual acceptation of 'part', are as existent and actual as he. Not so his implexive parts. Excepting only the trivial case of George himself, all his implexive parts are objects that are incomplete, and hence could not exist or be actual. Their only toehold on being is that vicarious participation, which Meinong calls "implexive being," that they derive from being embedded in George.

Meinong's incomplete objects owe something to Locke's abstract ideas, but he was not the first modern philosopher to transpose Locke's theory into an ontological key. In 1894 his younger contemporary Kazimierz

Twardowski sketched a theory of general objects, the objects of general ideas, which like Meinong's incomplete objects are both nonexistent and embedded in particulars. Thus, the general horse is embedded in all actual horses. Unlike Meinong, however, Twardowski allowed general objects to be literally parts of the individuals in which they are embedded: "The object of the general presentation is a part of the object of a subsumed presentation, a part which stands in the relation of equality to certain parts of objects of other individual presentations."[33] So the general horse, which does not exist, is nevertheless part of each actual individual horse.

There is a methodological issue raised by Meinong's concept of implexive containment, namely: what more has to be added to the formal principles of part–whole relations to insure that such a relation really is a part–whole relation in the usual acceptation, and not a strange relation like implexive containment? I confess I do not know, but consideration of this example suggests the answer may reside in the modal (i.e., non-extensional) properties of the relation.

6. Actual and Potential Parts

According to at least one interpretation,[34] Aristotle viewed a continuum such as a body, surface, or line, as not having actual proper parts that go to make it up. Nevertheless, Aristotle naturally could not view a continuum as a mereological atom, that is, an object without proper parts in any sense, like a Euclidean point. A possible *via media* is to see continua as being divisible, that is, as somehow containing potential objects, which could be actualized. For example, in the diagram

$$A \underline{\qquad\qquad} B \underline{\qquad\qquad} C$$

the line segment AC is a continuum that contains the line segments AB and BC as potential parts, and these could both come into actual existence by dividing AC at its midpoint B. AC is divisible without being actually divided, and in Aristotle's view each such continuum is infinitely divisible. There are difficulties both about interpreting Aristotle and about making sense of this doctrine, whether it is Aristotle's or not, but a reasonable first stab at an interpretation takes 'potential part' to mean 'part that is nonexistent'. This may be embodied in the definition

$$\text{Df}_{P<} \quad s_P < t \equiv \sim E!s \,\&\, s < t$$

Note that no sentence of the form $s_P < t$ can be true unless we reject

NEPT. Accepting TRAN, it trivially follows also that $<_P$ is transitive, and accepting REFL it follows that $<_P$ is reflexive on nonexistents.

An object could also be a potential whole in the same style:

Df $<_P$ $s <_P t \equiv \;\sim E!t \;\&\; s < t$

and similar remarks apply about its reflexivity and transitivity. An actual part would then be a part that exists, and likewise for an actual whole:

Df $_A<$ $s_A < t \equiv E!s \;\&\; s < t$
Df $<_A$ $s <_A t \equiv E!t \;\&\; s < t$

One can then combine these definitions to get potential parts of potential wholes, potential parts of actual wholes, actual parts of potential wholes, and actual parts of actual wholes, though on the Aristotelian view there would be none of the last sort. The metaphysical picture that this way of thinking invokes is that of an inner domain of actual or existent objects, connected by part–whole relations to an outer domain of potential or nonexistent objects, which also themselves stand in mereological relations. The part–whole relations would themselves be ontologically neutral. The Aristotelian picture of the line segments AB, BC, and AC would then be that only AC actually exists, though AB and BC are potential (i.e., nonexistent) parts of AC, whose sum indeed is AC. So (oddly) some sums of nonexistents are existent.

One of the various difficulties with this metaphysical picture is precisely the one mentioned before, that by accepting $s < s$ even when s does not exist, we cannot define o in the standard way if we want to insist that s o s even if s does not exist. The obvious way to get something like the standard definition of overlapping is to allow quantification over potential as well as actual objects, introducing quantifiers $\prod$ and $\sum$ that range over both existents and nonexistents, and for which the standard quantifier rules apply, then redefining overlapping by substituting '$\sum$' for '$\exists$'. But this solution is formally indistinguishable from that where we keep the classical quantifiers and introduce a distinction among existents between the actual and the nonactual. It is then, however, unclear what it means to be actual. On the other hand, allowing quantification over potential (but not necessarily existent) entities raises for these just the same questions that free logic raises for existents, namely, what do we do about singular terms that lack a denotation among either existents or nonexistents?

This brief discussion may serve to show that there are philosophical difficulties involved in Aristotle's doctrine. These difficulties may incline one to agree with Brentano,[35] that Aristotle's mereology (if it be his) is to be rejected rather than made the best of. But there may well be other work for the notion of a potential part and its kin that makes it worthwhile buying the concept despite its disadvantages, so it is not otiose to mention it within the framework of free mereology. The aspects of change and potentiality involved in the actual Aristotelian version are not analyzed in great depth, but it should be possible to do this along the lines of the temporal–modal extension of free mereology as I envisage it.

7. Parts of States of Affairs

There is other work that may be done by the notion of a part of a nonexistent whole that promises to appeal to a rather larger class of potential users. According to one way of thinking about states of affairs, there exist no states of affairs corresponding to false propositions, yet the objects that the proposition is about are nevertheless parts of the corresponding (nonexistent) state of affairs. I do not want to enter into a history of the many theories of states of affairs and related entities, nor discuss in detail whether any theory that accepts states of affairs is correct. I need only mention some "fairly standard" properties of states of affairs.

States of affairs are not to be confused with propositions. Propositions are typically contents of judgment, and/or meanings of sentences and bearers of truth and falsity. False propositions have the same mode of being as true ones. States of affairs, by contrast, are the objects of judgment, referents of sentences, and the things that go to make truths true and falsehoods false.[36] One might say judgments and sentences express propositions which in turn denote states of affairs.

The subjects of judgments and sentences are (typically) not parts of the propositions of which they are subjects, rather these propositions contain senses, meanings, concepts, or what-have-you, that denote the subjects. But these subjects are indeed parts of the states of affairs that such propositions denote.[37] States of affairs corresponding to true propositions exist. They are typically called 'facts'.[38] States of affairs corresponding to false propositions (Meinong called them 'unfacts') do not exist. One might say "there are not any," but for present purposes I take this to mean that there are (weak sense of 'are') some states of affairs that do not exist (subsist, obtain). The metaphysical picture is again that of an outer domain of objects without being.

Logical atomist theories of states of affairs hold that not every true judgment or sentence "corresponds to" a state of affairs. For example, there is no need for special disjunctive states of affairs. This is probably best taken to mean that some putative terms for states of affairs, for example, 'the state of affairs that Caesar conquered Gaul or Pompey conquered Britain' fail to designate either existing or nonexisting states of affairs: they are just empty.

Putting these ideas together, what can we say about parts of states of affairs? First, whether or not an object is part of a state of affairs has nothing to do with whether the state of affairs exists. Caesar is just as much part of the nonexistent state of affairs that he conquered Mexico as he is of the existent state of affairs that he conquered Gaul. And conversely (on some theories at least) Zeus is part of the (existent) state of affairs that the Greeks worshipped Zeus, though he does not exist. Caesar and Gaul are probably not the only parts of the state of affairs that Caesar conquered Gaul. Russell, for instance, would take the universal relation of conquering as a further, nonparticular part. Alternatively, one could think of the conquering as a particular moment, a complex event in this case, which is not conquering per se but a particular instance of conquering.[39]

To fix ideas, take any sentence A and represent 'the state of affairs that A' by '$[A]$'. Then on Russell's view, if we designate the universal relation R by the abstract $\lambda x \lambda y (x \, \mathsf{R} \, y)$, we have that

$$a < [a \, \mathsf{R} \, b] \ \& \ b < [a \, \mathsf{R} \, b] \ \& \ \lambda x \lambda y (x \, \mathsf{R} \, y) < [a \, \mathsf{R} \, b]$$

On the alternative view, instead of the last part we would have some particular instance r of the relation $\lambda x \lambda y (x \, \mathsf{R} \, y)$ as part of $[a \, \mathsf{R} \, b]$. All of this could proceed quite independently of the question whether or not $[a \, \mathsf{R} \, b]$ exists.

There are two prima facie points in favor of the second view, that the bit of the state of affairs corresponding to the predicate is particular rather than universal. One is simply that it does not of itself commit one to universals, so, other things being equal, is to be preferred over a theory that requires universals as well as particulars. The other is that, if an instance of a relation can only exist if those particular things that it relates are indeed related by it, then the existence of the individual relation entails the existence of the state of affairs that they are so related.

The universal is not so tied to the existence of states of affairs, since even on a moderate realist view according to which a universal can only exist if it is exemplified, the existence of John, loving, and Mary is not sufficient to entail that John loves Mary, since Hans may love Erna and so ensure that loving exists, yet John not love Mary. By contrast, if the particular individual that is the love of John for Mary exists, it follows that John loves Mary, and conversely, if John loves Mary, then there is some individual that is a love of John for Mary. Another point in favor of the particular love theory is that it is essential to this love that it is John's love for Mary, so from these essential relational properties we can recover the state of affairs as John's loving Mary rather than Mary's loving John, whereas a list of John, Mary, and loving will not enable us to say who loves whom. Relational particulars, we might say, come with their own case relation built in.[40]

Nevertheless, there are still problems for the part–whole view of states of affairs, which inclined some supporters of states of affairs to deny that their subjects are literally their parts. This was one point of difference between Russell and Meinong.[41] One problem is simply that of making sense of a whole composed of such categorially disparate objects as John, Mary, and love, or Tarski, admiration, and Aristotle's theory of truth.[42] Also, if we analyze subjecthood in terms of parthood, we get the absurd consequence that Caesar's big toe is a subject of the state of affairs that Caesar conquered Gaul. The natural response is to deny that the same sense of 'part' is involved when we say Caesar's toe is part of him and when we say he is part of the state of affairs, so we block the transitivity. As an expedient, this may work, but it is a name for the problem, not a solution to it. How do the two types of part differ?

One advantage of accepting nonexistent states of affairs is that it allows one to deal with negation in a way consonant with logical atomist principles. The big problem for logical atomism has always been to give an acceptable account of the falsehood of false atomic propositions without accepting negative states of affairs. van Fraassen[43] claims to avoid negative states of affairs, but he has a primitive predicate negation in addition to propositional negation, so his claim is not secure. van Fraassen cannot take some states of affairs to be nonexistent, because for him they are just certain tuples, which exist whenever their terms do. If we take states of affairs seriously as an ontological category in their own right, then we can simply say that an atomic proposition is true (on an interpretation, in a world) if its state of affairs (under that interpretation) exists (in that world) and false if its state of affairs does not exist.

A world can then be construed, in Wittgensteinian fashion, as a totality of states of affairs.[44] It would be nice to be able to replace much of what Wittgenstein loosely says about complexes, states of affairs and situations in the *Tractatus* by more manageable mereological vocabulary. At first sight, it would appear that nothing was further from Wittgenstein's mind when considering states of affairs, facts, and so on, than part and whole. A careful examination of what he says about complexes dispels this impression.[45] There is much mereology there. The problem is how to account for all of Wittgenstein's situations in a way that gives them a relatively harmless ontological status as complexes of states of affairs.[46] If this cannot be done, there is no mereological reconstruction of logical atomism. It is possible to do without conjunctive and disjunctive situations, provided we ignore negation. Let P be a set of atomic propositions, and for each $p \in P$ let $[p]$ be its corresponding atomic state of affairs. Then the conjunction of P may be taken to designate

$$\sigma x(\exists p \in P(x = [p]) \,\&\, \forall p \in P(E!\,[p]))$$

the sum of all P's states of affairs provided $E!\,[p]$ for all $p \in P$ (it is empty otherwise), whereas the disjunction of P designates

$$\sigma x(\exists p \in P(x = [p]))$$

the sum of those of P's states of affairs that exist, which is empty only if all propositions in P are false. If quantification is regarded as infinite conjunction and disjunction, this device takes care of general situations as well, always provided it is only atomic propositions that are conjoined or disjoined.

The difficulty is negation. I do not see how one can allow each situation to correspond to some sum of states of affairs and avoid postulating negative and complex states of affairs, which is to abandon logical atomism. Nevertheless, it is surprising how much of the theory of states of affairs can be put on a mereological footing, and I would claim that free mereology not only has its merits as a vehicle with which to interpret past theories of states; those who propose such theories today would do well to bear free mereology in mind.

My own view is that the truth-making role of states of affairs can be usurped by individual moments or tropes, things like John's love for Mary. In view of the problems sketched above, I prefer not to mix mereology and states of affairs. We cannot do without mereology, so what is to be done about states of affairs? My response is like that of the Scotsman who found out how to avoid the problems of mixing whisky with water: he did without water.

References

Bacon, J., "Four modal modelings," *Journal of Philosophical Logic*, vol. 17 (1988) pp. 91–114.
Bencivenga, E., "Free logics," in *Handbook of Philosophical Logic* (ed. Gabbay, D. and Guenthner, F.), vol. III, Reidel, Dordrecht (1986) pp. 373–426.
Brentano, F., *The Theory of Categories*, Nijhoff, The Hague (1981).

Chisholm, R. M., "Parts as essential to their wholes," *Review of Metaphysics*, vol. 26 (1973) pp. 581–603.
Chisholm, R. M., *Person and Object: A Metaphysical Study*, Allen & Unwin, London (1976).
Cocchiarella, N. B., "Some remarks on second order logic with existence predicates," *Noûs*, vol. 2 (1968) pp. 165–75.
Findlay, J. N., *Meinong's Theory of Objects and Values*, 2d edn., Clarendon Press, Oxford (1963).
Fine, K., "Modal theory for modal logic, Part III: Existence and predication," *Journal of Philosophical Logic*, vol. 10 (1981) pp. 293–307.
Fine, K., *Reasoning with Arbitary Objects*, Blackwell, Oxford (1985).
Goodman, N., *The Structure of Appearance*, 3d edn., Reidel, Dordrecht (1977).
Lambert, K., "On the philosophical foundations of free logic," *Inquiry*, vol. 24 (1981) pp. 147–203.
Lambert, K., "On the philosophical foundations of free description theory," *History and Philosophy of Logic*, vol. 8 (1987) pp. 57–66.
Lambert, K., "Incomplete objects: Meinong's theory and its applications," in *Essays on Meinong* (ed. Simon, P. M.), Philosophia, Munich (1990).
Leonard, H. S., "Essences, attributes, and predicates," *Proceedings and Addresses of the American Philosophical Association*, vol. 37 (1964) pp. 25–52.
Leonard, H. S. and Goodman, N., "The calculus of individuals and its uses," *Journal of Symbolic Logic*, vol. 5 (1940) pp. 45–55.
Leśniewski, S., "Podstawy ogólnej teoryi mnogości" ["Foundations of a general theory of manifolds"], *Prace Polskiego Koła Naukowe w Moskwie*, Sekcya matematiczno-przyrodnicza, 2. Moscow (1916).
Leśniewski, S., "O podstawach matematyki" ["On the foundations of mathematics"], *Przegląd Filozoficzny*, vol. 30 (1927) pp. 164–206; vol. 31 (1928) pp. 261–91; vol. 32 (1929) pp. 60–101; vol. 33 (1930) pp. 75–105, 142–70.
Leśniewski, S., "On the foundations of mathematics," *Topoi*, vol. 2 (1983) pp. 7–52. [Abridged English translation of Leśniewski 1927–30 by V. F. Sinisi.]
Leśniewski, S., *Stanislaw Leśniewski's Lecture Notes on Logic* (ed. Szrednicki, J. and Stachniak, Z.), Kluwer, Dordrecht (1988).
Meinong, A., *Über Moglichkeit und Wahrscheinlichkeit*, vol. 6 of *Alexius Meinong Gesamtausgabe*, Akademische Druck- u. Verlagsanstalt, Graz (1972).
Mulligan, K., Simons, P. M., and Smith, B., "Truth-makers," *Philosophy and Phenomenological Research*, vol. 44 (1984) pp. 278–321.
Parsons, T., *Nonexistent Objects*, Yale University Press, New Haven (1980).
Russell, B., *The Principles of Mathematics*, Allen & Unwin, London (1903).
Russell, B., *Our Knowledge of the External World*, Allen & Unwin, London (1914).
Simons, P. M., "On understanding Leśniewski," *History and Philosophy of Logic*, vol. 3 (1982) pp. 165–91.
Simons, P. M., "Leśniewski's logic and its relation to classical and free logics," in *Foundations of Logic and Linguistics* (ed. Dorn, G. and Weingartner, P.), Plenum, New York (1985) pp. 369–402.
Simons, P. M., "The old problem of complex and fact," *Teoria*, vol. 5 (1985a) pp. 205–26.
Simons, P. M., "Tractatus mereologico-philosophicus?" *Grazer Philosophische*

Studien, vol. 28 (1986) pp. 165–86.
Simons, P. M., *Parts. A Study in Ontology*, Clarendon Press, Oxford (1987).
Tarski, A., Appendix E in J. H. Woodger, *The Axiomatic Method in Biology*, Cambridge University Press (1937) pp. 161–72.
Tarski, A., "On the foundations of Boolean algebra," in A. Tarski, *Logic, Semantics, Metamathematics*, Clarendon Press, Oxford (1956) pp. 320–41. (German original published 1935.)
Twardowski, K., *On the Content and Object of Presentations*, Nijhoff, The Hague (1977).
van Fraassen, B. C., "Facts and tautological entailments," *Journal of Philosophy*, vol. 66 (1969) pp. 478–88.
van Fraassen, B. C. and Lambert, K., "On free description theory," *Zeitschrift für mathematische Logik und Grundlagen der Mathematik*, vol. 13 (1967) pp. 225–40.
Whitehead, A. N., *An Enquiry Concerning the Principles of Natural Knowledge*, Cambridge University Press (1919).

Notes

1. See the third investigation in Husserl (1970).
2. See Whitehead (1919, p. 74 ff., 101 ff). The theory is the basis for his famous "method of extensive abstraction." Whitehead may have started his formal work earlier than 1915, since Russell (1914, p. 92) offers axioms for the relation 'encloses', the converse to 'part'. He credits the ideas to Whitehead (p. v) and states that a formal treatment is to appear in the (unrealized) fourth volume of *Principia*.
3. See Leśniewski (1916, 1927–30, 1983).
4. Whitehead's *Nachlass* was destroyed on his instructions after his death.
5. Cf. the criticisms in Leśniewski (1988, pp. 171–8).
6. This is a pity. Cf. Simons (1982) and, for the application in mereology, Sections 1.3, 2.6, and 2.7 of Simons (1987).
7. See Tarski (1937, 1956).
8. See Leonard and Goodman (1940). Goodman varied the ideas, dropping the use of sets, in *The Structure of Appearance*; cf. Goodman (1977).
9. Whitehead's views diverged significantly from the majority and I shall not consider them further. Cf. Section 2.9 of Simons (1987) for some comments. I shall be publishing a more comprehensive account of Whitehead's mereology in a forthcoming paper.
10. Definitions of predicates are given using ':≡', read 'is defined to be equivalent with'; definitions of terms, term-forming functors, and operators are given using ':=', read 'is defined to be identical with'.
11. More particularly, the weak supplementation principle of Simons (1987, p. 28). In many contexts, this is interderivable with the strong supplementation principle (*ibid.*, p. 29):

$$\forall x \forall y (\sim(x < y) \supset \exists z(z < x \;\&\; z \mid y))$$

12. Tarski (1956, p. 333 n).

13. Cf. Simons (1985); Simons (1987, Section 2.8).
14. Cf. Simons (1987, especially Chapter 3).
15. See Simons (1987, Section 2.9) for a survey of nonclassical extensional mereologies.
16. Cf. Simons (1987, Chapter 3).
17. Simons (1987, Chapter 5), where temporal mereology is handled using temporally indexed predicates, although the tense-logical option is also worthy of consideration, and would most naturally be dealt with in a tensed free logic.
18. Leibniz's *Nouveaux essais sur l'entendement humaine*, Book II, chapter 27, sec. 4; Chisholm (1973); Chisholm (1976, Appendix B).
19. Simons (1987, Chapter 7f.). There is no theory of descriptions in the theory considered.
20. Useful survey articles are Lambert (1981, 1987), Bencivenga (1986). The best source for free description theories is van Fraassen and Lambert (1967).
21. Cf. Lambert (1987, p. 62). In the free description theory literature the resulting theory of definite descriptions is known as FD and is the minimal free description theory.
22. The idea is due to Leonard (1964, p. 30), and also, independently, to Cocchiarella (1968).
23. Fine (1981, p. 293f.), Simons (1987, p. 264). The idea behind the term is that if either object does not exist, the antecedent (in this case, that one is part of the other) is false.
24. Simons (1987, Section 2.5).
25. Cf. Meinong (1972, Section 25, pp. 168ff.). Incomplete objects promise to be the most useful of Meinong's menagerie, and seem set to attain even a measure of scientific respectability: the arbitrary objects of Fine (1985) show that incomplete objects are finding increasing use. For an appraisal of some uses, cf. Lambert (1989).
26. Cf. Parsons (1980, p. 19). In fact Meinong tends to say that an incomplete object neither has nor lacks P. Since I regard 'lacks' to means the same as 'does not have' I prefer to express the point using the contradictory or negative of a property (1972).
27. Cf. Meinong (1972), Findlay (1963, pp. 186ff.), Parsons (1980, pp. 21ff.).
28. Meinong (1972, p. 211).
29. Cf. Parsons (1980, p. 19).
30. Cf. Parsons (1980). By the Leibnizian principle, there is thus exactly one such object.
31. Cf. Findlay (1963, p. 168).
32. Cf. Meinong (1972, p. 210), also Findlay (1963, p. 168).
33. Twardowski (1977, p. 100).
34. The interpretation is that of Brentano. See Brentano (1981, *passim*) on Aristotle. I would not put my hand in a fire to vouch for the correctness of Brentano's interpretation, since there are passages in Aristotle's *Physics*, Book VI, 231 b 11–15 that imply that continua have other continua as parts. But whether or not this is definitely Aristotle's theory, it is a possible and interesting one.
35. Brentano (1981).
36. Cf. Mulligan, Simons, and Smith (1984).
37. So this theory is closer to that of Frege than that of Russell (1903), except that states of affairs rather than truth-values are the referents of sentences.

38. Cf. van Fraassen (1969) and Mulligan, Simons, and Smith (1984).
39. This would be the way consonant with Mulligan, Simons, and Smith (1984).
40. Some might use these as reasons to dispense with states of affairs in favor of these individual instances of attributes—cf. Mulligan, Simons, and Smith (1984). Another line would be that such instances (known down under as 'tropes') do not differ from states of affairs; cf. Bacon (1988). It makes life much easier for the theory of truth if there are nonexistent tropes, as Bacon is prepared to concede, but we three are not.
41. As we have seen, Meinong believes that the parts of existing objects themselves exist. Since the true state of affairs that Pegasus does not exist itself does exist (in the wide sense of 'exist' we are using), Pegasus could not on this view be part of this state of affairs, on pain of existence. One possible answer to Meinong is thus simply to revise the mereology and retain the idea that subjects of states of affairs are parts of them.
42. Again the particular instance theory scores here, since integrating particulars with particulars is not as problematic as integrating them with universals.
43. van Fraassen (1969).
44. In Wittgenstein's *Tractatus* there is a fixed fund of states of affairs that does not vary: a world is obtained by taking any selection of these as the existing ones. It is then determined which states of affairs do not exist at that world—they are just the remainder. But we may wish to allow the fund of states of affairs to vary from world to world, in which case we need to specify both which states of affairs exist and which do not exist at that world, and a world will be a totality not just of atomic facts (states of affairs that exist there) but also of atomic unfacts (those states of affairs that do not exist there).
45. For an account of Wittgenstein's confusion, cf. Simons (1985a); for some Tractarian mereology, cf. Simons (1986).
46. Wittgenstein calls situations (*Sachlagen*) what other people call states of affairs (*Sachverhalte*), reserving the latter term for the logically atomic case only. Roughly speaking, situations stand to states of affairs as truth-unctions of elementary propositions stand to elementary propositions.

Bibliography

Bencivenga, Ermanno, "Free semantics for indefinite descriptions," *Journal of Philosophical Logic*, vol. 7 (1978) pp. 389–405.
Bencivenga, Ermanno, "Free semantics," *Boston Studies in the Philosophy of Science*, vol. 47 (1980) pp. 31–48. [Revised version in this volume, Chapter 5.]
Bencivenga, Ermanno, "Truth, correspondence, and non-denoting singular terms," *Philosophia*, vol. 9 (1980) pp. 219–29.
Bencivenga, Ermanno, "Again on existence as a predicate," *Philosophical Studies*, vol. 37 (1980) pp. 125–38.
Bencivenga, Ermanno, "Free semantics for definite descriptions," *Logique et Analyse*, vol. 92 (1980) pp. 393–405.
Bencivenga, Ermanno, "Free logics," in *Handbook of Philosophical Logic*, vol. 3 (ed. Gabbay, D. and Guenthner, F.), Reidel, Dordrecht (1986).
Bencivenga, Ermanno, Lambert, Karel, and van Fraassen, Bas, *Logic, Bivalence and Denotation*, Ridgeview, Atascadero, Calif. (1986).
Burge, Tyler, "Truth and singular terms," *Noûs*, vol. 8 (1974) pp. 309–25. [Reprinted in this volume, Chapter 9.]
Cocchiarella, Nino, "A logic of possible and actual objects," *Journal of Symbolic Logic*, vol. 31 (1966) pp. 688–9.
Cocchiarella, Nino, "Quantification, time, and necessity," Chapter 13 in this volume.
Fine, Kit, "The problem of nonexistent objects," *Topoi*, vol. 1 (1982) pp. 97–140.
Garson, James, W., "Applications of free logic to quantified intensional logic," Chapter 6 in this volume.
Grandy, Richard, "A definition of truth for theories with intensional definite description operators," *Journal of Philosophical Logic*, vol. 1 (1972) pp. 137–55. [Reprinted in this volume, Chapter 8.]
Hintikka, Jaakko, "Existential presuppositions and existential commitments," *Journal of Philosophy*, vol. 56 (1959) pp. 125–37.
Hintikka, Jaakko, "Towards a theory of definite descriptions," *Analysis*, vol. 19 (1959) pp. 79–85.
Hintikka, Jaakko, "Cogito, ergo sum: Inference or performance?" *Philosophical Review*, vol. 71 (1962) pp. 3–32. [Reprinted in this volume, Chapter 7.]

Hintikka, Jaakko, "Studies in the logic of existence and necessity," *The Monist*, vol. 50 (1966) pp. 55–76.

Hintikka, Jaakko, "Existential and uniqueness presuppositions," in *Philosophical Problems in Logic* (ed. Lambert, Karel), Reidel, Dordrecht (1970).

Lambert, Karel, "The definition of E(xistence)! in free logic," in *Abstracts: The International Congress for Logic, Methodology and Philosophy of Science*, Stanford University Press (1960).

Lambert, Karel, "Existential import revisited," *Notre Dame Journal of Formal Logic*, vol. 4 (1963) pp. 288–92.

Lambert, Karel, "Notes on E! III: A theory of descriptions," *Philosophical Studies*, vol. 13 (1963) pp. 51–9.

Lambert, Karel, "Notes on E! IV: A reduction in free quantification with identity and definite descriptions," *Philosophical Studies*, vol. 15 (1964) pp. 85–8.

Lambert, Karel, "Free logic and the concept of existence," *Notre Dame Journal of Formal Logic*, vol. 8 (1967) pp. 133–44.

Lambert, Karel, "On the philosophical foundations of free logic," *Inquiry*, vol. 24 (1981) pp. 147–203.

Lambert, Karel, *Meinong and the Principle of Independence*, Cambridge University Press (1983).

Lambert, Karel, "Predication and ontological commitment," in *The Tasks of Contemporary Philosophy = Die Aufgaben der Philosophie in der Gegenwart* (ed. Leinfellner, Werner and Wuketits, Franz M.), Hölder-Pichler-Tempsky, Vienna (1986).

Lambert, Karel, "A theory of definite descriptions," Chapter 1 in this volume.

Lambert, Karel and van Fraassen, Bas, *Derivation and Counterexample*, Dickenson, Encino, Calif. (1972).

Leblanc, Hugues and Hailperin, Theodore, "Nondesignating singular terms," *Philosophical Review*, vol. 68 (1959) pp. 129–36.

Leblanc, Hugues and Meyer, Robert, "On prefacing $\forall x A \supset A(y/x)$ with (y): A free quantification theory without identity," *Zeitschrift für mathematische Logik und Grundlagen der Mathematik*, vol. 16 (1970) pp. 447–62.

Leblanc, Hugues and Thomason, Richard H., "Completeness theorems for some presupposition-free logics," *Fundamenta Mathematicae*, vol. 62 (1968) pp. 125–64.

Leonard, H. S., "The logic of existence," *Philosophical Studies*, vol. 7 (1956) pp. 49–64.

Mann, William E., "Definite descriptions and the ontological argument," *Theoria*, vol. 33 (1967) pp. 211–29. [Revised version in this volume, Chapter 14.]

McCall, Storrs, "Abstract individuals," *Dialogue*, vol. 5 (1966) pp. 217–31. [Reprinted in this volume, Chapter 12.]

Meyer, Robert and Lambert, Karel, "Universally free logic and standard quantification theory," *Journal of Symbolic Logic*, vol. 33 (1968) pp. 8–26.

Meyer, Robert, Bencivenga, Ermanno, and Lambert, Karel, "The ineliminability of E! in free quantification theory without identity," *Journal of Philosophical Logic*, vol. 11 (1982) pp. 229–31.

Posy, Carl J., "A free IPC is a natural logic: Strong completeness for some intuitionistic free logics," *Topoi*, vol. 1 (1982) pp. 30–43. [Reprinted in this volume, Chapter 3.]

Rescher, Nicholas, "On the logic of existence and denotation," *Philosophical Review*, vol. 69 (1959) pp. 157–80.

Rosser, J. B., "On the consistency of Quine's *New Foundations for Mathematical Logic*," *Journal of Symbolic Logic*, vol. 4 (1939) pp. 15-24.
Scales, Ronald, *Attribution and Reference*, University of Michigan Microfilms (1969).
Scales, Ronald, "A Russellian approach to truth," *Noûs*, vol. 11 (1977) pp. 169-74. [Reprinted in this volume, Chapter 11.]
Schock, Rolf, "Contributions to syntax, semantics and the philosophy of science," *Notre Dame Journal of Fomal Logic*, vol. 5 (1964) pp. 241-89.
Schock, Rolf, *Logics Without Existence Assumptions*, Almqvist and Wiksell, Stockholm (1968).
Scott, Dana, "Existence and description in formal logic," in *Bertrand Russell: Philosopher of the Century* (ed. Schoenman, Ralph), Allen and Unwin, London (1967). [In this volume, Chapter 2.]
Simons, Peter M., "Free part-whole theory," Chapter 16 in this volume.
Skyrms, Brian, "Supervaluations: Identity, existence and individual concepts," *Journal of Philosophy*, vol. 64 (1968) pp. 477-83.
Smiley, Timothy, "Sense without denotation," *Analysis*, vol. 20 (1960) pp. 125-35.
Thomason, Richmond H., "Modal logic and metaphysics," in *The Logical Way of Doing Things* (ed. Lambert, Karel), Yale University Press (1969) pp. 119-46.
van Fraassen, Bas, "The completeness of free logic," *Zeitschrift für mathematische Logik und Grundlagen der Mathematik*, vol. 12 (1966) pp. 219-34.
van Fraassen, Bas, "Singular terms, truthvalue gaps and free logic," *Journal of Philosophy*, vol. 67 (1966) pp. 481-95. [Reprinted in this volume, Chapter 4.]
van Fraassen, Bas, "Presupposition, supervaluations and free logic," in *The Logical Way of Doing Things* (ed. Lambert, Karel), Yale University Press (1969) pp. 67-91.
van Fraassen, Bas and Lambert, Karel, "On free description theory," *Zeitschrift für mathematische Logik und Grundlagen der Mathematik*, vol. 13 (1967) pp. 225-40.
Woodruff, Peter W., "On supervaluations in free logic," *Journal of Symbolic Logic*, vol. 49 (1984) pp. 943-50.